REGREENING THE BUILT ENVIRONMENT

Now in its second volume, *Regreening the Built Environment* provides an overview of physical and social environmental challenges that the planet is facing and presents solutions that restore ecological processes, reclaim open space, foster social equity, and facilitate a green economy.

Healing the planet requires a combination of strategies networked across multiple scales of development, including buildings, sites, communities, and regions. Case studies from a range of locations in the United States, Denmark, Vietnam, Germany, South Korea, Switzerland, France, and the United Kingdom, among others, demonstrate how existing gray infrastructure can be retrofitted with green infrastructure and low-impact development techniques. From this, the author shows how a building can be designed that creates greenspace or generates energy; likewise, a roadway can be a parkway, an alley can be a wildlife corridor, and a parking surface can be a garden. This new edition also includes case studies that have successfully reconnected communities that were fragmented by unjust planning practices and irresponsible patterns of development, resilient design solutions in response to natural disasters, passive design strategies that can make interior spaces more efficient and healthier, and expanded discussions on capturing carbon, renewable energy, agriculture, waste, public transit, and adaptive reuse, including innovative ideas on how to reimagine the shopping mall in the era of e-commerce.

The strategies presented in this book will stimulate discussions within the design profession and will be of great interest to students and practitioners of environmental studies, architecture, landscape architecture, and urban design.

Michael A. Richards, B.S., M.Arch., Ph.D., has spent much of his career exploring ways to integrate environmental sensibility into the built environment. He has worked as a planner, designer, researcher, and teacher at several colleges and universities, including the University at Buffalo, Buffalo State University, the University of Vermont, Vermont Technical College, Erie Community College, and the Community College of Vermont. In addition to this book series, he is the author of the science fiction novel, *Dream Makers*, which tells the story of a distraught architecture student who can alter the past, present, and future by manipulating his dreams.

REGREENING THE BUILT ENVIRONMENT

Nature, Green Space, and Sustainability

Second Edition

Michael A. Richards

Routledge
Taylor & Francis Group
LONDON AND NEW YORK

earthscan
from Routledge

Designed cover image: Michael A. Richards

Second edition published 2025
by Routledge
4 Park Square, Milton Park, Abingdon, Oxon, OX14 4RN

and by Routledge
605 Third Avenue, New York, NY 10158

Routledge is an imprint of the Taylor & Francis Group, an informa business

First edition published by Routledge 2018

British Library Cataloguing-in-Publication Data
A catalogue record for this book is available from the British Library

ISBN: 978-1-032-66865-9 (hbk)
ISBN: 978-1-032-66859-8 (pbk)
ISBN: 978-1-032-66866-6 (ebk)

DOI: 10.4324/9781032668666

Typeset in Times New Roman
by Apex CoVantage, LLC

In memory of my father, Michael A. Richards DDS (1938–2019), and my brother-in-law, Robert J. Staerker (1973–2020).

CONTENTS

PREFACE

It is difficult to believe that seven years have gone by since the first volume of this book was published. The older I become, Willie Nelson's song *Funny How Time Slips Away* seems to be always playing in the back of my mind. But given the pace of technological and societal change in recent times along with living through a global pandemic, I was delighted when the publisher asked me if I was interested in writing another volume, for there are many new challenges and opportunities that need to be discussed to keep the topics in this book relevant.

In the preface of the original volume, I talked about how the fragmentation of my neighborhood in Rochester, New York, caused by highway and roadway expansion was the impetus for me to embark on a career focused on regreening the built environment. Although the forest and ravine around my childhood neighborhood are still not restored, it seems the notion of mitigating ecological and social fragmentation caused by infrastructure – particularly roadways and highways – has gained considerable traction since 2017. For Chapter 7 (Roadways and Parking), I compiled a table of roadway removal, tunneling, and capping projects but didn't end up including it because it was getting too long (I stopped at about 120 projects!). As a graduate student in architecture in the late 1990s/early 2000s, my master's thesis proposed a 37-mile greenway between Lake Ontario and Lake Erie along the Niagara River that removed, reduced, and/or capped portions of the Interstate 190, the Scajaquada/198 corridor, the Kensington Expressway, and the Robert Moses Parkway. Since that time, I am pleased to see that portions of the northern Robert Moses Parkway have been removed, revegetated, and pedestrianized (see cover photo), and there are several projects in process to repurpose the Buffalo Skyway, put deck parks over the Kensington Expressway, and restore Frederick Law Olmsted's parkway system in the Scajaquada Creek corridor. As a result of the federal Reconnecting Communities and Neighborhoods Program, $185 million in grants have been allocated to 45 communities across the United States to reconnect neighborhoods that were victims of discriminatory zoning practices that often intentionally constructed urban highways as physical barriers to reinforce segregation. In my hometown of Rochester, New York, eastern portions of the "Inner Loop" section of Interstate 490 have recently been removed and replaced with pedestrian-friendly boulevards, mixed-use development, and mixed-income housing, and, as I write this book, plans are in process to remove the northern section. Given the heightened awareness of social equity as a critical tenet of sustainability, such projects can provide reparation and action for past injustices, empty platitudes, and performative action statements.

Despite these success stories, reversing existing sprawled patterns of development and preventing further sprawl still remain a challenge. In Chapter 6 (Transportation), a new case study is presented that proposes a regional transportation system in Vermont to alleviate traffic congestion that threatens the historic character of the small towns. The *Ski/Ride Vermont by Rail* concept has the potential to decrease auto congestion, decrease carbon emissions, revitalize struggling communities, prevent and/or curtail sprawled patterns of development, and provide equity by increasing transit options for those who cannot afford an automobile.

Although the COVID-19 pandemic altered our lifestyles, crippled the economy, and, worst of all, resulted in the loss of many lives, it has also presented planners and designers with new opportunities regarding the utilization of space. The increase in hybrid or fully remote work schedules along with the proliferation of e-commerce has caused an increase in vacant office space and brick-and-mortar retail stores. The adaptive reuse and/or repurposing of these spaces, along with making existing spaces more efficient, can provide alternatives to unnecessary new construction on undeveloped land. Multi-laned, auto-centric commercial strips can be retrofitted into pedestrianized boulevards with town squares in suburbs that never had defined village centers in the first place. Given these new trends, this volume has added discussions on "reimagining the shopping mall" and on the efficiency, comfort, and safety in interior spaces.

Many other personal factors also influenced the tone of this volume. In addition to the pandemic, in the last five years, my dad passed away, my brother-in-law passed away, we sold our house that I reference in the *Preface* of the first volume, and I have struggled with my own health issues, including a failed surgery on my lumbar spine. All these factors – combined with the escalating cost of real estate – forced me to leave the state of Vermont and move back to Rochester, New York. However, some good has come out of this. After all, we are resilient and can adapt (also themes in this volume). In addition to being closer to family, the move provided a different lens when discussing some of the strategies or, more specifically, the contrast between a rust-belt city already impacted by sprawl versus a mostly pastoral region attempting to balance its natural capital with increased growth and tourism.

This volume has also been slightly reorganized. Chapters 1–3 introduce basic principles of sustainability, resiliency, environmental justice, and green infrastructure and discuss the impetus for regreening the planet in relation to current environmental challenges. Chapters 4 and 5 discuss basic planning strategies, energy, carbon capture, food, and waste. Chapters 6–11 are organized by scale starting with regional- or community-scaled strategies (transit, roadways, and corridors), then proceeding to building and interior-scaled strategies. Chapter 12 zooms back out and looks at strategies for the distant and *very* distant future. Chapter 13 discusses ways to incorporate the regreening strategies into policy, practice, and education.

The personal factors in my life mentioned earlier also provided me with a much-needed perspective in a time – where despite increased globalization – we are also seeing increased polarization. All the regreening strategies discussed in this book are moot if they further divide us or if they sacrifice health, safety, wellness, quality of life, or equity by neglecting the specific needs and desires of people, despite whether we agree or not. Inclusivity, transparency, and equal representation of community members in decision-making are therefore imperative to truly embrace the principles of sustainability and resiliency. Differences of opinion and constructive criticism are vital in society and are at the core of our democracy, but somewhere along the line, differences have unfortunately made us just plain angry. We need to face the harsh reality that, on a planet as diverse as ours, it is impossible to satisfy everyone, but we can certainly strive to find some common ground, listen more/talk less, and/or agree to disagree. As a part-time musician, I am always amazed at music's ability to amalgamate people from different walks of life. It is my hope that, after reading this book, regreening and healing the planet will also become something that unites us, not divides us.

ACKNOWLEDGMENTS

Special thanks to the following people/groups for providing images:

Tor Anderson, True North Designworks
Udo Becker, Photographer
Gary Bentrup, USDA
Alex Bitterman, PhD
Karen Breider
Stefania Canta, Renzo Piano Building Workshop
Yu Chen, CX-Landscape
Peter Close
Jessica Culler, NASA
Daniel Dong, CX-Landscape
Susan Fell, WSDOT
Paul Fetters, Photographer
Jared Ficklin, Argodesign
Brenda Freije, Rethink 65–70
Magdalena Gruber, Photographer
Lily Hart, Confluence Project
Fritz Horton, F.W. Horton and Associates
Rasmus Hjortshoj, Photographer
Kariane Jeaurond, Dominique Perrault Architecture
Christina Joyce, MnDOT
Claudia Kalinowski, Landschaftspark Duisburg-Nord
Mario Kamer, Doppelmayr/Garaventa
Samuel Kim, University at Buffalo Libraries
Krzysztof Koniuszaniec, Deep Ocean Technology
Raymond Lee, Rafael Vinoly Architects
Vladimir Levin, Architect

Duncan Lewis, Duncan Lewis Scape Architecture
Meagan Lott, WSDOT
Al Mowbray, Oregon Metro
Amelia Fruergaard Lund, Bjarke Ingels Group
Mark Nowaczyk, Cannon Design
Koen Olthuis, Waterstudio
Jing Peng, CX-Landscape
Jochen Peschel, Coastersandmore.de
Zhi Ting Phua, Buffalo Olmsted Parks Conservancy
Gaëlle Lauriot-Prévost, Dominique Perrault Architecture
Travis Price, Travis Price Architects
Hafeez Raji, Hafeez Photography
James Ramsey, RAAD Studio
Sabrina Register, Seattle Public Utilities
Kimberly A. Richards, DDS
Kelly L. Richards
Phyllis A. Richards
Mark Ross, Portland Parks and Recreation
Joan Schumaker, Friends of the Genessee Valley Greenway
Lena Sieler, Landschaftspark Duisburg-Nord
Tracy Tackett, PE, Seattle Public Utilities
Christine Thuring, PhD
Laertis Antonious Ando Vassiliou, LAAV Architects
Kathrine Warren, town of Mountain Village
Silvia Winzap, Doppelmayr/Garaventa
Ken Wyner, Ken Wyner Photography
George Zhuo, CX-Landscape
Simon Zhao, CX-Landscape
Horst Zielske, Photographer
Special thanks to the following groups:
Audemars Piguet Hotel
Seattle Municipal Archives
Seattle Parks and Recreation
SolaRoad
Thyssen Krupp Transrapid/Stoiber Productions
University of Vermont, Planning, Design, and Construction
Vo Trong Nghia Architects (VTN) Architects

1

INTRODUCTION

A new paradigm for the built environment

According to the National Oceanic and Atmospheric Administration's (NOAA's) Global Monitoring Lab, carbon dioxide concentrations reached 419.3 parts per million in the atmosphere in 2023, setting a new record high and making it the warmest year among all years in NOAA's 1850–2023 climate record (Lindsey, 2024; NOAA, 2024). The International Panel on Climate Change (IPCC) Average Sixth Assessment Report showed that annual greenhouse gas emissions during 2010–2019 were higher than those in any previous decade on record (IPCC, 2023). Temperatures are rising, sea levels are increasing, polar ice is melting, and extreme weather events and natural disasters are becoming more commonplace around the world each year. Scientists overwhelmingly agree that humans and irresponsible patterns of development are responsible for this impact.

Starting around the time of the Industrial Revolution, people began moving away from their agrarian roots and symbiosis with the land toward a more machine-centric worldview that emphasized technology over nature. Throughout the nineteenth and early twentieth centuries, cities burgeoned, buildings eclipsed the skyline, and infrastructure intertwined the landscape, forever changing the scale and pace of society and human interaction with nature. Eventually, the same cities that once provided safe havens from the "natural" elements became overridden with congestion, crime, and pollution that ultimately forged an exodus of people away from urban centers toward the country in search of open space, clean air, and tranquility. The proliferation of the personal automobile and abundance of new, affordable housing in the suburbs expedited this exodus and centrifugal dispersal pattern, and with the car and subsequent gas tax came more roads, freeways, boulevards, parking surfaces, and development that further pillaged the landscape.

In the United States, such development was catalyzed by the Federal Highway Act of 1956, which procured funding for the construction of 41,000 miles of interstate highways.

For a short time, this strategy seemed to be both socially accepted and positive for the economy. Industry was still ample in the city centers, road construction created jobs, and the housing market boomed in the suburbs. However, as evident in recent times, this "American Dream" pattern of development did not prove to be sustainable in the long run. Industries and businesses eventually began to abandon urban centers, creating barren downtowns plagued with decrepit

DOI: 10.4324/9781032668666-1

infrastructure and plunging real estate values. Suburbs, which once offered a reprieve from urban life, have become plagued with the same congestion issues that the cities faced in the nineteenth and twentieth centuries, just with a more auto-centric look and feel. Farmland, wetlands, forests, and natural habitats have been paved over, clear-cut, and fragmented in favor of multi-laned boulevards flanked with billboards, strip malls, box retail, and large parking surfaces. To make matters worse, the increase in remote/hybrid work options and the proliferation of e-commerce have increased office and retail vacancy rates, which has further amplified the amorphous and unpedestrianized character of commercial strips in many suburban communities.

Sprawl does not discriminate – it can occur in old, new, liberal, conservative, small, medium, and large cities. For example, in older European cities such as London and Paris, the city fabric has engulfed so much of the peripheral land that it is difficult to distinguish what is "city" versus "suburb." In the United States, some of the denser metropolitan areas are on the same path to becoming one large megalopolis. In cities such as New York, Boston, and Philadelphia, one must travel further and further to escape the exponentially expanding urban growth ring. Eventually, Vermont, New Hampshire, and Maine will become the "suburbs" of New York and Boston – in many ways, this is already true. Steiner (2011, p. 3) describes such a landscape after driving from Austin, Texas, to Dayton, Ohio:

> The built environment is beyond dull; it is ugly and predictable. The interstate interchanges in Texas, Arkansas, Tennessee, Kentucky, and Ohio, or alternatively through Georgia, Alabama, Mississippi, and Louisiana, wound nature and culture. These assaults on our senses blend together with the same hotels, gas stations, fast food chains, and mega-churches. The edge-city malls combine the same shops. The skylines of Dallas, Little Rock, Memphis, Nashville, Louisville, and Cincinnati may be taller or shorter, but all look pretty similar.

The deleterious effects of sprawl are certainly more apparent in areas that have been impacted by loss of industry and economic blight. This is especially evident in what is known as the "inland north" region of the United States that surrounds the Great Lakes roughly from Michigan to central New York. Because of its proximity to the lakes and canals, this region became the heart of manufacturing in the late nineteenth and early twentieth centuries. However, the U.S. steel market eventually declined and forced many of the factories to scale down, move to locations where labor/taxes were cheaper, or close altogether. To make matters worse, interstate trucking started to take over the shipping industry from the canal, river, and rail routes, thus detracting commerce activity in the city centers. Cities became plagued with declining populations, urban decay, and crime. In the mid-1980s, this region grimly came to be known as the rust belt, with many areas experiencing a perfect storm of economic decline and regional/urban sprawl that continues to have devastating physical and financial impacts that, in many cases, are difficult to reverse.

The good news is that there seems to be a paradigm shift in the twenty-first century. Citizens are overwhelmingly accepting notions of global climate change and natural resource depletion and are increasingly exploring lifestyle changes that can, in some desperate attempts, reverse the damage we have afflicted upon our planet. We are beginning to re-examine patterns of development and starting to devise strategies to revitalize our urban centers, conserve existing greenspace, and "regreen" existing gray infrastructure. Politically and economically, more nations are investing in notions of sustainable development and resiliency to conserve natural resources, preserve open space, manage stormwater runoff, minimize pollution, and decrease fossil fuel consumption. These new standards have introduced new demands from society that question the types of projects that are needed in the built environment.

Since the twentieth century was about leaving our mark on the landscape, the twenty-first century needs to be about reducing our footprint on the landscape. Indeed, a paradigm shift needs to occur in the relationship between the natural and built environment. In the past, the two have frequently been treated as separate entities – infrastructure is merely placed on the landscape with no awareness given to the ecological or natural processes intrinsic to the site. A different approach would be to analyze the landscape and adjacent communities and formulate a design solution that minimizes destruction to habitat and open space and integrates with the ecological processes and natural elements of the site. As Ian McHarg (1969, p. 7) states in his book, *Design with Nature*,

> If one accepts the simple proposition that nature is the arena of life and that a modicum of knowledge of her processes is indispensable for survival rather more for existence, health, and delight, it is amazing how many apparently difficult problems present ready resolution.

A comprehensive land analysis might indicate that new infrastructure is not needed at all but rather an improvement of existing infrastructure at both the building and site level and design strategies that restore and protect the natural landscape, conserve resources, and create renewable energy. It is quite plausible that a building be designed that creates greenspace or generates more energy than it consumes; likewise, a roadway can be a parkway, an alley can be a wildlife corridor, and a parking surface can be reconfigured to reduce impervious surface by using the green stormwater infrastructure. Green infrastructure can be networked with alternative modes of transportation to reduce vehicle congestion, improve circulation, and mitigate social and ecological fragmentation. Removing barriers such as roadways and highways can reconnect communities that were fragmented by their construction, which was often done intentionally to segregate lower socio-economic groups, minorities, and people of color. Green space can also provide recreation opportunities in dense, urban areas that lack open space or parks. Comprehensive planning can control future growth patterns by establishing growth boundaries that foster compact development, land conservation, stormwater management, alternative transportation, and pedestrian-friendly communities. Zoning strategies, conservation easements, public transit, and the purchase or transfer of development rights also have the potential to reacquire land and reverse existing sprawled patterns of development. Existing infrastructure can be adaptively reused as an alternative to building on undeveloped land, and alternative sites for infrastructure can be explored on or under water, the ground, cliff edges, or brownfields to further conserve open land on the surface of the earth.

This book will discuss strategies on how "gray" infrastructure can be retrofitted with green infrastructure for corridors, roadways, transit networks, sidewalks, parking surfaces, buildings, rooftops, brownfields, and interiors. Although there are several completed case studies presented from around the world, some of the conceptual proposals were applied to several "test" sites near places where I have lived, particularly in the states of New York and Vermont. It should be noted that the conceptual proposals are intended to be used for scholastic purposes only and are *not* intended to be actual proposals for the given sites. However, this does not mean that the concepts presented cannot be applied to other sites or regions. Remember, as your introductory environmental science teacher probably already told you, "Think globally and act locally."

The ideas presented in this book are not intended to condemn the built environment, nor are they meant to insinuate the complete abandonment of new infrastructure or building development. The theories do, however, favor a more sustainable worldview that advocates for humans to be partners with nature rather than conquerors of nature and that architects, landscape architects, engineers, and urban/regional planners need to rethink their vocations and the methodologies behind their projects. It is hoped that the strategies presented in this book will spark a new dialogue in the design profession to further facilitate the restoration of ecosystems and natural habitats, unite fragmented communities, preserve cultural heritage, foster social equity, create a green economy, and promote sustainability and resiliency on global, regional, and local scales. Open land is such a fragile resource – in many cases, once it is developed, restoring it to its original state is often impossible. Ormsbee and Starke (2006, p. 36) eloquently describe the prominence and delicacy of the earth's natural resources:

> Land and the waters that lap its edges, flow across its surface, seep into its upper soil strata, and move within its deep aquifers are our ultimate resources. Mismanaged, they may be lost to us forever, and our national wealth and well-being proportionately diminished.

References

Intergovernmental Panel on Climate Change (IPCC). (2023). *Synthesis report of the IPCC sixth assessment report (AR6)*. IPCC. https://report.ipcc.ch/ar6syr/pdf/IPCC_AR6_SYR_SPM.pdf

Lindsey, R. (2024). *Climate change: Atmospheric carbon dioxide*. National Oceanic and Atmospheric Administration (NOAA). Climate.gov. www.climate.gov/news-features/understanding-climate/climate-change-atmospheric-carbon-dioxide

McHarg, I. (1969). *Design with nature* (p. 7). John Wiley & Sons.

National Oceanic and Atmospheric Administration (NOAA). (2024). *2023 was the world's warmest year on record, by far*. www.noaa.gov/news/2023-was-worlds-warmest-year-on-record-by-far#:~:text=Earth's%20average%20land%20and%20ocean,0.15%20of%20a%20degree%20C)

Ormsbee, J. S., & Starke, B. W. (2006). *Landscape architecture: A manual of environmental planning and design* (p. 36). McGraw-Hill.

Steiner, F. (2011). *Design for a vulnerable planet* (p. 3). University of Texas Press.

2

GREEN PRINCIPLES AND TERMINOLOGY

When I was an environmental studies student in the early 1990s, the environmental movement was still considered fringe, often confined to demonstration tables in the student union, where hippies in tide-eyes congregated to play a game of hacky-sack or trade their latest tapes from live concerts of Phish, the Grateful Dead, or in the University at Buffalo's case, the band *moe.* I had a classmate in college who was an engineering major who would frequently ridicule those of us majoring in environmental studies, calling us names like *tree-hugger*, *green-thumb*, or *granola*, and claiming our chosen degree path was useless and a waste of money and time. But then the new millennium arrived, and it seemed like the environmental movement finally became mainstream, with terms like *sustainability*, *resilience*, *green*, *net-zero*, *organic*, and *fair-trade* becoming commonplace. Although this section will describe the evolution of some of these terms, it is certainly not meant to be an exhaustive history of the green or sustainability movement, as this book's intent is to look forward more than backward.

Sustainability

A few years ago, I attended a panel discussion on the human impact of changes in our landscape. In posing a question, one of the audience members chastised a member of the panel for repeatedly using the word *sustainable*, claiming the word was passe and overused to a point that diluted its significance. While I do agree there has been a great deal of greenwashing in the past decade or so, I do not agree that the term *sustainable* has lost its relevancy. In fact, the term *sustainability* itself is inherently resilient and has – and will continue – to morph, adapt, and transform in response to changing ideals in society and changing issues on our planet.

Sustainability is hardly a new concept. I didn't invent it, you didn't invent it, Al Gore didn't invent it, Greta Thunberg didn't invent it (although I admire her commitment), and Leonardo DiCaprio didn't invent it. Although the terms *sustainable* and *sustainability* first appear in the Oxford English Dictionary in the second half of the twentieth century, notions of longevity and lastingness have been used and known for centuries (DuPisani, 2007). The earliest civilizations likely had relationships with their physical surroundings, and ecological awareness first appeared on record 5,000 years ago in the hymns of Vedic sages and in Taoist and Buddhist

DOI: 10.4324/9781032668666-2

philosophy (Weyler, 2018). In 1798, Thomas Robert Malthus wrote *An Essay on the Principle of Population*, which warned of future food shortages and ecological destruction given geometric progressions of population (Weyler, 2018). In the late nineteenth century, transcendentalists Ralph Waldo Emerson and Henry David Thoreau proposed the "intrinsic value of wilderness beyond its usefulness to humans" and botanist William Bartram and ornithologist James Audubon dedicated their work to the conservation of wildlife (Lallanilla, 2020; Weyler, 2018). In 1864, George Perkins Marsh, a Woodstock, Vermont, native, published *Man and Nature* (later published as *Physical Geography as Modified by Human Action*) that warned of the irreversible impacts of humans on the earth (Smithsonian, n.d.).

In the 1900s, the Industrial Revolution brought pollution, overcrowding, and a general dominance of the machine over nature that subsequently catalyzed further concerns about environmental degradation. Writer John Muir published over 300 articles and ten books that recounted his travels and established his passion for wilderness and concern over its destruction. He helped in establishing several National Parks, including Yosemite, Sequoia, Mount Ranier, Petrified, and Grand Canyon and thus became known as "the father of our National Park system" (Sierra Club, n.d.). In 1892, Muir and his supporters founded the Sierra Club, and in 1903, Muir worked with President Theodore Roosevelt to establish conservation programs (Sierra Club, n.d.). Gifford Pinchot became the first head of the Division of Forestry in 1898. While both Pinchot and Muir agreed that land had its limits, Muir favored a full preservation of natural resources whereas Pinchot favored more of a conservation management approach where natural resources could be utilized for human benefit. After World War II, the United Nations held the Scientific Conference on the Conservation and Utilization of Resources, and Congress enacted the Air Pollution Act in 1955.

Rachel Carson published Silent Spring in 1962 that culminated research on the impacts of agricultural pesticides on ecological and human health (IISD, n.d.). Many consider Carson's book a major catalyst for the *Green Movement* of the 1960s that popularized the concept of green design by re-evaluating development, buildings, and design on ecological, energy usage, and bioclimatic principles. Indeed, the 1960s and 1970s saw a surge in several environmental actions, including the establishment of the Environmental Defense Fund in 1967, the Intergovernmental Conference for Rational Use and Conservation of the Biosphere (UNESCO) in 1968, the formation of Friends of the Earth in 1969, the National Environmental Policy Act in 1969, the Environmental Protection Agency in 1970, the Natural Resource Defense Council in 1970, the first Earth Day in 1970, Greenpeace in 1971, the Clean Water Act in 1972, the Endangered Species Act in 1973, and the Worldwatch Institute in 1975, to name a few. The concept of sustainable development received international recognition at the United Nations Conference on the Human Environment in 1972 (SDC, n.d.). In 1984, the World Commission on Environment and Development convened and subsequently published *Our Common Ground* (also known as the Brundtland Report). The report established the classic definition of sustainable development: "development which meets the needs of the present without compromising the ability of future generations to meet their own needs" (SDC, n.d.).

In 1985, a hole in the ozone layer in Antarctica was discovered, and the World Meteorological Society, the United Nations Environment Program, and the International Council of Scientific Unions reported the buildup of carbon dioxide and greenhouse gases in the environment and subsequently predicted global warming. This paved the way for the International Panel on Climate Change in 1988, the Stockholm Environment Institute in 1989, the International

Institute for Sustainable Development (IISD) in 1990, and the United Nations Conference on Environment and Development in 1992. The UN also gathered in 2000 to establish Millennium Development Goals that addressed hunger, poverty, disease, environmental degradation, and discrimination, and again in 2002 at the World Summit on Sustainable Development.

As mentioned earlier, the notion of sustainability has been – and will continue – to adapt and transform. Some scholars have simplified the definition around the concept of perpetual prosperity, or more specifically, "the possibility that human and other forms of life will flourish on the planet forever" (Ehrenfeld, 2006). Others have taken a more ominous view and argue that no human activity is truly sustainable, for all our actions – whether minimized or not – will impact the natural state of our planet. Some are even more apocalyptic, suggesting that nothing in the universe is meant to last forever – our star (the sun) will eventually burn out and life in our solar system as we know it will cease to exist, so "live and let live."

Professor Robert Shibley (1991, pp. 3–4) and students at the University at Buffalo identified five general definitions of sustainability that were being perceived throughout the world as a part of a project for the Monteverde Conservation League and its Sustainable Development Commission:

> The definitions include the economic definition, which defines nature and human-made capital as assets serving economic functions; the human centered or homo-centric definition, which is based on an interpretation of the world which defines the earth and its resources for human use; the stewardship definition, which defines a balance or convergence of humans and nature; the bio-centric definition, which recognizes the protection of the environment above humans; and the eco-centric definition, where the primary objective is the conservation of species and the environment. One observation of the study stated is that the definition of sustainable development is best defined as that which is in the best interest of those offering the definition.

These five definitions bring into question the weighting of the three dimensions (or three Es) of sustainability: ecology (the physical environment), equity (the social environment), and economy. Many may be familiar with the Venn diagram of three overlapping circles that shows a common area where the three factors converge. The other common diagram is often referred to as a nested dependency model, where one of the three dimensions is the primary driver for the other two dimensions. Since this book is focused on "regreening," it tends to favor the eco-centric model of sustainability where mitigation of the physical environment can catalyze social equity and economic opportunities. Many comprehensive regional, municipal, and campus plans, on the other hand, may tend to favor the overlapping model. But this is often dependent on politics, leadership, community demand, and the composition of committee members. Since the first volume of this book was published, culture has continued to cultivate transformations of sustainability. Although I do not claim to be a sociologist, my informal observations indicate that Millennials and Generation Zs seem to be advocating more for the social dimension – including fostering environmental justice (discussed in the following text), diversity, inclusion, racial equity, human health, and wellness – and are also calling out older generations (Generation X and the Baby Boomers) for their historical practice of performative actions where they make hollow statements that are only meant to paint a narrative rather than taking specific action to forge a change.

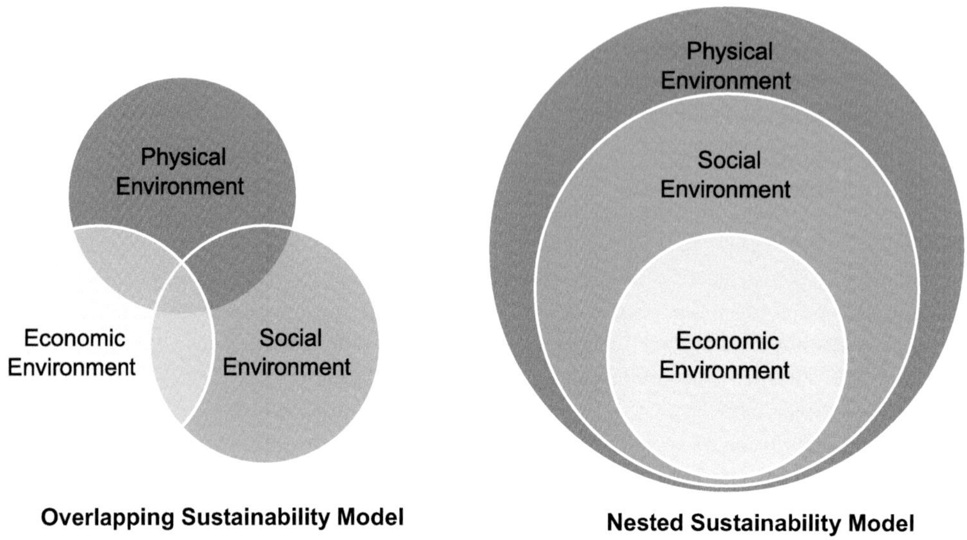

FIGURE 2.1 Sustainability models. Left: overlapping model; right: nested eco-centric model.

Source: Diagrams derived from Garcia, E. J., & Vale, B. (2017). *Unraveling sustainability and resiliency in the built environment* (p. 39, Figure 2.1). Routledge.

In 2015, the United Nations Sustainable Development Summit adopted 17 Sustainable Development Goals (SDGs), which have responded to immerging issues. The SDGs

> provide a shared blueprint for peace and prosperity for people and the planet, now and into the future . . . the goals recognize that ending poverty and other deprivations must go hand-in-hand with strategies that improve health and education, reduce inequality, and spur economic growth – all while tackling climate change and working to preserve our oceans and forests.
>
> *(UN, 2015)*

The 17 SDGs include:

- No poverty
- Zero hunger
- Good health and well-being
- Quality education
- Gender equality
- Clean water and sanitation
- Affordable and clean energy
- Decent work and economic growth
- Industry, innovation, and infrastructure
- Reduced inequalities
- Sustainable cities and communities

- Responsible consumption and production
- Climate action
- Life below water
- Life on land
- Peace, justice, and institutions
- Partnerships for the goals

The challenge with any definition or model of sustainability is that we have no idea what the future holds or what the needs of future generations might be – we can only make educated estimates. The term *needs* can also vary on the basis of worldview, geography, race, gender, religion, and socioeconomics. Sustainability, as a result, is relative to both spatial and temporal dimensions. A project that may have been considered sustainable in the twentieth century may not be so in the twenty-first century as technology improves, new problems arise, and standards and regulations become more stringent. For example, in the early 1900s, asbestos was considered one of the most durable and practical materials that offered solutions to a variety of construction complexities – we later learned of adverse health effects from asbestos that forced us to think otherwise. The World Trade Center's tube-tower design was considered cutting edge at the time it was designed. It offered an economically efficient method to build skyscrapers and provide open floor plans with ample space and light. The exterior framing provided structure to withstand lateral loads of the largest aircraft of its respective time. Who could have predicted what was to come in 2001? Before COVID-19, e-commerce may have been viewed as unsustainable with regard to the survival of small, local businesses. However, when put into the context of a global pandemic where people could not leave their homes, suddenly the virtual shopping business model was seen in a new light. And given the energy and material costs of brick-and-mortar retail buildings and the fuel consumption of consumers traveling to and from said retailers, it is easy to begin to question what truly is sustainable and to realize the sliding scale of sustainability relative to each moment in history.

Fortunately, sustainability recognizes human limits. A more accurate definition with "disclaimers" included might say,

> [D]evelopment that *strives* to meet the environmental, social, and economic *relative* needs of the present with the best available solutions without compromising the environmental, social, and economic well-being of future generations using the most accurate future prediction methods available in the present.

This factors out human differences and human limits – we can only do our best, and as a professor of mine used to say, "Everything can't be everything." Williams (2007, p. 17) recognizes this paradox and defines sustainability as a continuum where design and planning *approach* sustainability:

> A design is sustainable or it is not. If it is not sustainable, changes can be made to make it sustainable. If it is sustainable, by necessity it will change and evolve. Sustainability is not static – it is iteratively changing based on evolving knowledge that connects science and design . . . a design must be capable of functioning "unplugged" from the external nonrenewable energy sources and resources to be sustainable.

Courtney and Zencey (2012, p. 17) also state:

> Sustainability is an either-or goal – you either achieve it or you don't. And nature decides what is and isn't sustainable; it always has and it always will. We humans, with our limited vision, have to make our best guess as to what the limits imposed by nature must mean for us.

Indeed, a key aspect of sustainability is responsibility to the earth. We must question our projects, lifestyles, and activities. Is new development always necessary? If so, how can development work in harmony with nature? How can the impact on the natural landscape be minimized? How can development benefit diverse user groups? How can the project provide new opportunities for people? What is the payback of the project? If we continue to pose such questions and look outside the box, we will eventually discover that we are indeed engaging in sustainability.

Resiliency, resistance, and adaptation

Since the publication of the first volume of this book, there have been several publications, papers, discussions, and lectures emphasizing resilience and how it is related to sustainability. There seem to be five schools of thought in the literature: resiliency and sustainability are distinct strategies but complement each other; resiliency is a component of sustainability; sustainability is a component of resiliency; there is no difference between resiliency and sustainability; and those who feel resilience is the new sustainability. According to Google, searches for sustainability have tripled since 2004, and searches for resiliency have increased almost nine times during the same time frame (Carnow, 2022). Much of the debate stems from the various definitions used by institutions for both terms. Garcia and Vale (2017, p. 203) provide an extensive analysis of how the two concepts relate to each other but offer a complementary perspective:

> Learning about the resilience of the built environment as a way of understanding how cities work as complex adaptive systems will be of little use if it does not have sustainability as a goal . . . sustainability, therefore, can provide an ethical framework for designs that look toward enhancing the resilience of a city.

Other comparisons assume sustainability implies everlasting and maintains the same function, or as one writer states, "while sustainability looks at how current generations can meet their needs without compromising that ability for future generations, resilience considers a system's ability to prepare for threats, to absorb impacts, and to recover and adapt after disruptive events" (Carnow, 2022). Similarly, one college's website describing its program defines sustainability as "the ability to continue important functions indefinitely without a decline in quality and resilience as the ability to thrive in the face of change" (Drake, n.d.). Marchese et al. (2018) contrast the terms with time and scale:

> The primary objective of sustainability is to create desirable conditions for future generations. Thus, the effects of sustainability policies may not directly influence present conditions but may have substantial effects on future conditions. Resilience, on the other hand, is understood in many situations to apply to more immediate, temporal scales. Policies that increase the resilience of a system will protect the system in the short term from potential disturbances.

These either-or definitions seem to conclude that something can be designed as resilient in the short term but is not necessarily sustainable in the long term, and something can be designed as sustainable but may or may not be resilient to either an imminent and/or an unknown change.

Although they are often used interchangeably, there is also a distinction between resistance and resilience. Resistance refers to the ability of an ecosystem, community, or design to withstand the impacts of a disaster and retain its fundamental structure and function; resilience, on the other hand, refers to the ability of ecosystems, communities, and designs to recover from the stresses inflicted by disasters. The Resilient Design Institute (RDI, n.d.a) has defined ten general resilient design principles:

- Resilience transcends scales.
- Resilient systems provide for basic human needs.
- Diverse and redundant systems are inherently more resilient.
- Simple, passive, and flexible systems are more resilient.
- Durability strengthens resilience.
- Locally available, renewable, or reclaimed resources are more resilient.
- Resilience anticipates interruptions and a dynamic future.
- Find and promote resilience in nature.
- Social equity and community contribute to resilience.
- Resilience is not absolute.

The RDI also defines resilient design strategies across building, community, and regional/ecosystem scales. Although resilience can incorporate resistant design strategies, solutions are often examined more broadly to mitigate external threats. For example, at the building level, in addition to *resistant* structural systems to keep the structure operational in response to a disaster, other passive designs are often incorporated to make the structure more *resilient*, such as on-site renewable energy, green roofs to absorb rainfall, providing redundant water supplies, the use of vernacular design strategies for heating and cooling, and back-up systems for waste disposal. At the community scale, resilient strategies can include incorporating meeting hubs for education and/or gathering during emergencies, utilizing green stormwater infrastructure, providing space for local food production, utilizing advanced modeling technologies to predict the critical paths for future events, and providing compact development with pedestrian circulation networks that connect to critical services and regional transit systems. At the regional scale, resiliency can be fostered by protecting and restoring natural systems that provide ecosystem services, providing a stable, redundant energy grid, establishing a more diverse economy, and networking a variety of transit systems that can provide circulation, food, and other goods during times of emergency (RDI, n.d.b).

Personally, I do not believe the classification is as pertinent as understanding both concepts and where and how they might be applied given a specific project. Adaptability and flexibility in use are often considered when engaging sustainability in the design process, and despite some of the definitions above, a resilient design can consider the near and distant future. Using the rise of sea level as an example, resiliency and sustainability can be viewed as symbiotic or contradictory, depending on how you define them. For example, it may not be considered sustainable to construct new buildings along coastlines in the future but rather to reestablish riparian buffers or other landscape features that can diffuse storm surges and/or reconfigure themselves to adapt to different scenarios. The riparian zones/landscape features, therefore, are both a sustainable and resilient solution. As a comparison, let's say new buildings continue to be constructed in flood-prone areas but are built on hydraulic lifts that allow the structure to increase its height in response to rising water levels. Some could argue the design is both resistant given its ability to withstand disaster and resilient given its ability to adapt to the disaster but argue it is not sustainable because of the resources required to make it function and its impact on the ecosystem. Some may also argue it's not resilient because the hydraulic lifts provide only one solution to one known event: the rise of water. A truly resilient design can recover from what cannot be predicted. Let's take it a step further and say a meteor crashes into the site and destroys the building. The building is now no longer sustainable, resistant, or resilient – that is, unless the materials used in its construction can regenerate and/or reconfigure into another form that can generate new landforms or vegetation, create human or marine habitat, provide surge protection, or decompose into a material that can organically clean pollutants from the water.

Those that limit the definition of sustainability with definitive words such as *must*, *same*, or *indefinitely* as seen in one of the definitions referenced earlier are fueling the chasm between sustainability and resilience, which, in the long term, is ironically not sustainable. As mentioned earlier, sustainability itself is inherently a resilient concept. It is constantly evolving and in addition to ensuring viability for future generations, it strives to develop solutions that are flexible in use and can adapt to unpredictable changes. In this respect, adaptability is a strategy that enables persistence. In the same line of thought, it would be remiss for a resilient design to not consider the future. In essence, when we engage one, we are inevitably engaging the other. Sustainability and resilience should not be thought of as either-or concepts but rather symbiotic concepts that can complement or be components of each other, depending on the nature of the project and the external forces impacting the project.

Ecological design

If there ever is any confusion as to what ecological design encompasses or how to engage the ecological design process, there is a simple solution: take a walk outside in nature. Observe the plants, the animals, water movement, the wind, the sun, the temperature, the decomposition of waste, the flow of energy, and how wildlife creates a habitat that is symbiotic with nature. Ecology is the study of interactions between living organisms and their environment, and ecological design seeks to integrate design with the living processes. In the recent past, many designers interpreted this as merely blending form with natural features and/or copying natural forms (biomorphic design). While fusing built and natural forms is certainly desirable from an aesthetical perspective, it is only a small piece of the "holistic ecological design" pie. A comprehensive ecological design model strives to make the design function like a living organism, or as Van der Ryn and Cowan (2007, pp. 33–34) describe as a

> marriage of nature and technology, or as any form of design that minimizes environmentally destructive impacts by integrating itself with natural processes . . . and one that respects species diversity, minimizes resource depletion, preserves nutrient and water cycles, maintains habitat quality, and attends to all the other preconditions of human and ecosystem health.

Ian McHarg was one of the pioneers in ecological design. In 1969, he published *Design with Nature*, where he advocated sensitivity to place by declaring "that no human intervention into nature should be considered without a thorough analysis of the compatibility between a building and the ecosystem in question, including the specific topographical and microclimatic conditions of the site" (McHarg, 1969). The book demonstrates a comprehensive analysis of all the layers characteristic of a site to inform land use planning and the design of communities, including soils, hydrology, groundwater, topography, vegetation, wildlife, local weather patterns, wind, sun, and geological features, to name a few. This layer analysis (McHarg used stacked Mylar maps) formed the basis of the environmental impact statement process and was a precursor for modern Geographic Information Systems (GIS) software. Another concept that emerged from McHarg's book is the notion of bioregionalism, which is essentially understanding natural processes at a larger scale beyond the immediate project site. Regional analysis often reveals conflicts, history, and patterns of development that would otherwise not be apparent if the focus was confined to building or site scales.

McHarg's model has been adopted, transformed, and/or integrated into current landscape, planning, and environmental policy and has influenced much of the current thinking around the notion of sustainability in design. Van Der Ryn and Cowan (2007, pp. 72–73) lay out five general principles for ecological design: solutions grow from place, ecological accounting informs design, design with nature, everyone is a designer, and make nature visible. Steiner (2011) also describes five principles for twenty-first-century architecture: the location of a site, energy efficiency, water conservation, building materials, and beauty. A professor of mine in architecture school used to use the human body as an analogy to describe an "organic" approach to designing buildings. Like the human body, buildings need systems to think, breathe, circulate, filter pollutants, maintain temperature, and discard waste. Like the human body, there are efficient ways to use energy and not-so-efficient ways. And like the human body, buildings must not only adapt and be resilient but also get tired, age, and eventually die. I'm not sure if that professor coined this analogy, but it certainly works when trying to connect built and living systems and when attempting to engage in ecological design.

A thorough understanding of ecology at local and regional levels can reveal where and what types of development are or are not appropriate. Often, the analysis might reveal that no development is conducive for a site but rather the preservation, regeneration, and/or enhancement of original, predevelopment natural conditions. Such strategies include reforestation/revegetation, the conservation/enhancement of natural buffers along waterways, streambank and shoreline stabilization to protect against erosion and storm surge, wetland restitution, and bioremediation.

Where development is appropriate, ecological analysis reveals optimum bioclimatic design strategies that harness and utilize the natural forces (sun, wind, light, heat, and water), methods to treat and capture stormwater on-site, ways to treat and recycle waste, the use of native materials, strategies to decrease erosion and sedimentation in the construction process, and techniques to minimize the footprint of the infrastructure being constructed. The ecosystem model can also be applied to urban environments to understand human activities/behavior, social/cultural interaction, and the spatial relationships of humans, infrastructure, and nature. Indeed, an understanding of all aspects of "place," whether they be ecological, cultural, or historical, is imperative to inform the design process.

Biophilia, biomimicry, biomorphism, bioengineering

Biophilia is the tendency of humans to seek connections with nature and other living organisms. While biophilic design focuses more on relationships between humans and nature and biomorphic design is the design that is inspired by naturally occurring geometries and patterns found in nature and living organisms, biomimicry or biomimetic design learns from or mimics the systems, patterns, and ecological processes of nature to solve human design challenges. Biomimicry doesn't merely utilize nature for a specific benefit but rather functions like nature by emulating and learning from its systems, behaviors, and patterns (Biomimicry Institute, n.d.). These concepts are discussed further in Chapter 9.

Smart Growth and New Urbanism

Although Smart Growth and New Urbanism each share the objectives of fostering healthy, prosperous, and resilient places, each has its own set of principles. The U.S. Environmental Protection Agency created the Smart Growth Network in 1996 that formulated ten key principles that include creating a mix of land uses; taking advantage of compact design; creating a range of housing opportunities and choices; creating walkable neighborhoods; fostering distinctive, attractive communities with a strong sense of place; preserving open space, farmland, natural beauty, and critical environmental areas; directing development toward existing communities; providing a variety of transportation choices; making development decisions predictable, fair, and cost effective; and encouraging community and stakeholder collaboration in development decisions (SGA, n.d.). New Urbanism focuses more on urban design, or, more specifically, on creating a new urban coherency to mitigate the dispersed patterns of development in the mid-twentieth century. New urbanist principles include walkability, which includes positioning services within a ten-minute walk from home or work, creating pedestrian-friendly streets where buildings are close to the sidewalk, providing on-street parking, positioning parking behind buildings, and creating welcoming sidewalks with street lined trees; connectivity, which includes providing a network of dedicated walking/bicycling lanes, an interconnected and hierarchical street network; providing mixed-use and diversity for both the type of spaces and opportunities for different income ranges; mixed housing; design that is on the human scale and creates a sense of place; transect planning structures with the highest density at town centers and lower densities toward the edge; density and compact development; green transportation including networks of rail, bus, and pedestrian routes; sustainable development including renewable energy, eco-friendly design, and locally produced food and products; and enhancing well-being and quality of life (New Urbanism, n.d.).

Green infrastructure

Green infrastructure (GI) is a term that has evolved over time from an unconventional concept to a comprehensive solution. The 2019 Water Infrastructure Improvement Act defines GI as "the range of measures that use plant or soil systems, permeable pavement or other permeable surfaces or substrates, stormwater harvest and reuse, or landscaping to store, infiltrate, or evapotranspirate stormwater and reduce flows to sewer systems or to surface waters" (EPA, n.d.a). While this definition seems to focus more on green stormwater infrastructure (GSI), others have defined GI more as an integrated concept. The Green Infrastructure Ontario Coalition (n.d.) defines GI

> as the natural vegetative systems and green technologies that collectively provide society with a multitude of economic, environmental, health, and social benefits, including urban forests and woodlots; bioswales, engineered wetlands and stormwater ponds; green roofs

and green walls; parks, gardens and grassed areas; natural heritage systems (interconnected meadows, wetlands, ravines, waterways and riparian zones); and urban agriculture.

Tzoulas et al. (2007) utilize a fairly broad and inclusive definition: "It (GI) can be considered to comprise of all natural, semi-natural, and artificial networks of multifunctional ecological systems within, around, and between urban areas, at all spatial scales."

While I do prefer this all-encompassing definition, I would pull it back a step further and basically define GI as any natural, naturally simulated, or engineered technology that reduces the human footprint on the earth. Given this breadth, all of the concepts in this book are considered GI, including auto-alternative forms of transit; land conservation and preservation; clean energy; carbon capture; the reduction, reuse, recycling, and recovery of waste; riparian buffers and greenways; green belts; parks, open space, heritage sites; natural areas; sustainable and urban agriculture; impervious or "gray" infrastructure removal, reduction, and reconfiguration; wildlife and pedestrian over and underpasses; alternative sites; adaptive reuse; and GSI, which is discussed in the next section in more detail. It is important to reiterate that GSI is only one component of GI. For example, the Vermont Agency of Natural Resources (VT ANR) considers GSI a best management practice that is part of a larger series of eight low-impact development (LID) principles, which include conservation development, minimization of soil compaction, minimization of total disturbance, protection of natural water flows, protection of riparian buffers, protection of sensitive areas, reduction of impervious surfaces, and stormwater disconnection (VT ANR, n.d.).

Green stormwater infrastructure

GSI – which can also be referred to as sustainable urban drainage systems or "SUDS" – manages stormwater runoff by preserving and/or replicating the natural, predevelopment hydrologic processes of a site after development. In developed or "altered" landscapes, rainfall cannot infiltrate the ground due to soil compaction, impermeable surfaces, and the flattening of natural depressions in the landscape. The conventional design response to stormwater has been to connect impervious areas directly to storm sewers that subsequently channel the water to streams, lakes, rivers, or detention ponds. This method, also known as "pipe and pond," can damage water quality, increase flooding, and decrease the recharge of groundwater since water and sediment are not adequately filtered or absorbed by the soil or vegetation. By replicating natural hydrologic processes, water and sediment have adequate time to be filtered and absorbed, ultimately decreasing discharge and pollutants into natural bodies of water. GSI includes riparian buffers, street trees, green roofs, green walls, vegetated swales, filter strips, sand filters, infiltration trenches, rain barrels/cisterns, stormwater ponds/constructed wetlands, bioretention/rain gardens, permeable pavement, and planter boxes. GSI strategies can be applied to all gray infrastructure, including roadways, parking surfaces, sidewalks, plazas, buildings, and rooftops. These concepts will be reiterated in the succeeding chapters.

Disconnecting impervious surfaces

One of the most common practices utilized to control stormwater is to direct stormwater from impervious surfaces to a pervious surface that is sized and graded to reduce the volume of water and absorb pollutants. This can be achieved by using downspouts off rooftops that direct water to a vegetated area and/or capture water in a cistern or rain barrel rather than being connected directly to a sewer pipe. Although design standards will vary by municipality, most guidelines will define a minimum distance between the building and the vegetated area, dimensions of the vegetated area, the slope of the vegetated area, soil preparation, and the type of vegetation (which usually does not allow clumping species). Gravel areas around the base of the downspout near the building can prevent erosion and splash-back to the building. For residential applications, a standard 50-gallon (189-liter) rain barrel can be used but should be drained after large rainfall events. Above- or below-grade cisterns in the 200–10,000-gallon (757–37,854-liter) range can be used for larger impervious surfaces, in wetter climates, or for those interested in harvesting their rainwater for gray water use or irrigation needs. Water can also be routed to vegetated swales, infiltration trenches, rain gardens, or bioretention ponds (discussed in the next section).

Vegetated swales, sand filters, and infiltration trenches

Vegetated swales – also known as bioswales – are grass and/or herbaceous plant-lined trenches that can run parallel to roads, meridians, sidewalks, or parking lots and are designed to trap particulate pollutants, promote infiltration, and reduce stormwater runoff. In contrast to traditional concrete curbs and gutter systems, vegetated swales can pre-treat and absorb runoff more efficiently before it drains into water bodies. Water can be channeled into the swales either through inlets in the curb or by eliminating the curb and sloping the roads toward the swales. According to the U.S. Environmental Protection Agency's Green Streets Handbook (EPA, 2021), a parabolic or trapezoidal cross section with side slopes of 3:1 or flatter is recommended to maximize the wetted channel perimeter of the swale. The area draining into the swale should typically be no longer than 100 feet (30.5 meters) in length and no more than 1 acre (0.4 hectare), and the ponding depth should be a maximum of 6 inches (15 cm).

While swales can provide a cost-effective, simple solution for stormwater management, they are less efficient when the slope is extremely flat or extremely steep. Flat slopes often result in poor channeling and may lead to standing water that is a breeding ground for bugs and odors; steep slopes may lead to increased flows that erode the vegetation and decrease infiltration efficiency. This can be exacerbated by poorly drained soils under the swale. Sites with slopes greater than 5 percent can incorporate terracing, diversion berms, or check dams. Swales can also be coupled with other pretreatment systems such as sand filters, forebays, or filter strips to better remove particulate matter and/or can be used in conjunction with infiltration trenches, which are linear beds/basins filled with rock or with a sand/soil/loose aggregate mix that collect, store, and gradually release the water into the surrounding soil. The trenches are usually a minimum of 3 feet (~1 meter) in depth and can contain a perforated pipe submerged under the aggregate to increase infiltration efficiency. The trenches can be constructed at or below the surface and are often used under bioswales, sand filters, porous pavement, downspouts, or drip edges of buildings to improve drainage.

Bioretention cells and rain gardens

Bioretention cells, also sometimes referred to as rain gardens, are concave-shaped areas planted with native species that are used to capture stormwater from impermeable surfaces. The site is graded so water is channeled from impervious surfaces and/or swales, through grass filter strips, and then into the vegetated cells. Suspended solids, heavy metals, hydrocarbons, nutrients, and sediment are filtered by the mulch layer and the vegetation and root systems absorb the water/

FIGURE 2.2 Top left: Vegetated grass swale along the roadway with erosion control fabric; top right: stone-lined swale with overflow drain; bottom: stone-line swale along sloped road.

nutrients through transpiration. According to the U.S. EPA (1999, 2021), the amended planting soil should be sandy loam, loamy sand, or loam texture with clay content ranging from 10 to 25 percent and infiltration rates greater than .05 inches (0.13 cm) per hour. Vegetation should be both drought- and water-tolerant and should have density and diversity that mimic natural, native ecosystems as closely as possible to be able to endure the local climate and provide habitat to species. The cells are usually designed to allow ponding water up to about 6–12 inches (15–30 cm) and contain an overflow drain to prevent flooding in higher-volume rain events. Keeping the ponding depth below 6 inches allows the water to be slowly infiltrated into the ground and also ensures the water does not stand for extended periods of time. Ideally, the standing water will drain within 24 hours. Higher-volume flows and areas with fast-draining soils can be retrofitted with an aggregate base and a perforated underdrain to enhance infiltration. Bioretention areas are not recommended where the water table is within 6 feet (1.8 meters) of the ground surface, within 10 feet (3 meters) of foundations, where mature trees are located, above underground utilities, above septic systems, or on slopes that are greater than 20 percent.

Peak flows can be calculated using the Rational Formula of $Q = Cf \times C \times I \times A$, where Q equals peak flow in cubic feet per second; Cf equals runoff coefficient adjustment factor to account for the reduction of infiltration and other losses during high-intensity storms; C equals the runoff coefficient to reflect the ratio of rainfall to surface runoff; i equals rainfall intensity in inches per hour; and A equals drainage area in acres. The size of the bioretention area can generally be 5–7 percent (5 percent with a sand bed, 7 percent without a sand bed) of the drainage area multiplied by the rational runoff coefficient "C" (determined by the ground cover present at the site). The runoff coefficient is a variable that relates the amount of runoff to the amount of precipitation received; it is larger for impermeable, steep slopes with low infiltration and smaller for flat slopes and vegetated areas with higher infiltration. Table 2.1 shows different runoff coefficients for various types of ground cover. For example, a 20,000-square-foot parking lot constructed of asphalt would have a runoff coefficient of around .90, so the bioretention area with a sand bed would be 20,000 × .09 × .05, which would yield a size of about 900 square feet. The length of the cell should be twice the width and should not be less than 15 feet in width (4.6 meters) and 40 feet in length (12.2 meters). The width fosters the cell's ability to mimic nature and create a microclimate, while adequate length promotes the distribution of flow. Multiple, smaller cells are often more effective than one large cell due to increased surface area. Overall, there should not be more than one acre per rain garden.

Curb extensions and bump-outs

Curb extensions, also known as "bump-outs" or "bulb-outs," extend portions of the curb into the street. Such designs have traditionally been used at intersections as a means of traffic calming and to decrease crossing distances for pedestrians. However, bump-outs may also be placed along the

TABLE 2.1 Runoff coefficients for various land covers

Surface	*Flat slope, <2%*	*Rolling slope, 2–10%*	*Steep slope, >10%*
Pavement and roofs	0.90	0.90	0.90
Drives and walks	0.75	0.80	0.85
Gravel pavement	0.85	0.85	0.85
Lawns	0.17	0.22	0.35
Cultivated land, clay, and loam	0.50	0.55	0.60
Cultivated land, sand, and gravel	0.25	0.30	0.35
Meadows and pastureland	0.25	0.30	0.35
Woodlands and forests	0.10	0.15	0.20

Source: Oregon Department of Transportation Highway Division, Engineering and Asset Management Unit Geo-Environmental Section. (2014). *Hydraulics design manual*, Appendix 7-F, p. 3. www.oregon.gov/ODOT/GeoEnvironmental/Docs_Hydraulics_Manual/Hydraulics-07-F.pdf

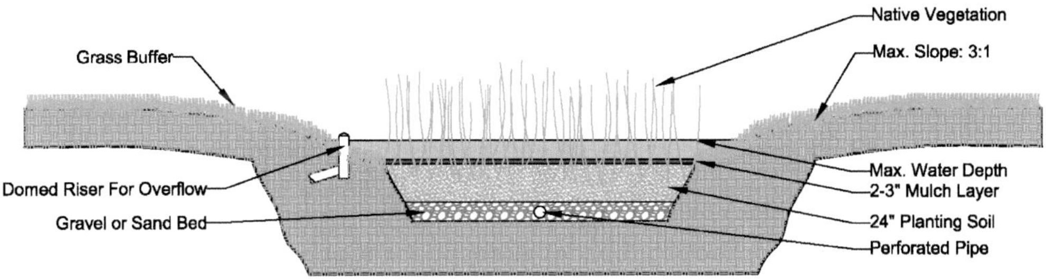

FIGURE 2.3 Typical cross section of a bioretention area.

Source: U.S. Environmental Protection Agency (EPA). (1999). *Stormwater technology fact sheet: Bioretention.* U.S. EPA Office of Water. EPA-832-F-99-012. https://nepis.epa.gov/Exe/ZyPDF.cgi/200044BE.PDF?Dockey=200044BE.PDF

road and can serve as catch areas for stormwater runoff. Water flows into the bump-out through an inlet that is set lower than the street's gutter elevation. Stormwater is temporarily stored in the bed until it infiltrates the root system, ultimately returning to the sewer or into an adjacent drain inlet. Similar to drainage swales, curb extensions can be combined with forebays, check dams, underdrains, infiltration trenches, or sand filters to improve sediment capture and infiltration.

Ribbon driveways

Ribbon driveways, also sometimes called Hollywood driveways, contain two parallel strips of hardscape (which would ideally be made from pervious pavement/pavers) separated by a strip of vegetation. Although such detail might seem minuscule, its effects on reducing impervious surfaces on a site can be significant, especially when the site is at or above the permitted impervious area lot coverage requirements.

Street trees

An anonymous source once stated, "The best time to plant a tree is 20 years ago; the second-best time is now." In addition to their aesthetic beauty and their ability to absorb carbon dioxide and provide oxygen, trees absorb odors, pollutant gases, and heat; provide natural cooling; absorb rainwater; stabilize soil; and provide wildlife habitat. The American Forestry Association estimates that a 50-year-old urban tree saves $75 a year in air conditioning, $75 a year in stormwater and erosion control, $75 a year in wildlife shelter, and $50 a year in air pollution control (Benedict & McMahon, 2006, p. 71). Climate Central (2023) analyzed data for 242 cities across the continental United States and estimated trees in these areas absorbed 110 billion gallons of stormwater a year, reduced four billion pounds of air pollution a year, and absorbed 150 million tons of carbon dioxide and greenhouse gases. Street trees can be planted individually in a planter or "tree box," in a linear tree trench below the surface that connects the individual tree planters, or in structural soil cells (Silva cells) that consist of a structural grid placed under the tree bulb that prevents soil compaction and root compression. When added to the parking lane, vegetation and street trees also provide a physical buffer between the road and pedestrians. The distance from the street curb to the sidewalk should be at least 10 feet (3 meters) to provide a safe distance between pedestrians and cars and to allow cars to turn without blocking sidewalks (see lower right image in Figure 7.2).

FIGURE 2.4 A bioretention area at Champlain College in Burlington, Vermont, was created at the base of a slope near the admissions building (Perry Hall). Stormwater is collected in tanks under the main lawn behind Perry Hall, channeled through a gravel field into the bioretention area, and connected back to the city's stormwater drainage system. Native plants, such as meadow sweet, red-osier dogwood, blue flag iris, and cardinal flower, filter pollutants and provide habitat for insects and birds.

FIGURE 2.5 Top: curb extension with bioretention; middle: tree box; bottom: ribbon driveway.

FIGURE 2.5 (Continued)

Wet ponds for larger-scale development

To provide water-quality and water-quantity treatment on larger sites, constructed wetlands or "wet ponds" that have a permanent pool of water throughout the year can be constructed to offset the runoff from development. The University of Vermont's stormwater treatment facility for the Southwest Campus Watershed is classified by the U.S. Environmental Protection Agency as a "wet extended pond" since water treatment is split between an upper detention storage (forebay) and the permanent micro-pool. Drainage pipes channel the stormwater into one of two forebays in the upper pond. The forebays are 4–6 feet (1.2–1.8 meters) deep and essentially pre-treat the water before it enters the permanent micro-pool chamber of the pond. During storm events, water is detained in the forebay and then slowly released over a 12–48-hour period into the micro-pool. Large, coarse sediments settle to the bottom of the forebay. Maintenance can therefore be done in the smaller forebay rather than having to dredge the main pool. Stone weirs between the forebay and the micro-pool ensure the sediment does not flow into the micro-pool. Water and sediment from the forebays then flow into the aquatic bench – the shallow area of the pond that contains aquatic plants that provide nutrient uptake – and ultimately into the main portion of the pond (the micro-pool), which provides detention and the settlement of smaller particles. The micro-pool geometry is longer than it is wide to provide more effective distribution and treatment of stormwater – the EPA recommends a ratio of at least 1.5:1. The lower pond provides redundancy that can decrease flow rates exiting the facility and provide additional

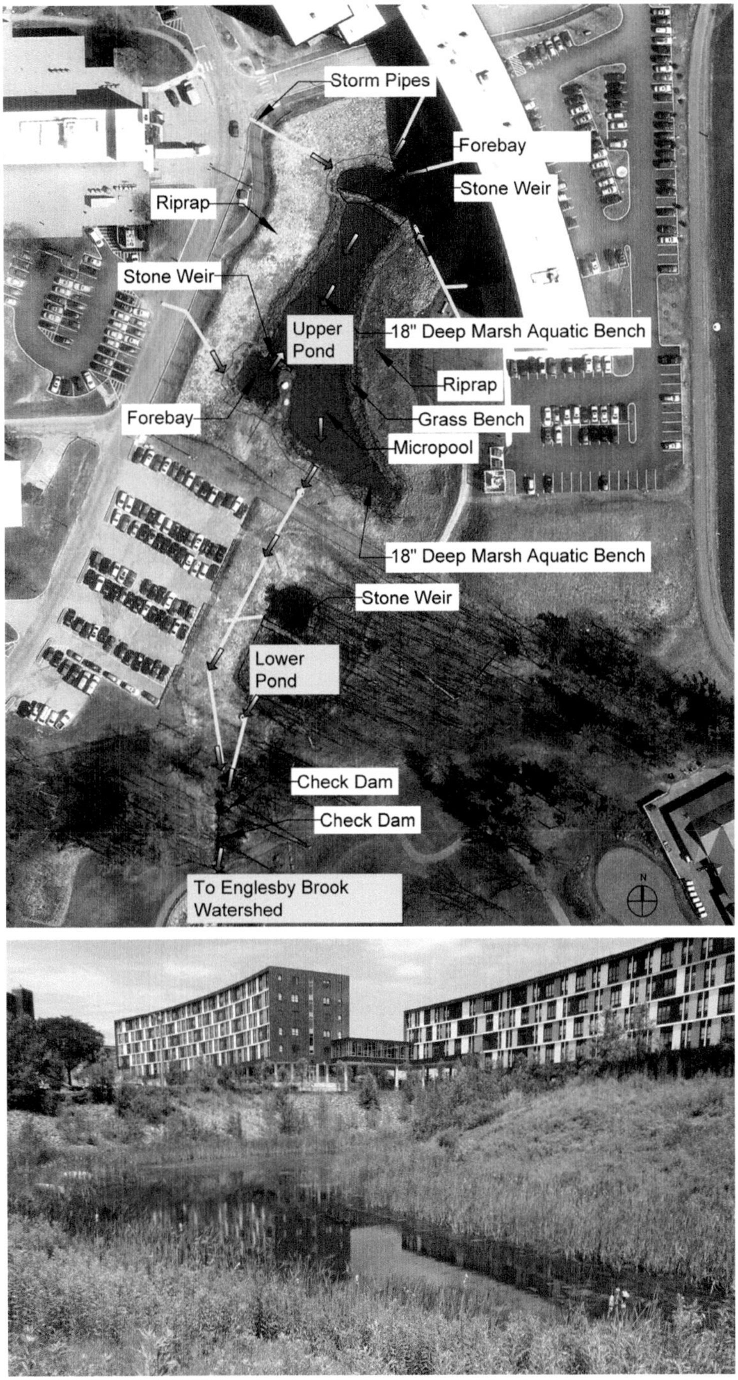

FIGURE 2.6 Top: University of Vermont Southwest Stormwater Treatment Facility. Bottom: close-up of the upper pond.

Source: Orthophotography courtesy of Esri, DigitalGlobe, GeoEye, i-cubed, USDA FSA, USGS, AEX, Getmapping, Aerogrid, IGN, IGP, swisstopo, and the GIS User Community storage/treatment in larger and/or extreme rainfall events. Check dams at the outlet end of the facility provide further control for discharge rates for water entering the adjacent Englesby Brook, which ultimately drains into Lake Champlain. The Southwest Stormwater Treatment Facility is one of five stormwater detention facilities on the campus that ensures the total capture of suspended solids, phosphorous loads, and stormwater from annual storm events complies with the Municipal Separate Storm Sewer System (MS4) General Permit, which is a federally required Clean Water Act permit that the state of Vermont has been authorized to administer.

Permeable, porous/pervious pavement, and turf-block systems

As previously mentioned, traditional forms of concrete and asphalt are impervious and therefore increase stormwater runoff and subsequent pollution and sedimentation into waterways.

FIGURE 2.7 Top left: interlocking permeable pavers; top right: porous asphalt in the parking area and pervious concrete on the pathway next to the parking area; bottom left: turf-block system with plantings on the grid lines around the pavers; bottom right: turf-block system with central plantings in void spaces of grid.

Permeable, pervious, and porous pavements, on the other hand, allow water to infiltrate the pavement and drain into the subsurface stone infiltration bed and/or filtrate to the soil and water table. Permeable pavement systems are interlocking or grid concrete pavers that allow water to pass through crushed aggregate in the seams between the pavers; porous asphalt and pervious concrete are manufactured to contain less fine aggregate so water can infiltrate the surface; and turf-block systems are a cellular grid constructed of plastic or concrete placed on the soil to prevent soil compaction in high traffic areas such as pathways, driveways, or parking surfaces. Dirt, sand, gravel, or vegetation can grow in the void spaces, which can be either between the grid lines or along the grid lines, depending on the configuration.

Several studies have shown that permeable pavement effectively reduces runoff and improves water quality. Collins et al. (2007) tested a parking lot at North Carolina State University that consisted of a control section of traditional impermeable asphalt, a section of pervious concrete, two types of permeable interlocking concrete pavers with different void ratios, and concrete grid pavers filled with sand and found that all four systems significantly reduced surface runoff volumes and peak flow rates when compared to standard asphalt. The pervious concrete had the highest surface infiltration rates, followed by the larger void interlocking pavers, the smaller void interlocking pavers, and the concrete grid pavers. Studies by Barrett et al. (2006), Bean et al. (2007), Clausen and Gilbert (2006), Rushton (2001), UNHSC (2012), and Van Seters (2007) also showed porous asphalt, pervious concrete, and permeable interlocking concrete pavers to effectively remove pollutants such as total suspended solids, metals, and nutrients.

Green roofs and green walls

Green roofs can be designed for both new and existing structures and include extensive, semi-intensive, and intensive systems based on the depth of the medium. Green walls can be applied vertically to both interior and exterior wall surfaces and include naturally occurring green walls, facade systems, and modular systems. These are discussed in more detail in Chapter 9.

LID workflow

Using LID, GI, and GSI principles, a typical design might have the following workflow:

1. Map all the natural features of the site, including soils, hydrology, vegetation, forest blocks, wildlife habitats, wetlands, streams, riparian buffers, and topography.
2. Identify sensitive areas to be preserved and/or protected.
3. Avoid development in a 50–100-foot buffer around riparian zones and waterbodies.
4. Plan development around preserved/protected areas.
5. Identify soil types, permeability, and depth of groundwater table.
6. Prioritize soil areas that should not be disturbed.
7. Minimize compaction and disturbance to soils during construction and remediate soils disturbed after construction.
8. Minimize total disturbance to the site by minimizing the size of the construction staging area, limiting areas where heavy equipment is used, and preserving natural topography, water flows, and vegetation.
9. Protect natural water flows by analyzing drainage patterns, optimizing flow patterns, and minimizing grading and clearing.

10 Utilize GSI, reduce impervious surfaces, and disconnect impervious areas to slow, absorb, and diffuse stormwater runoff.

GI strategies can be applied to all gray infrastructure – including roadways, parking surfaces, sidewalks, plazas, buildings, and rooftops – and can also be networked at multiple scales. A green roof on a house can be connected to a green wall which can be connected to a rain garden that sits adjacent to a pervious paved driveway. The rain garden can then tie into a vegetated garden lining the property that connects to a riparian buffer along a nearby stream that runs into a riparian zone along a river that flows through a forest into wetlands and ultimately into a lake. Continuously thinking between building, site, community, and regional scales is crucial in planning and designing efficient green networks.

Bioclimatic and passive design

Bioclimatic design is building or landscape design that bestows the local climatic conditions and natural laws of thermodynamics to provide optimal comfort, reduce consumption, and improve energy efficiency. It utilizes passive design strategies that do not rely on external electrical or mechanical systems for heating, cooling, and lighting. Bioclimatic and passive design consider building orientation, high-performance thermal insulation of the building envelope, window placement, daylighting, sun shading, high-performance glazing, passive solar heating and cooling of water and air, moisture and humidity control, and natural and balanced ventilation.

According to the International Passive House Association (n.d.), for a building to qualify as a passive house, it must meet the following criteria:

1 Space heating demand must not exceed 15 kilowatt hours annually or 10 watts peak demand per square meter of usable living space.
2 Space cooling demand matches heating demand with climate-dependent allowance for dehumidification.
3 Primary energy demand must not exceed 120 kilowatt hours annually for all domestic applications per square meter of usable living space.
4 Airtightness maximum of 0.6 air changes per hour at 50 Pascal pressure
5 Thermal comfort must be met for all living areas year-round with not more than 10 percent of the hours in any given year over 25 degrees Celsius.

Bioclimatic/passive design strategies – including passive solar air and water heating – are discussed in more detail in Chapters 5 and 11.

Universal and inclusive design

Universal design is defined as the "design of products and environments to be usable by all people, to the greatest extent possible, without the need for adaptation or specialized design" (NCSU, 1997). Steinfeld and Maisel (2012) at the Center for Inclusive Design and Environmental Access (IDEA Center) at the University at Buffalo define it as "a design process that enables and empowers a diverse population by improving human performance, health and wellness, and social participation." The IDEA Center also distinguishes universal design from accessible design, the latter referring to the minimum compliance with codes and standards for people with disabilities, including the 1968 Architectural Barriers Act, section 504 of the Rehabilitation Act of 1973, the Fair Housing Act Amendments of 1988, and the Americans with Disabilities Act of 1990 (IDEA, n.d.).

In 1997, a multi-disciplinary group compiled the Principles of Universal Design (NCSU, 1997; IDEA, n.d.), which include:

- Equitable use. The design does not disadvantage or stigmatize any group of users.
- Flexibility in use. The design accommodates a wide range of individual preferences and abilities.
- Simple and intuitive use. Use of the design is easy to understand, regardless of the user's experience, knowledge, language skills, or current concentration level.
- Perceptible information. The design communicates necessary information effectively to the user, regardless of ambient conditions or the user's sensory abilities.
- Tolerance for error. The design minimizes hazards and the adverse consequences of accidental or unintended actions.
- Low physical effort. The design can be used efficiently and comfortably and with a minimum of fatigue.
- Size and space for approach and use. Appropriate size and space is provided for approach, reach, manipulation, and use, regardless of the user's body size, posture, or mobility.

Although universal design embraces inclusivity, some scholars have distinguished inclusive design with a related but independent set of guiding principles. The Inclusive Design Research Centre (IDRC, n.d.) defines inclusive design as design that considers the full range of human diversity with respect to ability, language, culture, gender, age, sexual orientation, ethnicity, and other forms of human difference. They further define three dimensions, including recognizing diversity and uniqueness, inclusive processes and tools, and awareness of context and the broader beneficial impact. Inclusive design recognizes that individuals are complex and multi-faceted with unique needs that arise from numerous factors and seeks to create systems that can adapt or morph to such needs. In addition to accessibility and removing physical barriers, inclusive design considers ethics, equity, perspective, and behaviors and removing psychological barriers.

Environmental justice

As mentioned earlier, although the three-sector model of sustainability strives to find overlaps between the physical environment, society, and economy, social equity may sometimes be an afterthought that is often touted with empty platitudes and performative action statements. I also acknowledge that I may have had certain privileges that others did not have in my upbringing, education, or within my community. However, I am doing my best to understand the difficulties marginalized communities have and continue to face and am trying to learn how to effectively integrate environmental justice in the planning and design process. This is an oath that all planners, designers, and educators need to uphold in the twenty-first century to fully engage in sustainability.

In recent times, there has been a growing movement that seeks to mitigate unjust planning and development that adversely affects marginalized communities. Although the 17 Sustainable Development Goals have adopted more objectives surrounding social and racial equity, it is important to consider environmental justice not just as an addendum to sustainability but more as a comprehensive discipline with distinct areas of focus that need to be addressed in policy, planning, and development. The U.S. Environmental Protection Agency (EPA, n.d.b) defines environmental justice as

> the fair treatment and meaningful involvement of all people regardless of race, color, national origin, or income, with respect to the development, implementation, and enforcement of environmental laws, regulations, and policies. The goal can be achieved when everyone has the same degree of protection from environmental and health hazards and has equal access to the decision-making process to have a healthy environment to live, learn, and work.

The initiation of changes by lower-income citizens, marginalized groups, and people of color is key to transforming the top-down, reactionary planning process of the past.

If funding was made equally available to all neighborhoods regardless of socio-economic status, it is important to consider the financial impacts of projects on the community members. For example, GI networks have been shown to increase land values to adjacent properties, which could potentially price people out of their own neighborhoods and/or cause spatial-social restructuring and/or gentrification. Improvement projects in communities, therefore, are a delicate balance that requires transparency, community input, cost–benefit analyses, property assessment projections, anti-displacement policies, and/or subsidies to ensure that neighborhoods can become healthier, safer, and more livable without displacing people from their homes and/or businesses. The Urban Displacement Project established a framework of "three Ps of anti-displacement," including protection of tenants' rights, production of affordable housing, and preservation of existing affordable housing. Tenant protections include rent stabilization, safeguarding tenants from arbitrary evictions, providing tenants' rights resources, and guaranteeing tenants' right to counsel. Production of affordable housing can be accomplished through inclusionary zoning, land-banking parcels for future affordable housing, and through preference or right of return policies. Existing affordable housing can be preserved by establishing community land trusts, nominating buildings for acquisition by non-profits, policies that give right of refusal to tenants facing displacement, and policies that give tenants the ability to purchase or assign purchase rights to a non-profit, enabling cooperatives or associations that can provide ongoing ownership (Cash & Zuc, 2021).

In 1991, the National People of Color Environmental Leadership Summit gathered to establish principles of environmental justice. In their preamble, their mission is

> to build a national and international movement of all peoples of color to fight the destruction and taking of our lands and communities and to re-establish our spiritual interdependence to the sacredness of Mother Earth; to respect and celebrate each of our cultures, languages, and beliefs about the natural world and our roles in healing ourselves; to ensure environmental justice; to promote economic alternatives which would contribute to the development of environmentally safe livelihoods; and to secure our political, economic and cultural liberation that has been denied for over 500 years of colonization and oppression, resulting in the poisoning of our communities and land and the genocide of our peoples.
>
> *(EJN, 1996)*

The 17 principles of environmental justice that were adopted are listed next.

- affirms the sacredness of Mother Earth, ecological unity and the interdependence of all species, and the right to be free from ecological destruction.
- demands that public policy be based on mutual respect and justice for all peoples, free from any form of discrimination or bias.
- mandates the right to ethical, balanced, and responsible uses of land and renewable resources in the interest of a sustainable planet for humans and other living things.
- calls for universal protection from nuclear testing, extraction, production, and disposal of toxic/hazardous wastes and poisons and nuclear testing that threaten the fundamental right to clean air, land, water, and food.
- affirms the fundamental right to political, economic, cultural, and environmental self-determination of all peoples.
- demands the cessation of the production of all toxins, hazardous wastes, and radioactive materials, and that all past and current producers be held strictly accountable to the people for detoxification and containment at the point of production.
- demands the right to participate as equal partners at every level of decision-making, including needs assessment, planning, implementation, enforcement, and evaluation.
- affirms the right of all workers to a safe and healthy work environment without being forced to choose between an unsafe livelihood and unemployment. It also affirms the right of those who work at home to be free from environmental hazards.
- protects the right of victims of environmental injustice to receive full compensation and reparations for damages and quality health care.
- considers governmental acts of environmental injustice a violation of international law, the Universal Declaration on Human Rights, and the United Nations Convention on Genocide.
- must recognize a special legal and natural relationship of Native Peoples to the U.S. government through treaties, agreements, compacts, and covenants affirming sovereignty and self-determination.
- affirms the need for urban and rural ecological policies to clean up and rebuild our cities and rural areas in balance with nature, honoring the cultural integrity of all our communities, and providing fair access for all to the full range of resources.
- calls for the strict enforcement of principles of informed consent and a halt to the testing of experimental reproductive and medical procedures and vaccinations on people of color.
- opposes the destructive operations of multi-national corporations.
- opposes military occupation, repression, and exploitation of lands, peoples, and cultures, and other life forms.
- calls for the education of present and future generations which emphasizes social and environmental issues based on our experience and an appreciation of our diverse cultural perspectives.
- requires that we, as individuals, make personal and consumer choices to consume as little of Mother Earth's resources and to produce as little waste as possible; and make the conscious decision to challenge and reprioritize our lifestyles to ensure the health of the natural world for present and future generations.

Now that you have been introduced to these concepts, do you think you can design something that is sustainable, net-zero energy, resilient, does not disturb ecological processes or natural elements, reconnects fragmented communities, provides equity to marginalized groups, fosters diversity and inclusion, stimulates the economy, and makes the world a healthier place? Good luck! Hopefully, the succeeding chapters will provide inspiration and strategies to achieve these goals.

References

Barret, M. E., Kearfott, P., & Malina, J. F. (2006). Stormwater quality benefits of a porous friction course and its effect on pollutant removal by roadside shoulders. *Water Environment Research, 78*(11), 2177–2185. https://doi.org/10.2175/106143005X82217

Bean, E., Hunt, W., & Bidelspach, D. (2007). A field survey of permeable pavement surface infiltration rates. *ASCE Journal of Irrigation and Drainage Engineering, 133*(3), 249–255.

Benedict, M. A., & McMahon, E. T. (2006). *Green infrastructure: Linking landscapes and communities* (p. 71). Island Press.

Biomimicry Institute. (n.d.). *What is biomimicry*. https://biomimicry.org/what-is-biomimicry/?gclid=Cj0KCQjw0IGnBhDUARIsAMwFDLljBDcBg5O8DoRbDqqwHMhJMORnzyS-EGx8T0bjTrWH3IxBqS%E2%80%A6

Carnow, A. (2022, July 8). *Resilience and sustainability, the definitions, difference, and applicability in GIS*. ESRI. www.esri.com/en-us/industries/blog/articles/resilience-and-sustainability-the-definitions/

Cash, A., & Zuc, M. (2021). *Investment without displacement: From slogan to strategy*. Urban Displacement Project. www.urbandisplacement.org/blog/investment-without-displacement-from-slogan-to-strategy/

Center for Inclusive Design and Environmental Access (IDEA). (n.d.). *What is universal design*? University at Buffalo, School of Architecture and Planning, Center for Inclusive Design and Environmental Access (IDEA). https://idea.ap.buffalo.edu/about/universal-design/

Clausen, J. C., & Gilbert, J. K. (2006). Stormwater runoff quality and quantity from asphalt, paver, and crushed stone driveways in Connecticut. *Water Research, 40*, 826–832.

Climate Central. (2023). *The power of urban trees*. https://assets.ctfassets.net/cxgxgstp8r5d/32ijJSeRXMnSkG2OhjavzF/0e4e0fd63f67f00328e60f13280d55a6/FINAL_Power_of_Urban_Trees_2023__EN_.pdf

Collins, K. A., Hunt, W. F., & Hathaway, J. M. (2007, September 4). *Hydrologic and water quality comparison of four types of permeable pavement and standard asphalt in Eastern North Carolina*. North Carolina State University.

Courtney, E., & Zencey, E. (2012). *Greening vermont: The search for a sustainable state* (p. 17). Thistle Hill Publications.

Drake University. (n.d.). *What is sustainability and resilience*? Drake University College of Arts and Sciences, Environmental Science and Sustainability. www.drake.edu/enss/majorsminors/sandr/#:~:text=What

DuPisani, J. A. (2007). Sustainable development – historical roots of the concept. *Environmental Sciences, 3*(2), 83–96. http://doi.org/10.1080/15693430600688831

Ehrenfeld, J. (2006). Interview in big ideas for a little planet. *Architectural Record, 194*(8), 61.

Environmental Justice Network (EJN). (1996). *The principles of environmental justice*. www.ejnet.org/ej/

Garcia, E. J., & Vale, B. (2017). *Unraveling sustainability and resiliency in the built environment* (p. 203). Routledge.

Green Infrastructure Ontario Coalition. (n.d.). *What is green infrastructure*? https://greeninfrastructureontario.org/what-is-green-infrastructure/

Inclusive Design Research Centre (IDRC). (n.d.). *What is inclusive design?* https://idrc.ocadu.ca/about/philosophy/

International Institute for Sustainable Development (IISD). (n.d.). *Sustainable development timeline*. www.iisd.org/publications/guide/sustainable-development-timeline-2012

International Passive House Association. (n.d.). *Passive house certification criteria*. https://passivehouse-international.org/index.php?page_id=150

Lallanilla, M. (2020, October 30). *History of the green movement*. TreeHugger. www.treehugger.com/what-is-the-green-movement-1708810

Marchese, D., Reynolds, E., Bates, M. E., Morgan, H., Clark, S. S., & Linkov, I. (2018). Resilience and sustainability: Similarities and differences in environmental management applications. *Science of the Total Environment, 613–614*, 1275–1283. https://doi.org/10.1016/j.scitotenv.2017.09.086

McHarg, I. (1969). *Design with nature*. John Wiley & Sons.
New Urbanism. (n.d.). *Principles of new urbanism*. www.newurbanism.org/newurbanism/principles.html
North Carolina State University (NCSU). (1997). *The principles of Universal Design*. Center for Universal Design. https://design.ncsu.edu/research/center-for-universal-design/
Oregon Department of Transportation Highway Division, Engineering and Asset Management Unit Geo-Environmental Section. (2014). *Hydraulics design manual*, Appendix 7-F, p. 3. www.oregon.gov/ODOT/GeoEnvironmental/Docs_Hydraulics_Manual/Hydraulics-07-F.pdf
Resilient Design Institute (RDI). (n.d.a). *The resilient design principles*. www.resilientdesign.org/the-resilient-design-principles/
Resilient Design Institute (RDI). (n.d.b). *Resilient design strategies*. www.resilientdesign.org/resilient-design-strategies/
Rushton, B. (2001). Low impact parking lot design reduces runoff and pollutant loads. *Journal of Water Resources Planning and Management*, *127*(3), 172–179.
Shibley, R. (1991). *Sustainable design* (pp. 3–4). Planning, State University of New York at Buffalo, School of Architecture and Planning.
Sierra Club. (n.d.). *John Muir: A brief biography*. Sierra Club. https://vault.sierraclub.org/john_muir_exhibit/life/muir_biography.aspx
Smart Growth America (SGA). (n.d.). *What is smart growth*? https://smartgrowthamerica.org/what-is-smart-growth/#:~:text=Smart%20growth%20means%20protecting%20our,to%20live%2C%20helping%20everyone%20flourish
Smithsonian. (n.d.). *About George Perkins Marsh*. www.si.edu/spotlight/marsh-collection/about-george-perkins-marsh#:~:text=During
Steiner, F. (2011). *Design for a vulnerable planet* (pp. 32–34). University of Texas Press.
Steinfeld, E., & Maisel, J. (2012). *What is universal design*? University at Buffalo, School of Architecture and Planning, Center for Inclusive Design and Environmental Access (IDEA). https://idea.ap.buffalo.edu/about/universal-design/
Sustainable Development Commission (SDC). (n.d.). *History of SD*. www.sd-commission.org.uk/pages/history_sd.html#:~:text=The
Tzoulas, K., Korpela, K., Venn, S., Yli-Pelkonen, V., Kaźmierczak, A., Niemela, J., & James, P. (2007). Promoting ecosystem and human health in urban areas using Green Infrastructure: A literature review. *Landscape and Urban Planning*, *81*(3), 167–178. https://doi.org/10.1016/j.landurbplan.2007.02.001.
United Nations, Department of Economic and Social Affairs (UN) (2015). *The 17 goals*. United Nations. https://sdgs.un.org/goals
University of New Hampshire Stormwater Center (UNHSC). (2012). *University of New Hampshire Stormwater Center 2012 biennial report. https://scholars.unh.edu/stormwater/37/*
U.S. Environmental Protection Agency (EPA). (1999). *Stormwater technology fact sheet: Bioretention*. U.S. EPA Office of Water. EPA-832-F-99-012. https://nepis.epa.gov/Exe/ZyPDF.cgi/200044BE.PDF?Dockey=200044BE.PDF
U.S. Environmental Protection Agency (EPA). (2021). *Green streets handbook*. www.epa.gov/sites/default/files/2021-04/documents/green_streets_design_manual_feb_2021_web_res_small_508.pdf
U.S. Environmental Protection Agency (EPA). (n.d.a). *What is green infrastructure*? U.S. EPA. www.epa.gov/green-infrastructure/what-green-infrastructure
U.S. Environmental Protection Agency (EPA). (n.d.b). *Environmental justice*. www.epa.gov/environmentaljustice
Van der Ryn, S., & Cowan, S. (2007). *Ecological design* (10th anniversary ed., pp. 33–34, 72–73). Island Press.
Van Seters, T. (2007). *Performance evaluation of permeable pavement and a bioretention swale*. Seneca College, King City, Ontario (Interim Report #3). Toronto and Region Conservation Authority.
Vermont Agency of Natural Resources (ANR). (n.d.). *Low impact development*. https://dec.vermont.gov/sites/dec/files/wsm/erp/docs/sw_gi_1.0_LID_series.pdf
Weyler, R. (2018, January 5). *A brief history of environmentalism*. Greenpeace. www.greenpeace.org/international/story/11658/a-brief-history-of-environmentalism/
Williams, D. E. (2007). *Sustainable design: Ecology, architecture, and planning* (p. 17). John Wiley & Sons.

3

WHY REGREEN THE BUILT ENVIRONMENT?

Our planet is currently facing some of the most difficult environmental challenges in human history. The United Nations Framework Convention on Climate Change (UNFCCC, 2022a) defines the triple planetary crisis as three primary interlinked issues humanity is facing: climate change, pollution, and biodiversity loss. The non-profit organization Earth.org also recently published the 15 biggest environmental problems of 2023, including global warming from fossil fuels, poor governance, food waste, biodiversity loss, plastic pollution, deforestation, air pollution, melting ice caps and sea-level rise, ocean acidification, agricultural mismanagement, food and water insecurity, textile waste, overfishing, cobalt mining, and soil degradation (Robinson, 2023). As introduced in Chapter 2, the environmental crisis has also grown beyond physical problems and is now, more than ever, linked to economic stability and social equity. Indeed, implementing reparations that mitigate the impacts of human rights violations and discriminatory practices is imperative to fulfill the goals of sustainability by fostering health, wellness, justice, and equity. Given recent events, it may also be wise to add global pandemics and artificial intelligence (AI) to the list.

Although there will always be skeptics, political debate, and subjective spins on the climate change topic, the visual evidence is difficult to dispute. Sea levels are rising, polar ice is melting, and extreme weather events are intensifying and becoming more frequent. As pointed out by Time Inc. (2007, p. 64) in a special report on global warming, "If droughts and wildfires, floods and crop failures, collapsing climate-sensitive species, and images of drowning polar bears didn't quiet most of the global-warming doubters, then the destruction to New Orleans from Hurricane Katrina in 2005 did." Since this was written, similar destruction was also observed from Hurricane Irene in 2011, Sandy in 2012, and Harvey, Irma, and Maria in 2017.

However, despite the physical evidence, society remains divided on the issue. It seems there are five general opinions circulating in society today: those who believe in climate change and are actively engaged in reducing its impact; those who believe in climate change but are not being proactive due to disinterest, lack of motivation, financial constraints, or lack of incentive; those who don't believe climate change is as dire as the media portrays and believe this is just the earth's "natural course"; those who feel it is in the fate of the universe and/or in the hands of God; and those who flat out believe climate change is a hoax and actually advocate to continue

DOI: 10.4324/9781032668666-3

engaging activities that add more carbon to the environment. In recent times, climate change has also become politicized, which further polarizes proactive believers and adamant deniers. This polarization is dangerous since it creates an all-or-nothing, uncompromising culture with people on one side viewing the other side as self-righteous or sententious, and people on the other side viewing the other side as ignorant or uninformed. In the end, it is our planet that ultimately loses. Although it is a lofty goal, it is my hope that people from all corners of this debate will read this book and realize that our planet's health should not be a partisan issue that divides us, but rather a complex challenge that unites us. I often reference the movie *Jaws* when discussing climate change or facing any adversary, for that matter. Ultimately, the movie is really not about the shark, but rather about three people from very different walks of life forced to work together to address – as Blake Snyder describes in his book, *Save the Cat!* – the "monster in the house." In the twenty-first century, climate change is the monster and we all must come together to defeat it. Environmental issues in this chapter will be discussed with respect to the physical environment, the social environment, and the economic environment.

The physical environment

For the purposes of this book, the physical environment includes all the natural, ecological, biological, geological, and atmospheric factors that surround us, including air, water, soil, climate, vegetation, and ecosystems.

Carbon dioxide and global climate change

Land degradation, deforestation, the burning of fossil fuels, and various other anthropogenic activities have caused a surge of greenhouse gases in the atmosphere that has changed air quality and the global climate of our planet. Globally, carbon dioxide from fossil fuels and industrial processes accounts for most of the greenhouse gas emissions at 65 percent, followed by methane at 16 percent, carbon dioxide from deforestation, clearing for agriculture, and degradation of soils at 11 percent, nitrous oxide at 6 percent, and fluorinated gases at 2 percent (IPCC, 2014). According to the National Oceanic and Atmospheric Administration's (NOAA, 2024a) Global Monitoring Lab, carbon dioxide concentrations are currently at 423 parts per million. The International Panel on Climate Change (IPCC) Average Sixth Assessment Report showed annual GHG emissions during 2010–2019 were higher than in any previous decade on record (IPCC, 2023). For comparison, in 1900, roughly 2 billion metric tons of carbon dioxide were released into the atmosphere; this increased to 9 billion metric tons in 1960 and 35 billion metric tons in 2014; over half of the current total has been emitted since 1980 (Lindsey, 2022, June 23). China and the United States are currently the two biggest culprits, accounting for 30 percent and 14 percent of total global carbon dioxide emissions, respectively (Boden et al., 2017). The single largest source of emissions (25 percent of the total) comes from the burning of coal, natural gas, and oil for electricity and heat, followed by industrial processes at 24 percent, agriculture/deforestation at 24 percent, transportation at 14 percent, fuel extraction, processing, refinement, and transport at 10 percent, and buildings at 6 percent (IPCC, 2014).

Although the travel and economic restrictions of the COVID-19 pandemic resulted in a 5.4 percent decrease in carbon emissions, the amount of carbon dioxide in the atmosphere continued to grow at about the same rate as in preceding years, showing that temporary, short-term reductions are not enough to curb overall emissions (NASA, 2021). NOAA's temperature data show the ten warmest years on record have occurred since 2010, with 2023 having the highest globally averaged temperature over land and ocean surfaces since record keeping began in 1880, which is 2.12 degrees Fahrenheit (1.18 degrees Celsius) warmer than the twentieth-century average (NOAA, 2024b). In the near term (2021–2040), global average temperatures are more likely than not to reach 2.7 degrees Fahrenheit (1.5 degrees Celsius) above pre-industrial levels under the very low greenhouse gas emission scenario and likely or very likely to exceed 2.7 degrees Fahrenheit (1.5 degrees Celsius) above pre-industrial levels under higher emissions scenarios (IPCC, 2023). Although a two-degree Fahrenheit increase may not seem significant, it leads to a substantial amount of accumulating heat that causes temperature extremes, increased rainfall, decreased ice at the poles, and ecosystem disruption (Lindsey & Dahlman, 2024). Research conducted at Princeton University in 2013 predicts that even if we completely stopped carbon dioxide emissions today, the levels of carbon dioxide already in the earth's atmosphere could continue to warm the planet for hundreds of years (Kelly, 2013).

At the 2015 United Nations Climate Change Conference (COP21), 196 parties adopted the Paris Agreement, which is a legally binding international treaty on climate change with the goal of holding the increase in global average temperatures to below 2 degrees Celsius above pre-industrial levels and to pursue efforts to limit temperature increases to 1.5 degrees Celsius above pre-industrial levels (UNFCCC, n.d.). Since 2015, world leaders have recommended making the global warming target more stringent, limiting it to 1.5 degrees Celsius by the end of this century. To achieve this, greenhouse gas emissions must peak before 2025 and decline 43 percent by 2030 (UNFCCC, n.d.). Since 2020, members have been required to update their national climate plans – also known as nationally determined contributions (NDCs) – every five years. Since the deadline for the last update in October 2021, 39 parties have communicated new or updated plans, 74 percent of which have strengthened their commitments (UNFCCC, 2022c). In terms of greenhouse gas reduction commitments, the United Kingdom has committed to reducing economy-wide greenhouse gas emissions by at least 68 percent by 2030 compared to 1990 levels, the European Union has committed to a 55 percent reduction by 2020 compared to 1990 levels, Australia committed to a 43 percent reduction from 2005 levels by 2030 and net-zero emissions by 2050, and Brazil committed to reducing emissions 37 percent by 2025 and by 50 percent by 2050 compared to 2005 levels (UK SOS, 2022; European Commission, n.d.; Commonwealth of Australia, 2022; Federative Republic of Brazil, 2022). Despite the unfortunate sequence of events that led to the U.S. withdrawal from the Paris Agreement in 2017, Uncle Sam rejoined the treaty in 2021 and has set a goal to achieve a 50–52 percent reduction from 2005 levels in economy-wide net greenhouse gas pollution by 2030, including a goal to reach 100 percent carbon-pollution-free electricity by 2035, investing in energy retrofits and technology, reducing carbon pollution in the transportation sector by investing in renewable fuels, renewable energy, and auto-alternative transportation infrastructure, enhancing sustainable agriculture and forestry practices, and reducing non-carbon dioxide greenhouse gases (The White House, 2021). Ironically, China – the world's largest emitter of greenhouse gases – has yet to submit an update as of the writing of this book.

Although some skeptics claim greenhouse gas reductions will hurt the economy and increase the cost of energy, the literature seems to indicate otherwise. Studies by the World Resources Institute examining electric generation, electricity consumption, passenger vehicles, natural gas, and hydrofluorocarbons (HFCs) indicate that reducing emissions benefits the economy by saving consumers and businesses money, reducing energy bills, improving public health, and creating jobs (Igusky et al., 2014). The Global Commission on the Economy and Climate found that bold climate action could yield a direct economic gain of U.S. $26 trillion through to 2030 (The New Climate Economy, 2018). A study conducted by the Georgia Institute of Technology found that reducing greenhouse gas emissions from power plants can decrease electricity bills, increase the Gross Domestic Product, and significantly reduce sulfur dioxide, nitrogen oxides, and mercury emissions (Brown et al., 2021). As stated by Marilyn Brown, professor of Sustainable Systems in the School of Public Policy at the Georgia Institute of Technology,

> As energy is used more efficiently, non-competitive power plants can be retired, construction of new coal plants can be deferred, and transmission and distribution infrastructure investments can be delayed, all of which would lower rates and therefore lower the energy bills of all consumers.
>
> *(Phillips, 2015)*

Polar ice and sea levels

The warming of the planet is causing the polar land and sea ice in the Arctic Ocean and Greenland to melt and is subsequently causing sea levels to rise. Data from NASA's GRACE satellites (2020) show Antarctica is losing ice mass at an average rate of about 150 billion tons per year and Greenland is losing about 270 billion tons per year, accounting for one-third of the average global sea-level rise since 1993. According to Stager (2011, p. 26), we have added so much carbon that we have pushed the next ice age from naturally happening 50,000 years from now to 130,000 years from now. Until this natural cooling cycle happens, our planet will continue to warm, polar ice will continue to melt, and sea levels will continue to rise. The warmer sea temperatures will provide more thermal energy for tropical depressions that will ultimately produce stronger and more frequent hurricanes, high winds, torrential rains, storm surges, and larger ocean waves. Scientists have also found correlations between melting ice in the polar regions and the onset of earthquakes and volcanoes:

> The reason is that one cubic meter of ice weighs just over a ton, and glaciers can be hundreds of meters thick. When they melt and the water runs off, it is literally a weight off Earth's crust. The crust and mantle therefore bounce back, immediately as well as over thousands of years. That "isostatic rebound," according to studies of prehistoric and recent earthquakes and volcanoes, can make the planet's seismic plates slip catastrophically, and cause magma chambers that feed volcanoes to act like bottles of shaken seltzer.
>
> *(Begley, 2006)*

There is much debate on how fast and how high sea levels will rise in the future. Much of this uncertainty is due to different carbon emission scenarios, different melting scenarios, and how the ice sheets in Greenland and West and East Antarctica will respond. The International Panel on Climate Change estimates a 0.95–3.61-feet (0.29–1.1-meter) global sea-level rise by the end of this century and a report published by NOAA indicates a one-foot (.3 meter) increase in global mean average sea levels by 2100 compared to 2000 given a reduction of greenhouse gases and a 6.6-feet (2-meter) increase by 2100 compared to 2000 given a high greenhouse gas

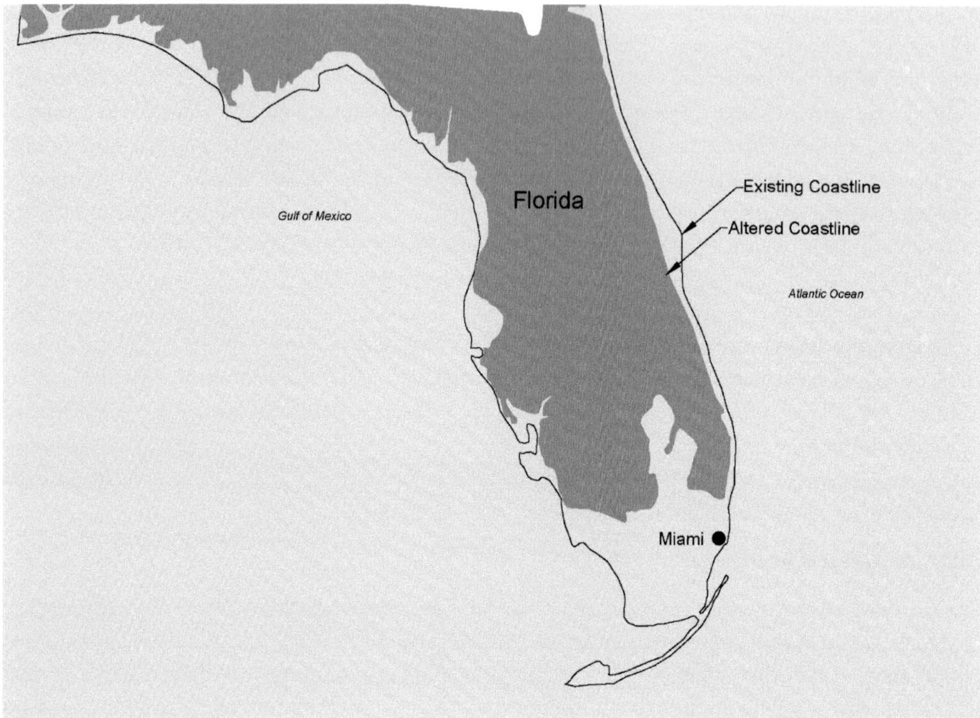

FIGURE 3.1 Approximate map of Florida given oceans rise 10 feet above global average sea levels.

Source: Map created by author; data derived from Climate Central (CC). (n.d.). *Surging seas: Risk zone map*. https://ss2.climatecentral.org/#8/27.665/-81.516?show=satellite&projections=0-K14_RCP85-SLR&level=5&unit=feet&pois=hide

emission scenario (IPCC, 2018; Lindsey, 2022, April 19). Some scientists, such as those from the University of Kansas, have also attempted to quantify sea-level rise in relation to land loss:

> The first 3 feet of sea level rise will cover nearly 400,000 square miles of coastal land and could displace more than 100 million people. . . Northwest Europe will lose about 13,000 square miles and uproot 12 million citizens, and southeastern United States could lose 24,000 square miles and displace more than 2.6 million people.
>
> *(Stager, 2011, pp. 130–131)*

Although there is still much to learn, one thing is certain: water is rising and coastlines and regions lying at or below sea level will be drastically altered in the distant future.

Natural resource depletion

According to the National Resources Inventory, about 44 million acres of land were newly developed in the United States between 1982 and 2017, bringing the total to about 116 million acres; that represents a 61 percent increase. Thus, 38 percent of developed land in the 48 conterminous states, Hawaii, Puerto Rico, and the U.S. Virgin Islands, was developed during the past 35 years (DOA, 2017). As Roxanne Warren (1998, p. 18) points out in her book *The Urban Oasis*,

> As woods and grasslands are sliced up into housing developments, malls, industrial parks, and highways, the whole complex fabric of the natural world has been gradually unravelling, leaving only isolated islands of nature preserves, and rendering untold numbers of plant and animal species vulnerable to extinction.

Soils, forests, wetlands, and natural habitats are being degraded and fragmented by urban sprawl, clear cutting, overgrazing, and agricultural mismanagement. As the surface of the earth becomes increasingly impermeable through the construction of buildings, roads, parking surfaces, sidewalks, and other forms of "gray" infrastructure, the natural hydrological cycle is interrupted by the increase of water runoff, decrease in infiltration and groundwater recharge, and decrease in evapotranspiration from vegetation. According to Earth Pledge (2005, p. 16), asphalt and concrete rooftops, roads, and parking lots cover up to 70 percent of land area in dense cities like New York, while open space in sprawling cities such as Phoenix, Arizona, is lost to development at a rate of 1.2 acres per hour. As a result, streams and water bodies become polluted and are subject to extreme flooding during wet cycles and severe drought during dry cycles. Conventional stormwater control measures – also known as curb and gutter systems – allow excess nutrients such as phosphorous, nitrogen, and organic particulate matter and pollutants such as heavy metals, hydrocarbons, pesticides, pathogens, and other waste to runoff over impervious surfaces and agricultural lands directly into streams, rivers, lakes, and oceans. This impacts water temperature, turbidity, pH, dissolved oxygen content, and the growth of phytoplankton, algae, and bacteria. The Clean Water Act of 1972 requires states to set designated uses for their water bodies, such as water supply, recreation, habitat, navigation, industrial, or agricultural, to name a few. Each designation has a set of specific standards and total maximum daily loads in determining whether the water quality is adequate for the specific use. If the standards are not met, the water source is considered impaired. According to the U.S. Environmental Protection Agency, there were more than 70,000 impaired water bodies identified and tracked as of 2014 in the United States (USEPA, 2022).

Fossil fuels

Since the Industrial Revolution, fossil fuels have been utilized to provide energy to the world. According to the International Energy Agency (IEA, 2021a), over 80 percent of the world's energy is produced from non-renewable sources, including 30.9 percent from oil, 26.8 percent from coal, and 23.2 percent from natural gas. As a comparison, biofuels and waste produce 9.4 percent, nuclear 5 percent, hydro 2.5 percent, and solar, wind, geothermal, tidal, and heat recovery 2.2 percent. As a result of the COVID-19 pandemic, oil demand was nine million barrels a day lower in 2020 than in 2019. However, as of 2023, oil demand reached 101.9 million barrels a day and is projected to reach 104.2 million barrels a day by 2026 (IEA, 2021b). The U.S. Energy Information Administration (EIA, 2023a) projects that world energy consumption will increase by 50 percent by 2050 compared with 2020. Although the U.S. EIA predicts that renewables will be the primary source of electrical generation in 2050, oil and natural gas production will continue to grow to support developing countries. The paramount question is, how long will the reserves of non-renewable energy sources last? This has been a topic with much debate and contention over the years.

In 1956, geologist M. King Hubbert devised a "peak oil" theory that described the point in time when the maximum rate of crude oil production was reached. Hubbert predicted the United States would reach peak oil in 1965 and the world would reach peak oil in 2000. His prediction was close for the United States when oil production peaked in the early 1970s and then declined for the subsequent 35 years. However, advances in technology – including hydraulic fracking of rock shale, offshore drilling, seismic mapping, drones, and hydrocarbon sniffing – led to an increase in oil production around 2010. While Hubbert primarily based his calculations on known (or "proven") reserves, modern technology has also improved the accuracy in predicting estimated reserves, which also introduces another variable in the equation. Based on the available studies, peak oil has been predicted to occur anywhere from 2019 to after 2050 (Fickling, 2022). As of 2021, the Organization for Petroleum Exporting Countries (OPEC, 2022) estimates that OPEC countries have 1,241.82 billion barrels of proven oil reserves left and non-OPEC countries have 3.25 billion barrels of proven oil reserves. Given these estimates along with the current rates of consumption mentioned earlier, that leaves about 40–45 more years until proven reserves are depleted. Supporters of big oil will read this as sensationalism and argue there are many more non-proven reserves; supporters of the green movement will tell you to stop driving and advocate for 100 percent renewable energy production; moderates will try and seek a balance of renewables and non-renewables. It seems plausible that all parties (even big oil) will concede that oil reserves will run out sometime in the twenty-first century. Oil is basically facing the same dilemma as film did in the last part of the twentieth century. Those who grew up in Rochester, New York, witnessed this first-hand. Eastman Kodak had around 145,000 employees and $19 billion in annual sales at its peak in 1990 when analog film was king. Kodak knew film was finite and did get into digital media, but their business model was simply over-vested in the old media. Today, they have about 8,000 employees and $2 billion in annual sales (Hardy, 2015). Even the executives in big oil have acknowledged oil is finite. In 2001, oil giant British Petroleum (BP) rebranded its name as "Beyond Petroleum," recognizing the need to reduce carbon emissions and invest in more renewable forms of energy. Although many might agree that BP is far from perfect (especially given the 2006 Alaskan Pipeline leak and the 2010 explosion/leak on its Deepwater Horizon vessel in the Gulf of Mexico), at least they are making some attempt – whether it is greenwashing or not – to rebrand their business model.

Natural gas is the second-most abundant and the "cleanest" of the non-renewables. The U.S. EIA estimates there are 445 trillion cubic feet (Tcf) of proved and 2,528 Tcf of unproved reserves of natural gas in the United States and 7,257 trillion cubic feet of proved dry natural gas reserves worldwide. At current U.S. consumption rates of 34.52 Tcf/year, our domestic supply of natural gas will last about 86 years. Coal is the largest emitter of carbon dioxide and most abundant of the non-renewables, with recoverable reserves in the United States estimated at 251 billion short tons and recoverable reserves worldwide estimated at 1.16 trillion short tons. Using the U.S. consumption rate of 0.577 billion short tons in 2021, recoverable coal reserves in the United States would last about 435 years (EIA, 2021a, 2022, 2023c).

Even if the earth had an infinite supply of fossil fuels, history has proven that their perils have outnumbered their benefits and that new, clean, renewable sources of energy would still have to be tapped. Mining accidents, air pollution, carbon dioxide, wars, trade embargos, and oil spills have led to unnecessary deaths and degradation of peace and well-being on our planet. One thing experts are sure of is that the supply of fossil fuels will drastically decline in the

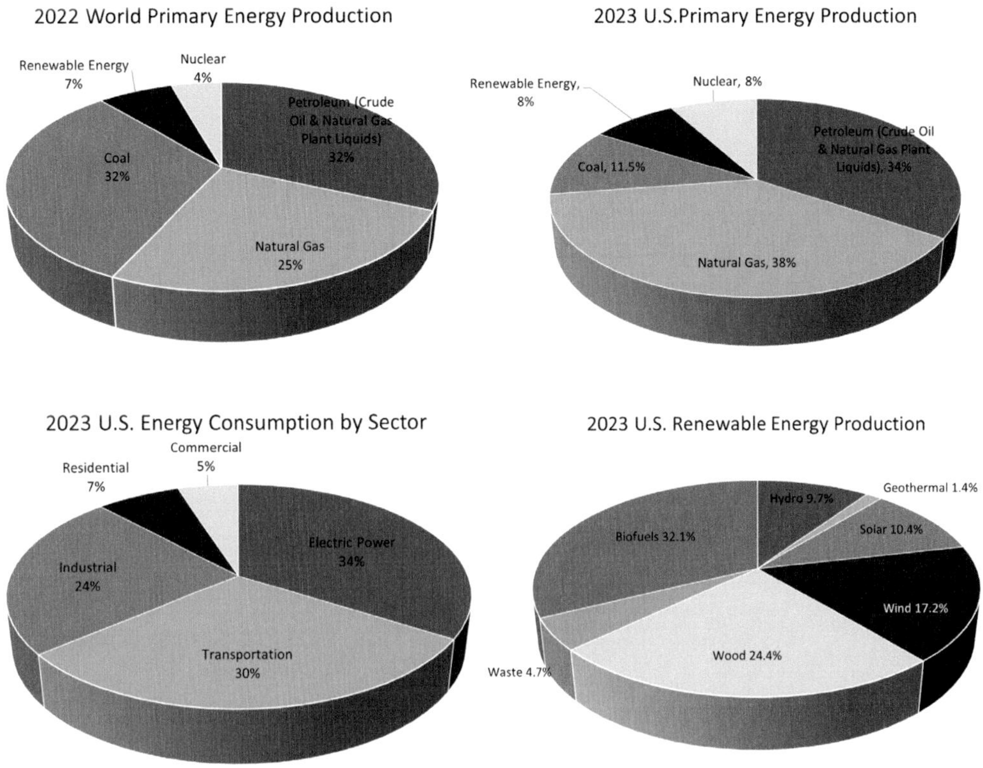

FIGURE 3.2 Top left: 2022 world energy production; top right: 2023 U.S. primary energy production; bottom left: 2023 U.S. energy consumption by sector; bottom right: 2023 U.S. renewable energy production.

Sources: U.S. Energy Information Administration (2023). U.S. total energy. https://www.eia.gov/totalenergy/data/monthly/pdf/sec1_5.pdf; By sector – https://www.eia.gov/energyexplained/us-energy-facts/data-and-statistics.php; International - https://www.eia.gov/international/data/world

twenty-first century and that a radical shift in our fuel and energy economy must be achieved – if we don't act now, we will most certainly be forced to act in the near future.

Decline of ecosystems loss of biodiversity

The scientific community estimates that the planet has undergone five major mass extinctions within the past 450 million years, and given current rates of human consumption and species decline, we may be in the midst of a sixth mass extinction (Lai, 2021). The 2005 Millennium Ecosystem Assessment (MA, 2005) determined that humans have altered ecosystems more intensely in the past 50 years compared to any period of history. The MA estimated that approximately 40 percent of the world's coral reefs have been lost or degraded, water withdrawals from rivers and lakes have doubled since 1960, and the global species extinction rate has increased as much as 1,000 times over the typical rate seen across the earth's history. The MA concluded that human activity is putting such a strain on the natural functions of the earth that the ability of the planet's ecosystems to sustain future generations can no longer be taken for granted.

The World Wildlife Federation, the United Nations Environment Program, and the Global Footprint Network also showed that humans have been living beyond their means since 1987. More specifically, at least 500 species native to the United States have become extinct since the early days of colonization, and at least 484 animal species and 654 plant species worldwide have become extinct since 1600 (WI, 2006, p. 92). A report by the Intergovernmental Science-Policy Platform on Biodiversity and Ecosystem Services (IPBES, 2021) showed that approximately one million animal and plant species are currently under the threat of extinction and 75 percent of the land-based environment and 66 percent of the marine environment have been significantly altered by human actions.

As discussed by Hellmund and Smith (2006, p. 55), ecological fragmentation on the landscape can be the result of six impacts: a decrease in the size and number of habitat patches; an increase in the distance between habitat patches; a change in the distribution of prey animals; an increase in productivity in habitat patches; a decrease of the quality of habitat patches; and the increase in the importance of edge effects. The authors go on to explain that these issues can be multiple and confounding, ultimately impacting the landscape on both spatial and temporal scales.

After reading the plethora of grim statistics above that can fill an entire book, one may begin to question if there is anything that can be done at all to save our planet. Why should professions – particularly designers, architects, environmental scientists, engineers, or land use planners – even care enough to change their methodologies to better adapt to climate change? As Curt Stager (2011, p. 12) eloquently states:

> The reason is simple. Although humans will survive as a species, we are faced today with the responsibility of determining the climatic future that our descendants will live in. It may well be a struggle to hold our carbon pollution to a minimum, but failing to take the heroic path and control our collective behavior is likely to drag us and our descendants into a realm of extreme warming, sea level rise, and ocean acidification the likes of which haven't been seen on Earth for millions of years.

Think about your children, nieces, and nephews –and their children, nieces, and nephews – will there be any clean air, water, and soil left? Will they be able to swim in the lake, walk through a forest, or ski down a mountain? As Brianna Cayo Cotter (2008, p. 27) of the Energy Action Coalition stated, "Climate change is our generation's civil rights movement." It is up to us to change our lifestyles, modify our vocations, and foster new thoughts and technology to mend our planet.

The social environment

The social environment includes the communities, neighborhoods, groups, and cultures to which we belong and the interactions and relationships we form both within our groups and between other groups. The health of the physical environment directly impacts the quality of life, public health, safety, behavior, and equity in the social environment. Over-consumption of the earth's natural resources and irresponsible patterns of development have fragmented communities, marginalized and segregated minorities, people of color, and those in lower socio-economic demographics altered cultural heritages, disrupted human-centric activities, and are the cause of physical and mental health problems. So many of our places are becoming utilitarian spaces where the design is dictated by non-human and non-environmental factors. As stated by Thayer (1994), "Evidence from the landscape would suggest that America is preoccupied by the utilitarian means for living at the expense of both the reasons and the context." Indeed, the balance between technology, humans, and nature seems to be weighted toward the end of technology, and like it or not, it is here to stay.

Population dynamics

Although the annual growth rate of the world population has started to slow in recent times, the fact remains that absolute numbers are still growing – the world population reached over eight billion in 2024, and with an annual average growth rate of 0.9 percent, it is projected to be 10.2 billion by 2060 (U.S. Census Bureau, 2024). Such numbers and subsequent demand for resources raise concern over the carrying capacity of the earth. It is quite possible sometime in this century that the population rate will exceed food supply, energy sources, and land area for habitation, agriculture, and waste disposal. According to Ebert (1993, p. 229):

> Human's relationship with the ecosphere proceeds along a J-curve; that is to say that at first humankind was an integral part of the living world, existing in harmony with the total system, and did not exert any excessive influence. But over time, humans assumed a different role and attained the position of changing the environment consciously, not instinctively.

Population dynamics also vary in different regions of the world and between developed and developing countries. For example, data from the United Nations UN75 Report (2020) shows that over half of the projected growth between now and 2050 will come from nine countries, including India, Nigeria, Pakistan, the Democratic Republic of the Congo, Ethiopia, Tanzania, Indonesia, Egypt, and the United States. Although older people (age 65 and older) are the largest growing demographic globally, sub-Saharan Africa and most countries in Asia, Latin America, and the Caribbean are seeing higher rates of growth in people ages 25–64. Around 55 percent of people live in urban areas worldwide; in the United States, about 80 percent live in urban areas, which the U.S. Census Bureau (2022) defines as densely developed commercial, residential, and non-residential areas. The UN75 report also showed that international migration increased from 2.8 percent in 2000 to 3.5 percent in 2019. In addition to relocating for family, work, or educational opportunities, this increase was attributed to people fleeing from violence, persecution, or poor environmental living conditions. In the United States, the net domestic migration in 2020–2021 was positive for the South and negative for the Northeast, West, and Midwest, with the Cape Coral–Fort Meyers–Naples, Florida, combined statistical area (CSA) having the largest increase and the New York–Newark, NY–NJ–CT–PA CSA having the largest decrease (U.S. Census Bureau, 2022).

In addition to environmental disparities within developed countries that tend to have a more significant impact on lower socio-economic communities than in upper socio-economic communities, there are global inequities regarding climate change. Although developed countries consume more energy and subsequently have higher emissions than developing countries, the impacts on the global climate disproportionately damage developing countries since their economies are dependent on activities such as agriculture, fishing, and forestry (Guivarch et al., 2021). A report by the World Bank Group (2020) estimates that climate change will drive 68 million to 135 million into poverty by 2030, with the largest concentrations in the countries of sub-Saharan Africa and South Asia. In terms of urbanization and sprawl, a report by the World Bank found cities with low productivity, low-income levels, and dysfunctional policies tend to grow horizontally at slow rates at the urban edge, which often results in crowding and poor living conditions as the population density grows; conversely, cities with higher productivity and responsive policies tend to grow more vertically and infill vacant lots, resulting in densification at the urban core (Lall et al., 2021). Due to these socio-economic disparities, it is important to recognize the relativity of environmental problems and how their impacts can vary between regions and across the globe and to understand that strategies that may be regarded as sustainable or resilient in one country/region may not be so in a different country/region.

Human behavior and public health

An examination of human history will show that we have spent more time living *with* nature than apart from it. As mentioned in Chapter 1, the Industrial Revolution sparked the first major shift from an agrarian/rural society to an urban society. Since that time, our lifestyles have continued to morph further away from the natural environment to more "artificial" environments. Built infrastructure, engineering marvels, and technological advances – while making many activities of life more convenient – have changed the dimension of the places in which we live, work, and recreate to an unnatural state that is out of sync with the human senses. Calthorpe (1993, p. 11) describes the aesthetics of "place" with four dimensions: scale, pace, pattern, and bounds:

> The scale of our environment is now set in proportion to institutions and bureaucracies rather than community and neighborhood; the pace is set by electronic sound bites and the auto rather than human breath and step; the pattern is established by mass production and discontinuous ownership rather than local craft and social continuity; and the bounds are set by wealth and power rather than proportion and nature.

As more of our land and natural resources are being consumed, we are being further disconnected from nature, which can lead to social dislocation, community fragmentation, and impairments to physical and mental health. The sounds, pollution, speed, and size of modern development have caused both tangible and intangible chasms in our social fabric. For example, in terms of sound, a typical human conversation is around 60 decibels (dB), rainfall is about 40–50 dB, city traffic is around 85 dB, an ambulance siren is around 125 dB, and a jet plane is around 135 dB. Some studies show that exposure to 85 dB can cause permanent damage after just eight hours, not to mention the mere nuisance that can increase stress levels and worsen mood (Dangerous Decibels, n.d.). Other studies show that people are less willing to help others and cast more extreme judgments of others under high noise conditions compared to low noise conditions (Gifford, 2002, pp. 278–279). Street noise can be so acutely deafening that it is difficult to have a conversation or listen to music, let alone hear the healing sounds of the natural world, such as the birds, the wind, or the rain. Sound, while not visually tangible, can disassociate people and/or serve as barriers just as divisive as visual obtrusions; more simply put, we tend to run away from noise.

The same intangible aversions can occur from air pollution. With industrialization and over-consumption of our natural resources, taking a step outside to get some fresh air can be a challenge seeing the air is often contaminated with primary pollutants such as carbon monoxide, sulfur dioxide, nitrogen oxides, volatile organic compounds, and PM-10 particulate matter from fuel combustion, transportation and industrial processes, and secondary pollutants such as ozone, peroxyacetyl nitrates, and sulfuric/nitric acids from chemical reactions between sunlight, nitrogen oxides, heavy metals, radon, and volatile organic carbons. High concentrations of air pollutants have been linked to a variety of health conditions, including bronchitis, cancer, lung disease, heart disease, impaired immune systems, learning disabilities, and even death in some cases. Organic and inorganic waste, PCBs, metals, and pathogens spread bacteria and viruses and deem water unusable for fishing, potable use, and recreation. Although legislation such as the Comprehensive Environmental Response, Compensation and Liability Act (CERCLA) in 1980, the Clean Air Act of 1970, and Clean Water Act in 1972 have mitigated some of the damage, there still have been over 47,000 waste sites identified in the United States, over 1,700 of which have been placed on the National Priorities List since 1982 (National Geographic, n.d.). The National Water Quality Inventory (EPA, 2017) indicates that 46 percent of rivers and stream miles, 18 percent of coastal waters and Great Lakes, and 32 percent of wetlands are in poor biological condition and 21 percent of the nation's lakes, ponds, and reservoirs are hypereutrophic. Worldwide, the World Health Organization states that 24 percent of global deaths and 28 percent of deaths of children under the age of five are due to modifiable environmental factors (WHO, 2018, September 13).

In addition to impacts on physical health, some studies have shown that bad odors negatively affect mood and attraction to others; other studies have shown a correlation between higher levels of photochemical oxidants in the air and instances of domestic disputes and psychiatric disturbance (Gifford, 2002, pp. 278–279). Similar to loud areas, people tend to avoid areas with poor air quality, creating dead pockets of space unsuitable for habitation or social interaction.

The sheer pace of development is also socially disengaging humans on both conscious and sub-conscious levels. As our environment continues to be designed around the automobile, we are physically and socially disconnecting people from their communities. To put speed in perspective, the average human walks about 3 miles per hour (mph) (4.8 km/h) and the average car speed in an urban area is about 30 mph (48 km/h). At these speeds, a car can travel a mile in 30 seconds, whereas the walker would travel the same distance in about 20 minutes. This may seem elementary to some, but such a difference has become so embedded in our culture, that we are – for want of a better expression – forgetting to stop to smell the roses. Roads, highways, and parking surfaces have fragmented our neighborhoods and have exploded the scale of development far beyond the dimensions of human movement. It often takes the same amount of time to cross a 20-foot-wide (6-meter) street as it does to walk a quarter of a mile (.4 km) that is unobstructed by traffic. The same infrastructure that once made our lives easier and expanded our horizons is now confining us to our homes, our electronic devices, and the inside of our cars. We are no longer designing spaces that are sensitive to human dimensions, but rather spaces that are centered on speed, efficiency, and maximizing profit. We are becoming too focused on the fastest way to get from point A to point B and in doing so, are losing touch with the journey along the way. This leads to social detachment and disengagement with the landscape that ultimately results in spaces that are deprived of function, meaning, or culture.

The scale of development has the same social disengagement effect as the pace of development. Buildings, interior spaces, and the distance between buildings are all simply becoming too large for effectual human interaction, mobility, safety, and physical/psychological comfort. For example, although high-rise construction does reduce land consumption to some degree, the visual impact on the landscape and the safety and personal comfort issues of constraining people high in the air with minimal escape options outweigh their benefits. If high rises were designed to mimic mountains and consisted of terraced external spaces allowing a person to descend the perimeter of the building to the ground, it would be a different story. But most buildings are vertical, confined spaces that are difficult to escape from in response to disaster, not to mention the fact that they isolate people from both nature and street-level interaction. Even before the tragedies at the World Trade Center in 2001, we should have been asking ourselves, is it healthy and safe to confine people thousands of feet in the air with no reasonable escape or connection to the outdoors? The pros and cons of high-rise construction are discussed in more detail in Chapter 9.

The digital revolution and the virtual environment

In the twenty-first century, the Digital Revolution and virtual environments are further decreasing human interaction both with each other and with natural environments. Even when people do venture outdoors, they are often glued to their cell phones, smartphones, tablets, laptops, headsets, earpieces, smart glasses, implants, or whatever new gadgets might come down the pipeline in the future. Many people who have had the experience of someone walking into them in a store, a mall, a college campus, or any public place (and a lack of acknowledgment from the said person who walks into you) will attest that the zombie apocalypse is here. In the workplace, technology has created a double-edged sword: we can process information faster and more efficiently than ever; however, the trade-off is that we spend more time in sedentary positions fixated on a computer screen. As a result of the COVID-19 pandemic, Microsoft Teams, Zoom, Facetime, and Skype enabled us to work and attend school remotely. Although fully remote and hybrid work schedules have enabled more autonomy and flexibility in schedules, many people may not be moving enough (including this author as I write this book). The sedentary aspect of the digital age is having adverse effects on our physical health. An article published in the *New York Times* suggests that when we sit all day, electrical activity in the muscles drops, leading to a cascade of harmful metabolic effects, and sadly even getting regular doses of exercise doesn't offset the damage (Vlahos, 2011). It seems sitting has become the new smoking. In the movie *WALL-E*, the future is depicted by overweight humans living in outer space (because their ancestors destroyed the earth) lounging in virtual reality chairs that essentially simulate any "real-life" experience. Although this was a fictional plot line in 2008, it might not be that far from where we are headed. Geo-location apps and devices have also allowed us to track each other in real time. While this is a good safety feature for monitoring youth, the elderly, or just someone you care about, others may feel George Orwell's prediction of *Big Brother* watching us is finally here. Combine this with AI and algorithms that track our internet searches and/or analyze our social media profiles, and it is easy to see the machines are taking over, and, unfortunately, we still cannot travel into the future and reprogram a cyborg terminator made of living tissue over a robotic endoskeleton to travel back to the present to protect us or alter the space-time continuum (yet, anyway).

Indeed, humans are spending more time indoors than at any point in history and this trend will only increase as technology continues to advance. Kripke (2019) measured the amount of time people spent outdoors in San Diego, California and found that although San Diego has one of the most temperate and sunny climates in North America, people were spending less than one hour outside on average. He also compared the San Diego results to other data collected in the more northerly climate of Rochester, Minnesota, and found people in the latter spent slightly more than an hour outside in summer months, but only about one-third as much as San Diego's residents in winter months.

Evidence is starting to accumulate regarding the link between physical/mental health and nature. Kuo and Taylor (2004) showed that children with attention-deficit disorder have fewer symptoms and enhanced ability to focus after outdoor activities such as camping and fishing when compared to indoor activities such as doing homework and playing video games. This data is consistent with a study that found a negative association between screen time and psychological well-being – including measures of self-control, relationships with caregivers, emotional stability, diagnoses of anxiety and depression, and mental health treatment – among children and adolescents (Twenge & Campbell, 2018). A Dutch study in 2009 examined 345,143 medical records and assessed health status for 24 conditions, including cardiovascular, respiratory, and neurological diseases, and found those who lived closer to green space experienced less anxiety and depression than those who lived further away from green space. Additionally, people living in urban environments had a higher prevalence of 15 of the 24 conditions, with the relationship strongest for anxiety disorder and depression. What specifically improves health could be any number of factors, including more natural sunlight/Vitamin D, clean air, and/or a general sense of tranquility invoked by nature (Maas et al., 2009). A meta-study of 1,252 people collected from ten different studies showed that just five minutes of daily exposure to nature improves mental health (ACS, 2010). The Max Planck Institute for Human Development in Germany looked at the effect of walking on working memory and found:

> The working memory performance of both age groups improved when walking at their chosen speed compared with when sitting or walking at a fixed speed set by the researchers. This was especially the case for more difficult versions of the working memory task and was more pronounced among the children than the adults. So, this would appear to be a clear case of mental performance being superior in a dual-task situation.
>
> *(Schaefer, 2014)*

There are various theories that attempt to explain nature's mechanism of restoring human well-being. Gifford (2002, pp. 389–390) describes three such theories, including the affective-arousal approach that asserts that nature's benefits derive from the positive emotions it elicits in us and that any calm, peaceful setting will have similar benefits; the attention restoration theory, where nature is said to compel our involuntary attention; and the evolutionary or biophilic approach that builds on the notion that humans are genetically more adapted to natural than built settings due to millions of years of evolving in natural settings. Although humans historically created a shelter for protection from the unwanted elements of nature, the research suggests that we may have taken this division too far. The human disconnect from the natural environment is causing physical and psychological impairments that further support the need to conserve, preserve, and restore the natural environment.

Community fragmentation and segregation

In addition to infrastructure physically fragmenting our communities, historical policies – some as recent as the mid-twentieth century – deliberately sought to segregate neighborhoods. For roughly 50 years after the abolishment of slavery, municipalities and local governments still enforced exclusionary zoning that prohibited the sale of property to black people. Many neighborhoods also enacted private covenants among themselves where they agreed to not sell their homes to certain minority groups. Although the Supreme Court ruled against race-based zoning regulations in 1917, municipalities found several nefarious ways to keep neighborhoods segregated without mentioning race, including banning apartment complexes or duplexes in certain neighborhood districts, requiring minimum lot sizes thereby precluding lower-income groups, banning certain types of businesses that catered to lower-income and/or minority groups, and the Federal Housing Administration (FHA) and the Home Owner's Loan Coalition's color-coding mapping practice that identified areas to issue or not issue mortgages. Four color codes were used, including *Green* zones, which consisted primarily of white, professional men and were considered the "best" neighborhoods; *Blue* zones, which were still considered "desirable" and had only modest infiltration on non-white groups; *Yellow* zones, which were "definitely declining" as a result of being bordered by or being infiltrated by non-whites; and *Red* zones, which were considered "hazardous" where non-white infiltration had already occurred (Lockwood, 2020). *Green* and *Blue* zones were considered good investments and were eligible for FHA loans, *Yellow* zones were considered risky, and *Red* zones were deemed ineligible for FHA loans. In addition, *Red* zones received less public investment for public works projects, infrastructure, and education, thereby further limiting opportunities for citizens in those neighborhoods seeking a better quality of life. One of the greatest aversions in current times cited as a reason to not invest in these neighborhoods is crime. This is ironic, seeing much of the crime is the direct result of the racist policies that segregated these neighborhoods in the first place.

Although the Fair Housing Act of 1969 ended the practice of color-coding neighborhoods, the divisions are still prevalent across many cities in the United States today. Former Red and Yellow neighborhoods are plagued with vacant buildings, decaying water and sewer infrastructure, high unemployment, high crime rates, and lack of green/park space. In many cities, redlining was combined with highway construction in the mid-twentieth century. In addition to physically removing homes and businesses for their construction, the highways that bisected cities served as a further physical barrier that literally walled off lower-income neighborhoods from the rest of the city. Several case studies of this are discussed in more detail in Chapter 7.

Sadly, the practice of segregation was hardly an isolated event in American history. As far back as 1830, President Andrew Jackson signed the Indian Removal Act that enabled the government to take native land east of the Mississippi River and forcibly relocate natives to territory west of the Mississippi, which is currently the state of Oklahoma. Tribes that did not willingly sign treaties giving the government the power to assist them in their move were violently assaulted, imprisoned, or murdered. Those who did survive went on a 1,000-mile exodus known as the Trail of Tears (LOC, n.d.). When America was first discovered by European settlers in 1492, there were roughly ten million indigenous people, and, by 1900, there were fewer than 300,000 (EJI, 2016). As of 2021, it is estimated there are between four million and seven million indigenous people and 574 native tribes in the United States, of which 20 percent live in American Indian areas or Alaska Native villages (IWGIA, 2022). A recent study estimates that indigenous nations in the United States have lost 98.9 percent of their historical land base since European settlers began colonizing the continent and that 42 percent of tribes have no federal or state-recognized land. The study also showed the land that is recognized is only 2.6 percent the size of the original area, is less likely to have economically valuable resources, and is more prone to extreme heat and low precipitation (Farrell et al., 2021).

In 1946, Congress created the Indian Claims Commission to compensate 176 tribes and bands for land that was seized by the federal government (Blakemore, 2019). The commission paid out $1.3 billion, which amounted to less than $1,000 for each Native American. In 1971, $962 million was awarded to Indians, Eskimos, and Aleuts to relinquish their aboriginal claims of 44 acres of land in Alaska. In both cases, the federal government held the money in trusts, denying Indigenous people direct control of their funds (Hassan & Healy, 2019). In addition to the inadequate compensation amounts and the mismanagement of the money in the trusts by the government, many tribes do not feel financial awards are adequate to mend the mistakes of the past. This is due to the United States still maintaining control of the lands and denying tribes their sovereign right to protect the natural and cultural resources that reside on the lands (Tsosie, 2007, pp. 47–68). To truly embrace reparative justice, an understanding and acknowledgment of historical injustice of the past and an understanding of the social, economic, and political inequalities that exist in the present are imperative in adjudicating claims of ownership, equal opportunity, and cultural rights (Tsosie, 2007, p. 48). In 1990, Congress passed the Native American Graves Protection and Repatriation Act that provides a process for federal agencies and museums that receive federal funds to transfer Native American Cultural items – including human remains, funerary objects, sacred objects, and objects of cultural patrimony – from their collections to Indian tribes, Alaska Native Corporations, and Native Hawaiian organizations (BLM, n.d.a). Tribal consultation is an essential component to understanding past, present, and future impacts on the cultural values, religions, and traditional practices that could be impacted by Bureau of Land Management (BLM) actions on public lands. According to the BLM, tribal consultation regarding public lands includes identifying tribal governing bodies and individuals from whom to seek input; transparent communications for BLM actions on lands and asking for input regarding impacts on activities, practices, and beliefs; using tribal information to define acceptable uses of the land; and maintaining a permanent record of consultations and decisions (BLM, n.d.b).

In 2018, the Land Back movement was initiated by Aaron Tailfeathers, a member of the Kainai Tribe of the Blackfeet Confederacy of Canada. Although the concepts of repatriation have been around since the colonizers first arrived 500 years ago, the Land Back movement reignited the goal to decolonize Indigenous land and return it to its original inhabitants. Even though returning land is one way to facilitate restitution, the Land Back Movement has a broader mission that ultimately seeks to restore sovereignty, preserve culture, and transfer more political and economic decision-making power to Indigenous communities. Therefore, one key element that is paramount in achieving restitution is for tribal governments to gain more parity at local, regional, and federal levels. Opposed to what many critics think, removing non-native communities from Indigenous land is *not* an objective of the movement. In fact, the movement seeks to catalyze non-Indigenous allies in the fight for environmental justice.

Ideally, Native Americans would not have to buy back land that was taken from them and is owned by private parties or municipalities. In recent times, there have been some success stories. As part of the Natural Resource Damage Assessment and Restoration Program regarding the contamination of Onondaga Lake, a settlement was reached between the Natural Resources Trustees and Honeywell International, Inc. that conveyed over 1,000 acres of land owned by Honeywell to the Onondaga Nation in the Tully Valley of central New York. In addition to signing over land, Honeywell was required to implement 18 restoration projects, including the restoration and preservation of wetlands and grasslands, improved recreational opportunities, and improved public access to Onondaga Lake, Ninemile Creek, and the Seneca River (DOI, 2022). The restoration of the land will utilize Indigenous management styles that evolved over time through symbiotic relationships with nature, also known as Traditional Ecological Knowledge.

In 2020, Congress passed legislation to restore ownership of 19,000 acres of the National Bison Range in Montana to the Salish and Kootenai tribes. Land trusts have also assisted in several land transfers, including the Western Rivers Conservancy's $4.5 million deal that returned 1,199 acres of land along the Little Sur River back to the Esselen Tribe of Monterey County, the reacquisition of a 150-acre island in Maine to the Wabanaki Confederacy, and the Nature Conservancy's Indigenous Peoples and Local Communities Program that transfers ecologically vital lands back to Indigenous people around the world (Robbins, 2021).

Voluntary land taxes have also been implemented for non-Native people living on Native land, including the Honor Tax in Humboldt County, California, where non-Native community members can pay a tax on the benefits they derive from lands occupied by the Wiyot Tribe; Real Rent Duwamish, which is a monthly income-based rent to benefit the Duwamish Tribe whose land once occupied Seattle; the Manna-hatta Fund Collective for non-native residents of New York to support Indigenous people in New York; the Honor Native Land Tax which supports Tiwa land in Albuquerque, New Mexico; and the Sogore Tea Land Trust, which uses a Shuumi tax where non-native community members in the San Francisco Bay area can pay an annual tax that is dependent on their home or their monthly rent (Maphalala, n.d.). While these voluntary programs are not full replacements for returning land, the revenue can go toward improving education, health, social services, and cultural preservation.

The economic environment

For a large part of the twentieth century, balancing economic growth with environmental protection was viewed as a paradox. Development – whether it was responsible or not – was often thought of as symbiotic with economic growth, and protecting the environment was thought to hinder development, ultimately decreasing economic growth. Hellmund and Smith (2006) refer to this as *creative destruction*: "The capitalist economy that has helped to produce a remarkable high standard of living for a minority of the world's inhabitants relies on the constant dissolution and reconstitution of nature and society into new forms." If you have ever played with Legos with a four-year-old, the first thing they do after you build something is knock it down – and then rebuild it – and then knock it down again. And when they get bored with that process, they just leave the mess in piles all over the floor and move on to something else. This is what keeps it interesting for them. This is essentially the same way we approached the built environment in much of the twentieth century and much of it was in the name of economic growth. Land developers thrive on this approach; little do they realize they are acting the same way as four-year-old toddlers playing with their toys.

In the twenty-first century, this paradigm is starting to change. It is now being recognized that global climate change, environmental restoration, and sustainable development are no longer alternatives that need to be sacrificed for economic growth, but are mandatory issues that must be addressed that can actually drive the economy. The United Nations defines a *green economy* where growth in employment and income are driven by public and private investments into activities, infrastructure, and assets that reduce carbon emissions and pollution, enhance energy and resource efficiency, and prevent the loss of biodiversity and ecosystem service (UNEP, n.d.).

Indeed, notions of conserving land, restoring nature, and building sustainable, net-zero carbon buildings are starting to overshadow the "develop as much as we can for as cheap as we can" approach in terms of long-term economic growth.

Ecosystem services

Ecological economics poses the idea that a sustainable society doesn't draw down any of the capital stocks on which its operation depends. Capital stocks can be built/economic capital, natural capital, social capital, and cultural capital (Courtney & Zencey, 2012, p. 28). Ecosystem services are defined as the direct or indirect benefits that ecosystem functions provide to society. The Millennium Ecosystem Assessment (MA) identified four major categories of ecosystem services, including provisioning services, which are any benefits to people that can be extracted from nature, such as food, drinking water, fuel, clothing, and medicine, to name a few; regulating services, which are benefits provided by ecosystem process from regulating natural processes, such as pollination, carbon capture, climate regulation, water purification, and flood control; cultural services, which is a benefit that preserves or enhances cultural history, identity, and sense of place; and supporting services, which are natural processes that allow the earth to sustain basic life forms, such as photosynthesis, nutrient cycling, the creation of soils, and the water cycle. Supporting services are necessary for the other three ecosystem services (NWF, n.d.).

Ecosystem services and natural capital are often not captured or adequately quantified compared to capital from consumption, manufacturing, or investment. Costanza et al. (1997) estimated the economic value of 17 ecosystem services, including gas regulation, climate regulation, disturbance regulation, water regulation, water supply, erosion control and sediment retention, soil formation, nutrient cycling, waste treatment, pollination, biological control, refugia, food production, raw materials, genetic resources, recreation, and culture. Although it is challenging to quantify and/or put values on human life, health, well-being, aesthetics, and ecological function, the researchers calculated the value per unit area of each ecosystem service using either the sum of consumer and producer surplus, the net rent, or price times quantity and estimated that ecosystems provided at least U.S.$33 trillion worth of services annually, with U.S.$1.3 trillion a year in gas regulation, U.S.$1.8 trillion a year in disturbance regulation, U.S.$2.3 trillion waste regulation, and U.S.$17 trillion a year in nutrient cycling – 63 percent of the value came from marine systems and 38 percent from terrestrial systems, including U.S.$4.9 trillion a year from wetlands and U.S.$4.7 trillion a year from forests.

Green infrastructure, greenways, and parks have been shown to increase land value, save money, and generate new capital by attracting and/or creating new businesses, residents, and recreational opportunities. Preserving watersheds and marsh areas can drastically save dollars by preventing flood damage and providing a natural sink for pollutants, thereby eliminating the need to implement non-natural solutions. Planting trees alone has been shown to benefit the economy. The American Forestry Association estimates that a 50-year-old urban tree saves $75 a year in air conditioning, $75 a year in stormwater and erosion control, $75 a year in wildlife shelter, and $50 a year in air pollution control (Benedict & McMahon, 2006, p. 71). And then of course there is the benefit of tourism. Believe it or not, people still do like nature. For example, the National Park Service reported close to 312 million annual recreational visits in 2022 in their 407 areas, the Vermont Ski Areas Association reported close to four million skier/rider visits for 2021–2022, and over three million people visit the Appalachian Trail per year (NPS, n.d.; Rivard, 2022; ATC, n.d.).

Investing in green infrastructure, transportation, and renewable energy creates jobs on our soil and can ultimately decrease our dependency on foreign oil, which can in turn further save the taxpayers millions – if not billions of dollars funding our military presence overseas to protect said oil interests. Economic geographer Roger Stern of Princeton University estimated that the United States has misallocated $8 trillion since 1976 protecting the oil flow from the Persian Gulf that fuels much of the global economy (Thompson, 2011). Investing in auto-alternative transportation would save money in the long run based on energy consumption alone. Data reveals that urban car travel is nearly two times as energy consumptive as average bus travel, six times more energy intensive than an average urban electric train, and 3.7 times more than a typical light rail or tram system travel (Newman & Kenworthy, 1999, p. 78). Money can also be saved by divesting funds away from highway maintenance and expansion. According to Newman and Kenworthy (1999, p. 56), a ten-year, $100 billion transit investment would yield improved worker output valued at $521 billion; a comparable expenditure on highways would yield $237 billion. The highest annual level of net benefits from such an increase in transit would gain $15 billion in ten years compared to only $7 billion in the same time frame for road construction. Investment in rail can also influence urban form by upzoning nodes around transit stations to facilitate compact, mixed-use, transit-oriented development (TODs). Tax incentives can be implemented to foster the movement of businesses into TODs and land in between the TODs can be re-acquired and rezoned for open space, parks, or riparian zones.

The relationship between economic growth and development patterns is also constantly evolving. In the mid-twentieth century, highway expansion from city centers to suburbs allowed people to live in different locations to where they work. The construction of highways, new suburban homes, and retail infrastructure in suburbs had positive impacts on economic growth. However, over time, the exodus to the suburbs began to hurt some city centers, particularly medium-sized and rust-belt cities that experienced industrial decline and outsourcing of manufacturing. Such cities were plagued with vacant businesses and buildings, population declines, declining tax revenues, decaying infrastructure, and high unemployment and crime. While the exodus away from the city initially provided citizens with larger lots, more open space, and less congestion, recent trends show that more of the population is centralizing around suburbs, essentially urbanizing suburbia and causing auto congestion that is on-par or even worse than the city congestion they were fleeing from in the first place. According to the Pew Research Center (2020), 31 percent of the U.S. population lives in urban cores, 25 percent live in large suburban counties with over one million people, 30 percent live in small metropolitan counties with fewer than one million residents, and 14 percent live in rural areas.

The 2020 U.S. Census also showed suburbs are becoming more diverse. Minorities represent more than one-third of the population in 31 of the 56 major metropolitan areas, which is more than double the 2000 numbers (Frey, 2022). Although younger adults (25–34) tend to gravitate more toward urban centers, a larger portion of older adults (early Millennials, Generation X, and Baby Boomers) are living in the suburbs (Lee, 2022). Data by the Brookings Institute collected between 2000 and 2015 showed the highest growth of poverty rates in small metropolitan areas; in 2015, poverty rates were 19.6 percent in large cities, 17.2 percent in rural areas, 16 percent in small metro areas, and 11.2 percent in the suburbs (Kneebone, 2017). Given these shifts in demographics, the stereotype of suburbs as predominately white, middle to upper-income citizens and the city as lower socio-economic groups and people of color is becoming antiquated. In their book, *Retrofitting Suburbia: Urban Design Solutions for Redesigning Suburbs*, the authors state "it is no longer useful to talk about cities versus suburbs. The 'suburbs' are behaving more and more like center cities and the metropolises embrace both as they become more polycentric" (Dunham-Jones & Williamson, 2011, p. xxvii). As a result, the redevelopment of one should complement the other. Although the urban form may look physically different from the suburban form, they both benefit from similar design strategies, including infill development, corridor redesign, the removal of highways, adaptive reuse, removing and/or reconfiguring buildings and large parking surfaces, redesigning road networks, improving transit, improving pedestrian movement, clustering development, and creating more compact, pedestrian-centric village centers where there are none.

And just when we thought we had the economy and development patterns figured out, we were hit with a global pandemic. In an already digitally enabled economy, the pandemic further pushed our dependence on microchips and screens. We were (and some still are) confined to the inside of our homes where we worked, learned, and utilized e-commerce to a point that has questioned the relevance of physical space and location to the economy. In the aftermath of the pandemic, both cities and suburbs have witnessed an increase in vacant office and retail space, failed businesses, and a decline in available staffing and labor. This represents both new challenges and new opportunities to re-evaluate, retrofit, and/or regreen the built environment.

References

American Chemical Society (ACS). (2010, May 21). *In the green of health: Just 5 minutes of 'green exercise' optimal for good mental health.* Science Daily. www.sciencedaily.com/releases/2010/05/100502080414.htm.

Appalachian Trail Conservancy (ATC). (n.d.). *About the trail.* www.appalachiantrail.org/about-the-trail

Begley, S. (2006, June 9). How melting glaciers alter Earth's surface, spur quakes, volcanoes. *Wall Street Journal Science Journal.* www.wsj.com/articles/SB114981650181275742

Benedict, M. A., & McMahon, E. T. (2006). *Green infrastructure: Linking landscapes and communities* (p. 71). Island Press.

Blakemore, E. (2019, August 29). *The thorny history of reparations in the United States.* History. www.history.com/news/reparations-slavery-native-americans-japanese-internment

Boden, T. A., Marland, G., & Andres, R. J. (2017). *National co2 emissions from fossil-fuel burning, cement manufacture, and gas flaring: 1751–2014.* Carbon Dioxide Information Analysis Center, Oak Ridge National Laboratory, U.S. Department of Energy. http://doi.org/10.3334/CDIAC/00001_V2017

Brown, M. A., & Herrera, V. S. (2021, December 15). Combined heat and power as a platform for clean energy systems. *Applied Energy, 304.* https://doi.org/10.1016/j.apenergy.2021.117686

Calthorpe, P. (1993). *The next American metropolis: Ecology, community, and the American dream* (p. 11). Princeton Architectural Press.

Climate Central (CC). (n.d.). *Surging seas: Risk zone map.* https://ss2.climatecentral.org/#8/27.665/-81.516?show=satellite&projections=0-K14_RCP85-SLR&level=5&unit=feet&pois=hide

Commonwealth of Australia. (2022). *Australia's nationally determined contribution.* https://unfccc.int/sites/default/files/NDC/2022-06/Australias%20NDC%20June%202022%20Update%20%283%29.pdf

Costanza, R., d'Arge, R., de Groot, R., Farber, S., Grasso, M., Hannon, B., Limburg, K., Naeem, S., O'Neill, R. V., Paruelo, J., Raskin, R. G., Sutton, P., & van den Belt, M. (1997, May 15). The value of the world's ecosystem services and natural capital. *Nature, 387,* 253–260. https://doi.org/10.1038/387253a0

Cotter, B. C. (2008, April). The new green U. *College Planning and Management, 4*(4), 27.

Courtney, E., & Zencey, E. (2012). *Greening vermont: The search for a sustainable state* (p. 28). Thistle Hill Publications.

Dangerous Decibels. (n.d.). *Noise induced hearing loss.* www.dangerousdecibels.org/education/information-center/noise-induced-hearing-loss/

Dunham-Jones, E., & Williamson, J. (2011). *Retrofitting suburbia: Urban design solutions for redesigning suburbs.* John Wiley and Sons.

Earth Pledge. (2005). *Green roofs: Ecological design & construction* (p. 16). Schiffer Publishing Ltd.

Ebert, C. H. V. (1993). *Disasters: Violence of nature, threats by man* (2nd ed., p. 229). Kendall/Hunt Publishing Company.

Equal Justice Initiative (EJI). (2016, July 1). *Forced removal of Native Americans.* https://eji.org/news/history-racial-injustice-forced-removal-native-americans/

European Commission. (n.d.). *Paris agreement.* https://climate.ec.europa.eu/eu-action/international-action-climate-change/climate-negotiations/paris-agreement_en#:~:text=The%20EU's%20initial%20nationally%20determined,by%20the%20end%20of%202018

Farrell, J., Burow, P. B., McConnell, K., Bayham, J., Whyte, K., & Koss, G. (2021, October 29). Effects of land disposition and forced migration in indigenous Peoples in North America. *Science*, *374*(6567). http://doi.org/10.1126/science.abe4943

Federative Republic of Brazil. (2022, March 21). *Nationally determined contribution*. https://unfccc.int/sites/default/files/NDC/2022-06/Updated%20-%20First%20NDC%20-%20%20FINAL%20-%20PDF.pdf

Fickling, D. (2022, September 29). *Peak oil has finally arrived. No, really*. Bloomberg, L.P. Washington Post. www.washingtonpost.com/business/energy/peak-oil-has-finally-arrivedno-really/2022/09/28/f67f2f0a-3f68-11ed-8c6e-9386bd7cd826_story.html

Frey, W. H. (2022, June 15). *Today's suburbs are symbolic of America's rising diversity: A 2020 census portrait*. Brookings Institution. www.brookings.edu/research/todays-suburbs-are-symbolic-of-americas-rising-diversity-a-2020-census-portrait/#:~:text=The%202020%20census%20showed%20that,San%20Antonio%2C%20and%20San%20Diego

Gifford, R. (2002). *Environmental psychology: Principles and practice* (3rd ed., pp. 278–279, 389–390). Optimal Books.

Guivarch, C., Taconet, N., & Mejeane, A. (2021, September). *Linking climate and inequality*. International Monetary Fund. www.imf.org/en/Publications/fandd/issues/2021/09/climate-change-and-inequality-guivarch-mejean-taconet

Hardy, Q. (2015, March 20). At Kodak, clinging to a future beyond film. *The New York Times*. www.nytimes.com/2015/03/22/business/at-kodak-clinging-to-a-future-beyond-film.html#:~:text=It

Hassan, A., & Healy, J. (2019, June 19). America has tried reparations before. Here is how it went. *The New York Times*. www.nytimes.com/2019/06/19/us/reparations-slavery.html

Hellmund, P. C., & Smith, D. S. (2006). *Designing greenways: Sustainable landscapes for nature and people* (pp. xi, 55). Island Press.

Igusky, K., Gasper, R., Obeiter, M., Aden, N., & Forbes, S. (2014, October 10). *By the numbers: How the U.S. economy can benefit from reducing greenhouse gas emissions*. World Resources Institute. www.wri.org/insights/numbers-how-us-economy-can-benefit-reducing-greenhouse-gas-emissions#:~:text=Reducing

Intergovernmental Panel on Climate Change (IPCC). (2014). *Climate change 2014: Mitigation of climate change*. Cambridge University Press. www.epa.gov/ghgemissions/global-greenhouse-gas-emissions-data.

Intergovernmental Panel on Climate Change (IPCC). (2018). *Special report on the ocean and cryosphere in a changing climate*. www.ipcc.ch/site/assets/uploads/sites/3/2022/03/01_SROCC_SPM_FINAL.pdf

Intergovernmental Panel on Climate Change (IPCC). (2023). *Synthesis report of the IPCC sixth assessment report (AR6)*. IPCC. https://report.ipcc.ch/ar6syr/pdf/IPCC_AR6_SYR_SPM.pdf

Intergovernmental Science-Policy Platform on Biodiversity and Ecosystem Services (IPBES). (2021). *UN report: Nature's dangerous decline unprecedented; Species extinction rates accelerating*. United Nations. www.un.org/sustainabledevelopment/blog/2019/05/nature-decline-unprecedented-report/

International Energy Agency (IEA). (2021a). *Key world energy statistics 2021*. www.iea.org/reports/key-world-energy-statistics-2021

International Energy Agency (IEA). (2021b). *Oil 2021: Analysis and forecast to 2026*. www.iea.org/reports/oil-2021

International Work Group for Indigenous Affairs (IWGIA). (2022, April 1). *The indigenous world: Indigenous peoples in the United States*. www.iwgia.org/en/resources/indigenous-world.html

Kelly, M. (2013, November 24). *Even if emissions stop, carbon dioxide could warm Earth for centuries*. Princeton University Press. www.princeton.edu/news/2013/11/24/even-if-emissions-stop-carbon-dioxide-could-warm-earth-centuries#:~:text=Nov.,the%20journal%20Nature%20Climate%20Change

Kneebone, E. (2017, February 15). *The changing geography of US poverty*. The Brookings Institute. www.brookings.edu/testimonies/the-changing-geography-of-us-poverty/

Kripke, D. (2019, November). *Brighten your life*. www.brightenyourlife.info/

Kuo, F. E., & Taylor, A. F. (2004). A potential natural treatment for attention-deficit/hyperactivity disorder: Evidence from a National Study 2004. *American Journal of Public Health*, *94*(9). www.niu.edu/~carter/courses/526/articles/Kuo_and_Taylor.pdf

Lai, O. (2021). *The world's mass extinction events, explained*. Earth.org. https://earth.org/what-and-when-were-the-mass-extinction-events/?gad_source=1&gclid=Cj0KCQiA5-uuBhDzARIsAAa21T8VIbjXMnu6GEUwMbJfZL_qMlwhum_VOLXFzcwhLcFWe3qMwWpqVzoaAlONEALw_wcB

Lall, S., Lebrand, M., & Park, H. (2021, July 29). *Slums, sprawl, and skyscrapers*. World Bank. https://blogs.worldbank.org/sustainablecities/slums-sprawl-and-skyscrapers

Lee, H. (2022). Are Millennials leaving town? Reconciling peak Millennials and youthification hypotheses. *International Journal of Urban Sciences*, *26*(1), 68–86. http://doi.org/10.1080/12265934.2020.1871061

Library of Congress (LoC). (n.d.). *Immigration and relocation in U.S. history: Removing native Americans from their land*. www.loc.gov/classroom-materials/immigration/native-american/removing-native-americans-from-their-land/

Lindsey, R. (2022, April 19). *Climate change: Global sea level*. NOAA. www.climate.gov/news-features/understanding-climate/climate-change-global-sea-level

Lindsey, R., & Dahlman, L. (2024). *Climate change: Global temperature*. https://www.energy.gov/sites/default/files/2024-02/093.%20Rebecca%20Lindsey%20and%20Luann%20Dahlman%2C%20NOAA%2C%20Climate%20Change_%20Global%20Temperature.pdf

Lockwood, B. (2020, June 16). *The history of redlining*. ThoughtCo. www.thoughtco.com/redlining-definition-4157858

Maas, J., Verheij, R. A., de Vries, S., Spreeuwenberg, P., Schellevis, F. G., & Groenewegen, P. P. (2009). Morbidity is related to a green living environment. *Journal of Epidemiology and Community Health*, *63*(12), 967–973. www.jstor.org/stable/20721102

Maphalala, K. (n.d.). *Native land taxes reclaim rights, power, and the environment*. Natural Investments, LLC. https://naturalinvestments.com/native-land-taxes/

Millennium Ecosystem Assessment. (2005). *Ecosystems and human well-being: synthesis*. Island Press.

National Aeronautics and Space Administration (NASA). (2020). *Vital signs: Ice sheets*. https://climate.nasa.gov/vital-signs/ice-sheets/#:~:text=Antarctica%20is%20losing%20ice%20mass,adding%20to%20sea%20level%20rise

National Aeronautics and Space Administration (NASA). (2021, November 9). *Emission reductions from pandemic had unexpected consequences on atmosphere*. NASA Jet Propulsion Laboratory, California Institute of Technology. www.jpl.nasa.gov/news/emission-reductions-from-pandemic-had-unexpected-effects-on-atmosphere

National Geographic (NG). (n.d.). *How close are you to a Superfund site*? www.nationalgeographic.com/superfund/

National Oceanic and Atmospheric Administration (NOAA). (2024a). *Trends in atmospheric carbon dioxide*. https://gml.noaa.gov/ccgg/trends/

National Oceanic and Atmospheric Administration (NOAA). (2024b). *2023 was the world's warmest year on record*. www.noaa.gov/news/2023-was-worlds-warmest-year-on-record-by-far#:~:text=Earth's%20average%20land%20and%20ocean,0.15%20of%20a%20degree%20C)

National Park Service (NPS). (n.d.). *Frequently asked questions*. www.nps.gov/faqs.htm

National Wildlife Federation (NWF). (n.d.). *Ecosystem services*. www.nwf.org/Educational-Resources/Wildlife-Guide/Understanding-Conservation/Ecosystem-Services#:~:text=Four%20Types%20of%20Ecosystem%20Services,regulating%2C%20cultural%20and%20supporting%20services

The New Climate Economy. (2018). *Unlocking the inclusive growth story of the 21st century: Accelerating climate change in urgent times*. The Global Commission on Economy and Climate. https://newclimateeconomy.report/2018/wp-content/uploads/sites/6/2018/09/NCE_2018_FULL-REPORT.pdf

Newman, P., & Kenworthy, J. (1999). *Sustainability and cities: Overcoming automobile dependence* (pp. 56, 78). Island Press.

Organization for Petroleum Exporting Countries (OPEC). (2022). *OPEC share of world crude oil reserves, 2022*. https://www.opec.org/opec_web/en/data_graphs/330.htm

Pew Research Center. (2020, June 27). *A growing share of the population lives in the suburban counties of large metro areas*. www.pewresearch.org/social-trends/2020/07/29/prior-to-covid-19-urban-core-counties-in-the-u-s-were-gaining-vitality-on-key-measures/psdt_07-28-20_suburban-urban-00-2/

Phillips, A. (2015, July 30). *Obama's clean power plan will actually lower your energy bill, according to new study*. Think Progress. https://archive.thinkprogress.org/obamas-clean-power-plan-will-actually-lower-your-energy-bill-according-to-new-study-141cad0314cd/

Rivard, B. (2022, June 13). *Vermont ski areas bounce back for the 2021–2022 season*. Ski Vermont. https://skivermont.com/vermont-ski-areas-bounce-back-for-2021-22-winter-season

Robbins, J. (2021, June 3). *How returning lands to native tribes in helping protect nature*. Yale Environment 360. https://e360.yale.edu/features/how-returning-lands-to-native-tribes-is-helping-protect-nature

Robinson, D. (2023). *15 biggest environmental problems of 2023.* Earth.org. https://earth.org/the-biggest-environmental-problems-of-our-lifetime/
Schaefer, S. (2014, October). The influence of cognitive load and walking speed on gait regularity in children and young adults. *Gait and Posture*, *41*(1).
Stager, C. (2011). *Deep future* (pp. 12, 26, 130–131). St. Martin's Press.
Thayer, R. L. (1994). *Gray world, green heart: Technology, nature, and the sustainable landscape*. John Wiley & Sons.
Thompson, M. (2011, April 24). *Has the U.S. wasted $8 trillion defending oil from the Persian Gulf.* Time. http://nation.time.com/2011/04/24/a-question-for-the-obama-administration/
Time, Inc. (2007). *Global warming: The causes, the perils, the solutions, the actions: What you can do* (p. 64). Time Books, Time Inc.
Tsosie, R. (2007). Acknowledging the past to heal the future: The role of reparations for native nations. In J. Miller & W. Kymlicka (Eds.), *Reparations* (pp. 47–68). Oxford University Press.
Twenge, J. M., & Campbell, W. K. (2018, October 18). Associations between screen time and lower psychological well-being among children and adolescents: Evidence from a population-based study. *Preventive Medicine Reports*, *12*, 271–283. http://doi.org/10.1016/j.pmedr.2018.10.003
UK Secretary of State for Business, Energy, and Industrial Strategy (UK SOS). (2022, September). *United Kingdom of Great Britain and Northern Ireland's nationally determined contribution.* https://unfccc.int/sites/default/files/NDC/2022-09/UK%20NDC%20ICTU%202022.pdf
United Nations (UN). (2020). *UN75 2020 and beyond*: *Shifting demographics.* www.un.org/en/un75/shifting-demographics
United Nations Environment Programme (UNEP). (n.d.). *Green economy.* www.unep.org/regions/asia-and-pacific/regional-initiatives/supporting-resource-efficiency/green-economy
United Nations Framework Convention for Climate Change (UNFCCC). (2022a). *What is the triple planetary crisis*? https://unfccc.int/blog/what-is-the-triple-planetary-crisis
United Nations Framework Convention on Climate Change (UNFCCC). (2022b, October 26). *2022 NDC synthesis report.* https://unfccc.int/ndc-synthesis-report-2022
United Nations Framework Convention for Climate Change (UNFCCC). (2022c). *2022 NDC synthesis report.* https://unfccc.int/ndc-synthesis-report-2022
United Nations Framework Convention on Climate Change (UNFCCC). (n.d.). *The Paris agreement.* UNFCCC. https://unfccc.int/process-and-meetings/the-paris-agreement
U.S. Census Bureau. (2022). *Nation's urban and rural populations shift following 2020 census.* www.census.gov/newsroom/press-releases/2022/urban-rural-populations.html#:~:text=Despite%20the%20increase%20in%20the,down%20from%2080.7%25%20in%202010
U.S. Census Bureau. (2024). *International database.* www.census.gov/data-tools/demo/idb/#/country?COUNTRY_YEAR=2023&COUNTRY_YR_ANIM=2023
U.S. Department of Agriculture (DOA). (2017). *Summary report: 2017 natural resources inventory.* Natural Resource Conservation Service, Center for Survey Statistics and Methodology, Iowa State University. www.nrcs.usda.gov/sites/default/files/2022-10/2017NRISummary_Final.pdf
U.S. Department of the Interior (DOI). (2022, June 29). *Secretary Haaland applauds return of traditional homelands to Onondaga nation.* www.doi.gov/pressreleases/secretary-haaland-applauds-return-traditional-homelands-onondaga-nation
U.S. Department of the Interior, Bureau of Land Management (BLM). (n.d.a). *Native American graves protection and repatriation act.* www.blm.gov/NAGPRA#:~:text=The
U.S. Department of the Interior, Bureau of Land Management (BLM). (n.d.b). *Tribal consultation.* www.blm.gov/tribalconsultation
U.S. Energy Information Administration (EIA). (2023c). *How much natural gas does the United States have and how long will it last*? www.eia.gov/tools/faqs/faq.php?id=58&t=8
U.S. Energy Information Agency (EIA). (2021a, April 13). *Frequently asked questions: What is the volume of world natural gas reserves*? www.eia.gov/tools/faqs/faq.php?id=52&t=8#:~:text=As%20of%20January%201%2C%202020,proved%20reserves%20of%20natural%20gas
U.S. Energy Information Agency (EIA). (2022, October). *Coal explained: How much coal is left*? www.eia.gov/energyexplained/coal/how-much-coal-is-left.php#:~:text=EIA%20estimates%20U.S.%20recoverable%20coal,58%25%20is%20underground%20mineable%20coal
U.S. Energy Information Agency (EIA). (2023a). *Annual energy outlook 2023.* www.eia.gov/outlooks/aeo/narrative/index.php

U.S. Environmental Protection Agency (EPA). (2017). *National water quality inventory: Report to Congress*. https://www.epa.gov/sites/default/files/2017-12/documents/305brtc_finalowow_08302017.pdf

U.S. Environmental Protection Agency (EPA). (2022). *Impaired waters restoration process: Listing*. U.S. EPA. www.epa.gov/tmdl/impaired-waters-restoration-process-listing

Vlahos, J. (2011, April 14). Is sitting a lethal activity? *The New York Times*. www.nytimes.com/2011/04/17/magazine/mag-17sitting-t.html?_r=0

Warren, R. (1998). *The urban oasis* (p. 18). McGraw Hill.

The White House. (2021, April 22). *Fact sheet: President Biden sets 2030 greenhouse gas pollution reduction target aimed at creating good-paying union jobs and securing U.S. leadership on clean energy technologies*. www.whitehouse.gov/briefing-room/statements-releases/2021/04/22/fact-sheet-president-biden-sets-2030-greenhouse-gas-pollution-reduction-target-aimed-at-creating-good-paying-union-jobs-and-securing-u-s-leadership-on-clean-energy-technologies/

World Bank Group. (2020, October 7). *Global action urgently needed to halt historic threats to poverty reduction*. www.worldbank.org/en/news/feature/2020/10/07/global-action-urgently-needed-to-halt-historic-threats-to-poverty-reduction

World Health Organization (WHO). (2018, September 13). *Preventing disease through healthy environments: A global assessment of the burden of disease from environmental risks*. www.who.int/publications/i/item/9789241565196

Worldwatch Institute. (2006). *Vital signs 2006–2007* (p. 92). W.W. Norton & Company.

4

PLANNING STRATEGIES AND TOOLS

Comprehensive plans

Comprehensive plans – also sometimes referred to as master plans, general plans, campus plans, or land use plans – are used to guide long-term development (~10–20 years) for cities/towns, college campuses, counties, or entire regions. Comprehensive plans usually define a set of working principles, strategies, or goals that address broad concepts such as sustainability, resiliency, environmental justice, sense of place, and diversity, equity, and inclusion and more granular concepts addressing health, public safety, circulation, transportation, accessibility, efficient use of facilities, historic preservation, adaptive reuse, land conservation/preservation, and land-banking sites for future development. The process usually has several phases, including a research or assessment phase where a core group investigates other relevant comprehensive plans and current affairs/trends and collects and analyses data; a public outreach phase that can refine goals and principles through community input from committees, charettes, meetings, and surveys; a phase where the plan is drafted; a phase of review and revision; and a phase of approval where the plan is adopted by stakeholders, committees, and/or governing boards.

Public participation and transparency in the process are imperative to produce an effective plan. This includes formulating a variety of communication methods to ensure diverse representation from all demographics that reside in a given community. In recent times, technology such as video conferencing, interactive mapping with crowd-sourcing data collection abilities, online surveys like Mentimeter, and social media have made it easier to receive more community input. While this can be advantageous, it can also be overwhelming given the tendency of people to be more unfiltered when commenting virtually than when they are in person. Although the planning staff that are managing the plan may have their own biases and preferences, they must do their best to stay objective by listening, respecting, and entertaining ideas from all corners of the community.

In addition to public involvement, the success of the comprehensive plan is dependent on various other political, economic, and social factors that can alter specific visions, recommendations, and guiding principles. For example, a college campus may designate a portion of the land as a "no-build" zone since the space is used by students for recreation. However, increased

DOI: 10.4324/9781032668666-4

enrollment and a shortage of housing on the campus and in the surrounding community may force decision-makers and/or developers to consider the site for more housing since it is the only site that can accommodate the amount of space needed that does not violate lot coverage limits as defined by the municipalities zoning ordinances. Likewise, an investor may want to develop a community solar project on land designated for agricultural use in the comprehensive plan. The prospect of increased renewable energy may satisfy a carbon-neutrality goal defined in the plan but may contradict a goal of preserving local agricultural land to reduce food miles traveled. Natural disasters and global pandemics may also shift goals in the plan. As a result of the COVID-19 pandemic, many plans are now looking to make existing spaces more efficient, adaptable, and resilient rather than expanding or building new facilities. Given these scenarios, it is important to acknowledge that the comprehensive plan isn't exactly written in stone but is more organic with moving targets.

Subdivision regulations and zoning ordinances

While zoning ordinances establish specific uses or types of development across different zones, subdivision regulations guide the pattern of development on the land by establishing lots/parcels that are safe and compatible with the physical characteristics of the site. Often, public capital investment in infrastructure can influence the parcels that are created. Subdivision regulations can ensure that lots are adequately sized, can be accessed by roads, driveways, and sidewalks, and have access to potable water, sewer, and utilities. In addition, the regulations can protect open space and sensitive ecological or cultural areas and can ensure development is in harmony with existing settlement patterns. Subdivision regulations can be used solely or in conjunction with zoning ordinances.

Although architects and developers often regard zoning ordinances as nuisances that interfere with their design concepts, such regulations directly influence the look, feel, and function of space in communities. Zoning ordinances can specify site layout characteristics, including lot area, setbacks (the distance from front, rear, and side boundaries), lot coverage (amount of impervious surface), stormwater flows/detention, parking requirements, curb cuts, and roadway/driveway placement; structural requirements, including height, the number of stories, floor area ratios, materials, or structural requirements in response to specific climatic conditions of the region (i.e., hurricanes, tornadoes, flooding, snow loads, etc.); and of course permitted uses and exceptions. There are various types of zoning ordinances that can be used individually or in combination with each other, including functional zoning, incentive zoning, intensity zoning, cluster zoning, performance zoning, and form-based zoning.

Functional zoning

Functional zoning – also referred to as Euclidean zoning that was named after a land use case in Euclid, Ohio, in 1926 – is the most common type of zoning that separates land use by type (commercial, industrial, residential, recreational, green space, etc.). One common criticism of functional zoning is that it limits a mix of uses and/or designs, resulting in insular, drab districts with dead spaces. People are forced to live in one area and work or shop in another with few pedestrian connections between the zones. Given its restrictive use requirements, functional zoning can be made more flexible by replacing and/or enhancing it with other zoning strategies discussed in the following text.

Cluster, incentive, and inclusionary zoning

Cluster zoning or subdivision allows the minimal lot size to be reduced and structures to be placed closer together given the space saved is used for open space or public purposes. Incentive zoning permits certain allowances in exchange for something needed by the community. For example, residentially zoned districts might allow developers to exceed the permitted lot coverage or a minimum number of units per acre if they agree to build a specific percentage of low or moderately priced housing units or housing for the elderly. Likewise, a developer may be allowed to exceed minimum heights in a commercially zoned district if the building will have net-zero emissions. Other towns offer incentives in the form of tax incentives if developers agree to retrofit an existing structure or build on an existing developed site rather than build a new one or build on an undeveloped site. In contrast to incentivizing a specific type of use, some municipalities build inclusionary zoning requirements into their ordinances so they are not optional, which can be more effective in promoting social equity since many developers may or may not feel the incentives are adequate.

Intensity zoning

Intensity zoning divides land use by permitted density. This can be defined as a permitted number of units per acre or in the form of impervious lot coverage. For example, a city may allow 200 units per acre in the urban center, 50 units per acre on the perimeter, and 5 units per acre on the outskirts. Likewise, a city may permit a total impervious lot coverage of 75 percent per parcel in the central business district but only 40 percent per parcel in the outskirts. Similar to performance zoning and form-based zoning discussed in the next section, intensity zoning can allow more flexibility in the type of development since the developers may not necessarily be restricted to a specific use – that is unless it is combined with functional zoning. For example, some municipalities have multiple residential districts (R1, R2, R3, etc.) or multiple commercial districts (C1, C2, C3, etc.), each with its own set of density requirements.

Performance zoning

Performance zoning establishes standards a development must meet rather than a specific use. This allows for a mix of uses if the development fulfills the standards outlined in the ordinance. Typical operational standards that may be specified include noise, vehicle trips, commercial hours, pervious pavement, stormwater capture, or location of driveways and parking. As with intensity zoning, however, performance zoning can also be used in conjunction with functional zoning whereby both use and performance measures are specified.

Although it is a state-wide law that developers must conform to in addition to the zoning ordinances stipulated by local municipalities, Vermont's Act 250 Land Use and Development Act can be considered an "environmental performance-based" regulating measure. Signed into law by Governor Deane Davis in 1970, the law created nine District Environmental Commissions to administer the law and provide a public, quasi-judicial process for reviewing and managing the environmental, social, and fiscal consequences of major subdivisions and developments in Vermont. Before granting a permit, the District Commission must ensure that the development or subdivision meets the following ten criteria:

- Will not result in undue water or air pollution.
- Has sufficient water available for the needs of the subdivision or development.
- Will not unreasonably burden any existing water supply.
- Will not cause unreasonable soil erosion or affect the capacity of the land to hold water.
- Will not cause unreasonably dangerous or congested conditions with respect to highways or other means of transportation. As appropriate, will incorporate transportation demand management strategies and provide safe access and connections to adjacent lands and facilities and to existing and planned pedestrian, bicycle, and transit networks and services.
- Will not create an unreasonable burden on the educational facilities of the municipality.
- Will not create an unreasonable burden on the municipality in providing governmental services.
- Will not have an undue adverse effect on aesthetics, scenic beauty, historic sites, or natural areas, and will not imperil necessary wildlife habitat or endangered species in the immediate area.
- Conforms with the Capability and Development Plan, including the following considerations: the impact the project will have on the growth of the town or region, primary agricultural soils, productive forest soils, earth resources, extraction of earth resources, energy conservation, private utility services, costs of scattered developments, public utility services, development affecting public investments, settlement patterns.
- Conforms with any local or regional plan or capital facilities program (NRB, n.d.).

According to Courtney and Zencey (2012, p. 47),

> Over time, the Act 250 review process has approved 98 percent of the permits that have come before it – though the vast majority of those permits were for projects whose design was amended and improved in the permitting process. The law didn't put a halt to development, but it gave the people of the state, through its processes and criteria, a strong voice in setting limits and conditions.

Form-based zoning

Form-based code defines zones by their physical characteristics and has a stronger emphasis on physical aesthetics, context, and building form rather than by use stipulated in functional zoning. Andres Duany – architect, urban planner, and founder of the New Congress for Urbanism – devised an urban-to-rural transect system that categorizes elements of the built environment along a line from the rural periphery to the urban core. The transect, which provides the basis for the Smart-Code form-based zoning model, consists of six zones: T1, the natural zone, includes wilderness, preserved lands, and lands not suitable for development; T2, the rural zone, includes open fields and/or cultivated areas with sparse development; T3, the suburban zone, includes low-density residential areas with large setbacks and cul-de-sac streets; T4, the general urban zone, includes mixed-use and residential with a variety of styles; T5, the urban center zone, includes dense mixed-use with shallow building setbacks; and T6, the urban core, which has the highest density and mix of uses. An additional DA category is assigned to airports, college campuses, amusement parks, or large industrial parks that do not fit into the transect. Although form-based code allows for more of a mix of uses than functional zoning, some designers might still feel constricted by the transect. From a greening perspective, form-based code seems more concerned with preserving land in the T1 and T2 sectors and may neglect to soften the T5 or T6 areas. In some ways, contrary to claims of flexibility, it still seems the transects may tend to "type-cast" specific design aesthetics that may or may not be compatible or even desired in a particular community.

Relative to the "regreening" theme of this book, it seems that a combination of incentive, inclusionary, performance, cluster, and intensity zoning is needed to preserve natural features and open space and foster social equity. Although performance measures, clustering development, and density restrictions may preserve natural elements and protect a community's character, they may make development more difficult, which ultimately leads to escalating real estate prices, countering any benefit to social equity. Combining performance, intensity, or cluster zoning with inclusionary and/or incentive-based zoning may be needed to provide adequate and affordable housing opportunities to prevent the displacement of lower- and middle-income groups from communities or even entire regions.

The aforementioned subdivision regulations and zoning ordinances are more effective before development occurs. Once development sprawls across the landscape, it is difficult to reobtain the land and/or re-cluster development into compact nodes. The ordinances are also privy to politics, legalities, budgets, socioeconomics, and tax base. Some communities are apprehensive about applying too many zoning restrictions for fear they will lose out on tax income. In amending zoning, it all comes down to the makeup of the planning boards – some will certainly have a more capitalistic view, while others will take a more ecological or social-centric view. Variance within and between communities also creates chasms in the contiguity of community or regional comprehensive plans. Regional governance can mitigate this fragmentation if regulations are imposed on the individual jurisdictions that require them to conform to a regional comprehensive plan. Portland's metropolitan governance (discussed later) is a good example of a regional body coordinating with local communities to guide cohesive, future development.

Development agreements

Some municipalities also have legislation enabling them to enter directly into a development agreement with a developer, which allows developers to bypass the existing ordinances if they are in compliance with the terms outlined in the agreement. This can be useful if the project is complex and might require rezoning, conditional use permits, or variances. However, such agreements have also been criticized for their lack of transparency and/or as being undemocratic since they can circumvent the planning and review process. Some municipalities often create zoning overlays to redefine uses and allowances more than what is stated in the zoning ordinances. Landowners developing a planned unit development (PUD) may also get the benefit of being able to apply a mix of different uses otherwise not allowed in the zoning ordinance. For example, a townhouse or condominium development that has a homeowner's association might be in a residential zone, but if it is approved as a PUD, it might allow for a mix of residential, commercial, and recreational uses.

Land conservation and preservation

It would be remiss to not discuss land conservation and preservation at this point, as both should always be considered prior to any project and/or should be integrated into both local and regional plans and zoning regulations to ensure open space is not consumed by irresponsible patterns of development. Conservation is also a "preventative" strategy in protecting land and natural resources as opposed to the "regenerative" strategies discussed in the succeeding chapters of this book. Conservation and preservation are slightly different strategies but often work together to achieve the common goal of sustaining species and ecosystems. Land conservation balances the needs of people and nature by simultaneously protecting unused or underused lands while also providing benefits to people. Therefore, conservation does not prohibit human interaction with or use of the land but rather strives to manage the use so it does not exceed the capacity of the land. Preserving land, on the other hand, focuses on ensuring the permanence of a species or system regardless of its potential functionality. Such preserved land is designated "wilderness," as defined by the Wilderness Act of 1964, which prohibits commercial enterprise, permanent roads, motor vehicles, motorized equipment or motorboats, aircraft, and permanent structures. The act does permit activities that existed on the land before it was designated a wilderness, such as grazing, mineral extraction, and timber harvesting, but bans such activities in newly designated areas. When the act was initiated, 54 areas comprising 9.1 million acres in 13 states were designated wilderness and integrated into the National Wilderness Preservation System; currently, there are 762 areas comprising roughly 111 million acres managed by the National Park Service, the Bureau of Land Management, the U.S. Fish and Wildlife Service, and the U.S. Forest Service (The Wilderness Society, n.d.).

In 1965, Congress also created the Land and Water Conservation Fund (LWCF) with a bipartisan commitment to safeguard natural areas, national parks and forests, water resources, cultural heritage, and recreation opportunities. The LWCF uses revenue from offshore oil and gas drilling to support the fund. The LWCF also provides matching grants to state and tribal governments for the acquisition and development of public parks and recreation areas. Since 1965, there have been more than 45,000 projects, $5.2 billion in funding, and over eight million acres protected (DOI, n.d.; NRPA, n.d.). The Great American Outdoors Act was signed into law in 2020 authorizing $900 million annually in permanent funding for the LWCF and the Outdoor Recreation Legacy Partnership is funded by the LWCF and enables communities to create outdoor recreation spaces, parks, and connections between people and the outdoors in economically underserved communities. Although the fund has been susceptible to partisan politics and diverted in times of economic recession, the data shows that the land conserved has had a positive economic impact in nearly every county in the United States. It is estimated that the LWCF generates $646 billion annually and supports more than six million jobs (NRPA, n.d.).

The debate between conservationists, preservationists, and private landowners over the most effective way to manage land can be dated back to the early twentieth century, when forester Gifford Pinchot led the conservation movement and naturalist John Muir led the preservation movement. Both agreed that land had its limits, but Pinchot favored an approach that generated more economic benefit – for the greatest good for the greatest number over the greatest period of time – while Muir favored an approach that fully preserved all of the natural elements of an ecosystem with minimal human impact. This debate continues today, and some groups feel neither approach is truly sustainable. For example, national parks and national wildlife refuges, which fall somewhere between conservation and preservation, have the intent to protect areas of scenic or ecological significance while also providing public access. However, as the number of visits to such areas continues to increase, these goals are often seen as antagonistic. As Bill Bryson (1998, p. 47) states in his book *A Walk in the Woods*:

> The National Park Service has somewhat of a tradition of making things extinct . . . and, in fact, mostly what the Forest Service does is build roads. There are 378,000 miles of roads in America's National Forests with the main purpose of allowing private timber companies to get to previously inaccessible stands of trees.

The U.S. Forest Service did adopt a policy known as "ecosystem management" in the 1990s based on newer, more sustainable approaches to forestry management that emphasize ecological health over log harvesting. However, it will take many years to reverse the damage already done during a long history of irresponsible forestry management.

TDRs and PDRs

Conservation and preservation initiatives require either the direct acquisition of land through outright purchase or donation or the negotiation of easements with landowners. Acquisitions are usually conducted by state or federal governments, universities, or local or national non-profit land trusts. Since public funding is usually limited, land trusts are often better suited and more prepared to negotiate with private landowners. For example, the Land Trust Alliance currently represents more than 950 land trusts in 93 percent of U.S. counties; 80 percent of land trusts provide public access to their protected lands, amounting to 1.2 million acres of land and 9,761 miles of trails open to the public (LTA, n.d.).

If the outright purchase of a property is out of the question, conservation easements and development rights can be negotiated with landowners. In a conservation easement, the landowner voluntarily gives up specific rights with respect to how the land will be developed but retains ownership and use of the land. Most conservation easements are donated by the landowner, making them eligible for tax credits. Easements are also permanently attached to the property in perpetuity, which protects the land even if it is sold in the future. If the land is not donated, states, communities, and land trusts may attempt to negotiate a purchase-of-development rights (PDR) or transfer of development rights (TDR) with the landowner, where public or private funds are used to purchase the land's development rights. As with conservation easements, the ownership remains with the private landowner. The cost of development rights is based on an appraisal and is the difference between the fair market value of the land without the easement and its value as restricted by the easement. This situation can be a win–win for both communities and landowners – the latter are compensated but retain ownership of their land, while the former are able obtain open space to protect wildlife, preserve agriculture, and manage growth. Funds are often generated by private donations, grants, state/federal appropriations, lotto revenue, or bonds. Bonds that are voted on for open space/land conservation have shown high success rates in recent times. According to Planners Web, between 1999 and 2002, 544 successful state and local ballot measures generated approximately $16.7 billion in funds for land conservation programs. In 2003, voters approved an additional 99 measures with a total value of $1.3 billion. Overall, between 75 and 80 percent of initiatives and referendums that raise taxes or fees for land conservation have passed (Miller & Krieger, 2004).

The state of Vermont has been a leader in conservation and preservation efforts since the 1960s. In 1960, the Vermont chapter of the Nature Conservancy was started by two professors from the University of Vermont – James Marvin and Hub Vogelmann – who began purchasing and preserving environmentally sensitive land across the state. The Vermont Natural Resources Council was established in 1963 and the Vermont Land Trust in 1977. Since its establishment, the latter has permanently conserved more than 2,200 parcels of land covering more than 620,000 acres, or about 11 percent of the private, undeveloped land in the state. This conserved land includes more than 775 working farms, hundreds of thousands of acres of productive forest land, and numerous parcels of community land (VLT, n.d.).

Bolton Valley Nordic Center

The Bolton Valley Resort in northern Vermont contains 1,144 acres of forested land that provides habitat for countless plant and animal species and houses, an alpine, lift-serviced ski/snowboard area, and over 60 miles (100 km) of Nordic and backcountry ski trails for recreational use. The trails were cleared over 40 years ago by Gardiner Lane and a group of friends who are now referred to as the "Old Goats." In 2011, news surfaced that the land was going to be sold to a private developer and closed to the public. Members and friends of the Old Goats, along with many other outdoor enthusiasts, were shocked to hear that this pristine wilderness might be lost forever. Luckily, one of the Old Goats was (and still is) working part-time on the ski patrol for Bolton Valley. Ann Gotham, a full-time nurse practitioner, became the leading voice on this issue and spearheaded the formation of the Friends of Bolton Valley Nordic and Backcountry. She wrote to the potential buyer and explained the value of the wilderness and the connection people had with the land and eventually convinced them to withdraw their proposal.

Working with the Vermont Land Trust (VLT), the Friends of Bolton Valley Nordic and Backcountry then negotiated a price with Redstone Commercial Group (the owners of Bolton Valley) of just over $1.8 million. The VLT would be the buyer and would eventually turn the land over to the state to be added to Mount Mansfield State Forest. The challenge, of course, was raising the funds. The first major donation came from an $800,000 grant from the Vermont Housing and Conservation Board. Then, over the next 15 months, the VLT received over 1,200 donations, ranging from $1 raffle tickets to three anonymous $100,000 gifts. Just before the deadline, the Conservation Alliance came in with $50,000 to close the funding gap. As Ann Gotham said of this extraordinary accomplishment, "The passion to conserve this much-loved parcel ignited the hearts, minds, and passions of a large community of people of all ages extending far past the boarders of Vermont."

Elise Annes, Vice-President of Community Relations for the VLT, added:

> Every so often a conservation opportunity comes along that is at the heart of what Vermonters want to hold onto . . . [B]y helping to protect the Bolton Nordic and backcountry land, hundreds of volunteers, donors, and supporters have made their wishes known – public recreation and natural habitat are a part of what makes Vermont special.

The VLT and the Department of Forests, Parks, and Recreation created a fund for the future care of this land. The state also issued a license agreement to Bolton Valley Resort to maintain the Nordic and backcountry trails; the Department of Forests, Parks, and Recreation manages the rest of the property (FOBVNB, 2016).

Urban growth boundaries

In Portland, Oregon, a metropolitan form of governance known as Metro has initiated progressive land use strategies and environmental legislation that has proven to be successful in fostering smart growth and sustainability. Metro works with local governments, businesses, environmental interests, and citizens to develop future settlement plans. Prior to the establishment of Metro, Governor Tom McCall signed Senate Bill 100 into law in 1973 that created the Land Conservation and Development Commission and the Department of Land Conservation and Development. The bill required urban growth boundaries, wise use of urban land, and the

FIGURE 4.1 Land conserved at the Bolton Valley Nordic Center.

protection of natural resources. The first urban growth boundary was proposed for Portland in 1977 and eventually inherited by Metro when it was established by voters in 1978. Since this time, all the cities and metropolitan areas in the state of Oregon have established urban growth boundaries to control urban expansion onto farm and forest lands.

Metro has also worked with citizens to establish a 50-year growth strategy known as the 2040 Growth Concept. The policies in the 2040 Growth Concept are to encourage safe and stable neighborhoods for families, compact development that uses land and money efficiently, a healthy economy that generates jobs and business opportunities, protection of farms, forests, rivers, streams, and natural areas, a balanced transportation system to move people and goods, and housing for people of all incomes in every community (Metro, n.d.).

Such regional cooperation and sustainable planning have also resulted in the prevention of highway construction through the city and contributed to the construction of the 15-mile (24-km) Metropolitan Area Extension (MAX) light rail system, which connects downtown Portland to the suburban community of Gresham, the replacement of a downtown freeway with a riverfront park (discussed further in Chapter 7), the introduction of trees, seats, flower planters, sculptures, lighting, and artwork at street level, the development of new urban residences, and extensive traffic calming schemes, including exclusive rights of way and the city-sponsored Reclaim Your Street program. Suburbs have also witnessed a reversal of urbanization through the Skinny Streets program, where several roads have been reduced to 18-foot (5.5-meter) widths.

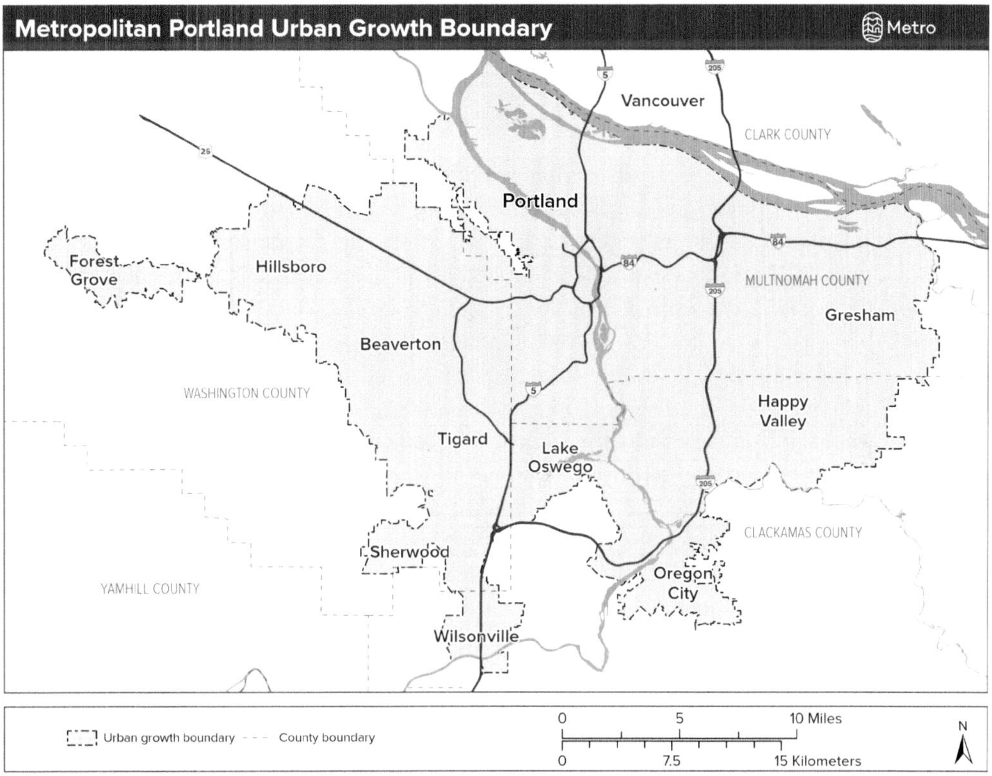

FIGURE 4.2 Portland's urban growth boundary.

Source: Courtesy of Metro

Future planning identifies nine regional centers interconnected by light rail links. Since 1990, regional voters have agreed to tax themselves to fund a fourfold expansion of the MAX system that will create a 58-mile (93 km) network that serves all quadrants of the region (Metro, n.d.). In addition to Metro planning, proactive public–private master planning is under way for the western corridor of Portland. This initiative has been orchestrated by the 1,000 Friends of Oregon, an independent environmental interest group that carried out a project known as LUTRAQ (Land Use, Transportation, and Air Quality Study) to identify and demonstrate alternatives to a freeway planned through the west side. The study rearranged land use into four types of development, including mixed-use centers to urbanize existing downtown areas along planned light rail lines, urban transit-oriented developments (TODs) at station areas, neighborhood TODs within a short feeder bus ride of the light rail, and secondary areas within a mile of each center. Overall, LUTRAQ has influenced west-side progress since 1990, including the construction of more than 6,000 dwellings and $230 million of mixed-use development (Metro, n.d.).

Portland has been gradually reinvented around the idea that it should be a human, livable, environment in which people can walk, sit, dine alfresco, be entertained, and find residence in housing set in attractive public environments – overall, the combination of enhanced transit access, human attractions, and extra housing, rather than more road space and parking, has improved the local economy by increasing from 5 to 30 percent of the metropolitan area's total retail turnover (Cervero, 1998, p. 421).

Geospatial analysis

As mentioned in Chapter 2, Scottish landscape architect Ian McHarg demonstrated a comprehensive method of site analysis in his book *Design with Nature* that formed the basis of the environmental impact statement process and was a precursor for modern Geographic Information Systems (GIS) software. McHarg utilized mylar overlays to examine the ecological processes inherent in a site, including soils, hydrology, groundwater, watersheds, viewsheds, topography, vegetation, wildlife, local weather patterns, wind, sun, and geological features. This analysis introduced ecological awareness and environmental sensibility into the field of design that was – up to that point – more preoccupied with aesthetic and human dominance over nature.

With advancements in GIS software, Light Detection and Ranging systems, aerial photography (historic and current), drones, and planimetric mapping along with the centralization of online data by many municipalities and regions, site analysis has become much more sophisticated. Like everything else in the digital world, it has made analysis both easier due to the accessibility of information and more difficult due to sheer the amount of data that can be analyzed. In the state of Vermont, the Agency of Natural Resources (VT ANR) has developed a state-wide interactive Bio-Finder map that identifies landscape scale components, including interior forest blocks, connectivity blocks, riparian wildlife connectivity, surface water and riparian areas, physical landscape blocks, and physical landscape diversity; and community and species scaled components, including natural communities, aquatic habitats, wetlands, terrestrial wildlife crossings, riparian wildlife crossings, and rare and uncommon species. Overall, the various layers are used to identify the highest priorities for maintaining Vermont's ecological function and have been a useful tool for identifying major natural-built conflict areas. When utilizing the regreening strategies discussed throughout this book, these layers can be useful for prioritizing conflict areas that need mitigation. They are also useful in networking green infrastructure at multiple scales, including connecting smaller "stepping-stone" green infrastructure at the site level to larger green infrastructure at community and regional levels. For example, a green roof on a building can be connected to a green wall, which can be connected to a bioretention area, which can be connected to a border garden that connects to a riparian buffer along a stream that flows into a river that ultimately flows into a lake. Using a neighborhood in Shelburne, Vermont, as an example, Figure 4.3 illustrates an analysis of a site using a sample of the VT ANR layers and how they connect to priority natural areas in the community and the region.

Orthophotography courtesy of Esri, DigitalGlobe, GeoEye, i-cubed, USDA FSA, USGS, AEX, Getmapping, Aerogrid, IGN, IGP, swisstopo, and the GIS User Community; data layers courtesy of Vermont Agency of Natural Resources (ANR, n.d.).

Sprawl indexes

In addition to ecological fragmentation, geospatial analysis can provide insight into sprawl-induced social fragmentation by identifying conflicts with pedestrian-scale movement in a community. Although visual analysis can reveal aggregate patterns, measuring the degree

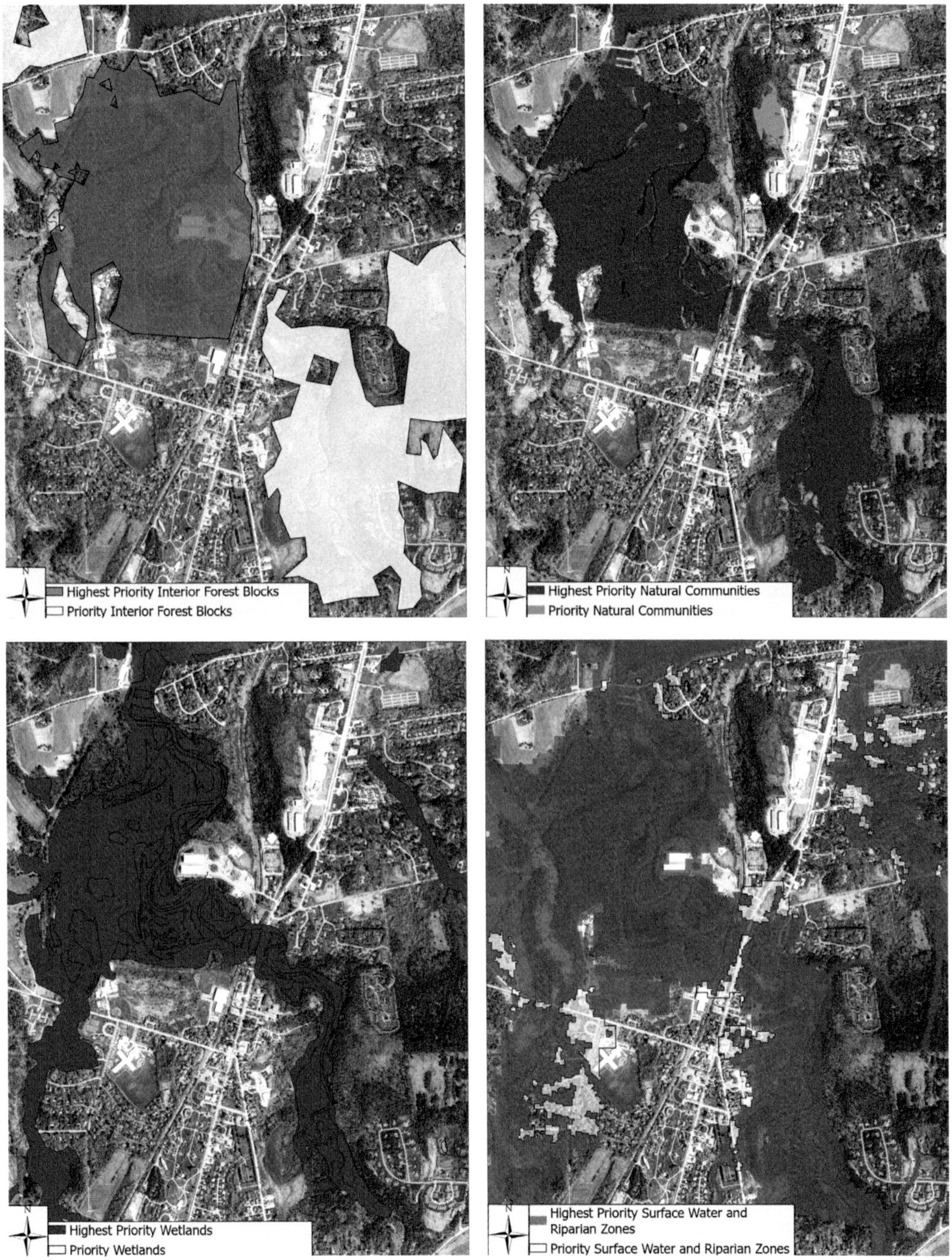

FIGURE 4.3 GIS analysis using sample data layers from the Vermont Agency of Natural Resource's Biofinder. Top left: interior forest blocks; top right: natural communities; bottom left: wetlands; bottom right: surface water and riparian zones.

of existing sprawl and how a particular design solution impacts that metric can also enhance the poignancy of a project. Smart Growth America (2014) measured the costs and benefits of sprawling development in 221 metropolitan areas and 994 counties in the United States using four primary factors: residential and employment density; neighborhood mix of homes, jobs, and services; strength of activity centers and downtowns; and accessibility of the street network. Each of the factors had specific components that were scored to create a sprawl index. Each of the four primary factors was equally weighted creating an average index of 100. Areas with scores over 100 were considered to be more compact and connected, while areas scoring under 100 were considered more sprawling. Using two of the regions that are used for several case studies throughout this book, Rochester, New York, has a sprawl index of 74.50 and ranks 189 out of 221 metropolitan areas analyzed by Smart Growth America, while Burlington/South Burlington, Vermont, has a score of 135.06 and ranks 14 out of the 221 areas examined. Even though these regions have a large gap between their scores, both still have ample opportunities for regreening and restoration projects that could contribute to sustainability. However, Rochester may require more strategies to reverse existing sprawled patterns of development while Burlington may require more strategies to prevent further sprawl. Some models also use more spatial methods to calculate sprawl, such as using land cover data or lot coverage to calculate percentages of undeveloped land in defined grid cells, while other models focus on population or housing density as their primary indicators. Although many of these indexes may seem approximate, they still can provide an effective methodology for quantifying the impact of a particular design solution and giving the field of design more grounding in science rather than the "because it looks good" reasoning.

Segregation indexes

As discussed in Chapter 3, the impacts of color-coding zoning practices in the early to mid-twentieth century are still prevalent in many neighborhoods today. Although many metropolitan areas still see stark divisions between urban and suburban communities, disparities in prices between housing developments are also yielding distinct enclaves within suburban communities. Before implementing any new urban or suburban retrofit, it is important to understand the demographic and geographical distribution of people within a community. Massey and Denton (1988) conceptualized residential segregation among five distributional characteristics, including evenness, or the degree to which groups are distributed proportionately across different areas; exposure, or the extent to which members of different groups share common residential areas within a city; concentration, or the degree of a group's agglomeration in a space; centralization, or the extent to which a group resides toward the center of a city; and clustering, or the degree to which specific groups are adjacent to one another. The authors also present several indices for each of the characteristics, including the index of dissimilarity, the Gini coefficient, the entropy index, and Atkinson index for evenness, the interaction and isolation indexes for exposure, the delta index for concentration, a proportion of the total and relative centralization index for centralization, the index of absolute clustering, the index of spatial proximity, and distance decay index for clustering.

Although it is limited to comparing only two variables, the index of dissimilarity is the most common formula used to compare the geographical distribution of one population subgroup versus another. With improvements in U.S. Census Data collection, the distribution by race, ethnicity, income, gender, age, migration patterns, life expectancy, income, and employment/

unemployment, to name a few, can be analyzed. The index of dissimilarity can be calculated as follows:

$$D = 100*0.5*\sum \left| Pxi/Pxt - Pyi/Pyt \right|$$

D = index of dissimilarity being compared for two groups
∑ = sum of all tracts being compared
Pxi = the population of group x in tract i
Pxt = total population of group x across all tracts being compared
Pyi = the population of group y in tract i
Pyt = total population of group y across all tracts being compared

The closer the index is to 0 means more integration of the two subgroups and an index closer to 1.0 means less integration of the two subgroups. Using the four neighborhood quadrants in the city of Rochester, New York, as an example and comparing the white and black populations, the index is 0.44, meaning 44 percent of blacks or whites would have to move to achieve an even distribution of population by race throughout the entire city.

Understanding how the data informs planning and design decisions can be more complex than the formulas themselves. Although several studies show that exposure to diversity is positive for all people by broadening social networks, increasing tolerance and fair-mindedness, and decreasing racial and ethnic prejudice, many communities may be resistant to change (Turner & Rawlings, 2009). For example, some lower-income communities may be apprehensive about infrastructure improvements and/or new housing projects out of fear of an increase in taxes and/or assessed values, which could drive up the cost of real estate and subsequently cause gentrification that further displaces citizens to even more desolate, fringe locations. Those who do have the means to relocate may not do so for fear of being outcasts or being labeled as "tokens" in predominately white neighborhoods. Some middle- or upper-income neighborhoods may adopt the "not in my backyard" (NIMBY) mentality regarding diversity in their neighborhoods out of cultural bias and/or misconceived stereotypes that they feel will "tarnish" their community. When I lived in Buffalo, New York, while attending college in the 1990s, I remember many community members in the suburbs expressing concern that the "city element" would infiltrate their neighborhoods if the light rail was extended. Even more recently in 2020, my family sold the house that we lived in for 45 years. One of the neighbors – whom we never even met up to that point – rang our doorbell and asked us, "Are the people who bought your house black?" I have also had discussions with people in upper-income brackets who argue, "I worked hard to

TABLE 4.1 Segregation index across four neighborhoods in the city of Rochester

Quadrant	*Total*	*White*	*Black*	*Wi/Wt*	*Bi/Bt*	*Abs. value: Wi/Wt–Bi/Bt*	*Wi/Wt-Bi/Bt*2*
NE	64,731	12,467	28,616	0.14	0.32	0.18	0.37
NW	54,727	17,711	19,962	0.20	0.23	0.03	0.06
SE	64,692	48,446	8,617	0.54	0.10	0.44	0.88
SW	49,054	11,153	31,361	0.12	0.35	0.23	0.46
Total	**233,204**	**89,777**	**88,556**			**0.88**	
Average							**0.44**

get to where I am, why should someone who didn't work hard get to live as well as me?" In the twenty-first century, these types of attitudes seem contemptible, but discrimination and fear of difference are both sadly still prevalent.

The reality is there are people who work hard and succeed, people who don't work hard and succeed, people who work hard and don't succeed, people who don't work hard and don't succeed, people who are lucky, people who are unlucky, and everything in between. Wealth does not always make you "good" and poorness does not always make you "bad." And of course, everyone's definitions of "success" and what constitutes "hard work" and "worthiness" are subjective. It is important to keep in mind the various struggles people may have endured – including racism, sexism, discrimination, mental/physical abuse, bullying, trauma, physical assault, sexual assault, mental/physical health problems, addiction, and exposure to unhealthy environments – and the subsequent uneven distribution of opportunities and privileges that are generated from said struggles. All because someone does not achieve an arbitrary measure of success (whatever that may be) doesn't mean they should be cast aside and/or denied access to decent living conditions and healthy communities. Society functions better when the opportunity is equalized and people can live, work, and recreate in places of their choosing.

Despite ongoing discrimination, younger generations (Millennials and Generation Z) tend to be more progressive than their predecessors (Generation X, Baby Boomers, Silent Generation). Data from the Pew Research Center shows 62 percent of Generation Z and 61 percent of Millennials believe diversity is good for society compared to 52 percent of Generation X, 48 percent of Boomer, and 42 percent of Silent Generation (Parker et al., 2019). As a result, younger generations may be more inclined to make existing neighborhoods more diverse and inclusive rather than gravitating toward a specific cultural or ethnic enclave. This can influence the types of homes, businesses, and urban forms that can be implemented when retrofitting existing or planning for new communities. In response to the changing culture, it is also imperative to have diversity, equity, and inclusion in the decision-making process by comprising boards, forums, and planning staff with representative citizens from their respective communities. This can enhance transparency and can provide much-needed qualitative data to the quantitative indexes that can reveal the needs, wants, desires, and fears of the community.

Homeownership ranks high as a metric for achieving the American Dream. Since historically single-use zoned neighborhoods have catered to middle- to upper-income households, providing mixed-income housing is one strategy that can foster diversity. Creating new multifamily housing options and/or rezoning existing single-family zoned neighborhoods to include multifamily housing can provide apartments, duplexes, in-law apartments, and townhouses across a mix of price ranges that can foster more variety and flexibility in where people can and choose to live. The keywords here are *mixed* and *choice*. Gating off exclusively subsidized communities doesn't provide the parity that results from integrating different incomes and styles of homes. Large, subsidized housing projects often stigmatize lower socio-economic citizens and create inter-community divisions of haves and have-nots. In addition to altering zoning regulations, housing developers can be incentivized to provide a more diverse range of housing styles in a development. Such incentives include expedited permitting, waivers for building permit fees, and density bonuses that allow a builder to construct more units if a specific portion of them are considered affordable. The Federal Housing Administration, the U.S. Department of Housing and Urban Development, the U.S. Department of Veteran Affairs, and the U.S. Department of Agriculture (USDA) all offer government assistance, grants, and loans for first-time buyers and lower-income households that can also increase options for lower-income citizens. Certainly, not all suburban neighborhoods have to be rezoned or redesigned, nor should everything always be 100 percent mixed-income/use. Finding the sweet spot of variety and placement of different housing styles is unique for every community and can vary due to demographics, geography, existing infrastructure, tax base, income, and politics.

References

Bryson, B. (1998). *A walk in the woods: Rediscovering America on the Appalachian trail* (p. 47). Broadway Books.

Cervero, R. (1998). *The transit metropolis*. (p. 421). Island Press.

Courtney, E., & Zencey, E. (2012). *Greening Vermont: The search for a sustainable state* (p. 47). Thistle Hill Publications and Vermont Natural Resources Council.

Friends of the Bolton Valley Nordic & Backcountry (FOBVNB). (2016). *Who are we*? https://bolton-friends.org/friends-of-bolton-valley-nordic-and-backcountry-fobvnbc-who-are-we-2/

Land Trust Alliance (LTA). (n.d.). *What is a land trust*? https://landtrustalliance.org/why-land-matters/land-conservation/about-land-trusts

Massey, D. S., & Denton, N. A. (1988). The dimensions of residential segregation. *Social Forces*, *67*(2), 281–315. https://doi.org/10.2307/2579183

Metro. (n.d.). *2040 Growth concept*. www.oregonmetro.gov/2040-growth-concept

Miller, G., & Krieger, D. (2004, January 23). *Purchase of open development rights: Preserving farm land and open space*. Planners Web. https://plannersweb.com/2004/01/purchase-of-development-rights-preserving-farmland-and-open-space/

National Recreation and Park Association (NRPA). (n.d.). *The land and water conservation fund*. www.nrpa.org/our-work/advocacy/the-land-and-water-conservation-fund-lwcf/#:~:text=LWCF%20funding%20has%20protected%208%20million%20acres%20across%20all%2050%20states

Parker, K., Graf, N., & Igielnik, R. (2019, January 17). *Generation Z looks a lot like Millennials on key social and political issues*. Pew Research Center. www.pewresearch.org/social-trends/2019/01/17/generation-z-looks-a-lot-like-millennials-on-key-social-and-political-issues/

Smart Growth America. (2014). *Measuring Sprawl 2014*. https://smartgrowthamerica.org/resources/measuring-sprawl-2014/

State of Vermont Natural Resources Board (NRB). (n.d.). *Act 250 program*. https://nrb.vermont.gov/act250-program

Turner, M. A., & Rawlings, L. (2009). *Promoting neighborhood diversity: Benefits, barriers, and strategies*. The Urban Institute. www.urban.org/sites/default/files/publication/30631/411955-Promoting-Neighborhood-Diversity-Benefits-Barriers-and-Strategies.PDF

U.S. Department of the Interior (DOI). (n.d.). *Land and water conservation fund*. www.doi.gov/lwcf

Vermont Agency of Natural Resources (ANR). (n.d.). *BioFinder*. https://anr.vermont.gov/maps-and-mapping/biofinder

Vermont Land Trust (VLT). (n.d.). *About us*. https://vlt.org/about/

The Wilderness Society. (n.d.). *National wilderness preservation system*. www.wilderness.org/articles/article/national-wilderness-preservation-system

5

ENERGY, CARBON CAPTURE, FOOD, WASTE

Energy

Energy sources are classified as being renewable or non-renewable. Non-renewable energy sources – which consist of fossil fuels (coal, oil, and natural gas) and nuclear energy – cannot be replenished at rates fast enough to keep up with consumption. Fossil fuels formed some 300–360 million years ago during the Carboniferous Period when plants, algae, animals, and plankton died, drifted to the bottom of the seas and lakes and/or beneath the crust, and were crushed with sediment that created anaerobic processes, high heat, and pressure that transformed them into crude oil, natural gas, or coal. The specific organic material buried, the amount of time, and the environmental conditions of the region determined which type of fossil fuel was formed. Crude oil is the residual sludge of the decomposed organic material, natural gas is the trapped methane in between the layers of sediment, and coal is compressed organic matter that doesn't decompose into residual sludge. Although fossil fuels are more energy-dense than renewable sources of energy, their high emission of carbon dioxide into the atmosphere is altering the climate of the planet. The emissions produce hazardous air pollution, including sulfur dioxide, nitrogen oxides, mercury, and particulate matter that can cause acid rain, eutrophication, impairments to human health, and damage to plants, wildlife crops, and trees. As evident with British Petroleum's (BP's) Deepwater Horizon Spill in the Gulf of Mexico in 2010, the Ixtoc-1 spill near Mexico in 1979, the Exxon-Valdez spill near Alaska in 1989, and countless others, the extraction and transport of oil has led to devastating oil spills that have damaged coastlines, polluted fisheries, injured or killed wildlife, and led to economic losses in cleanup costs and losses in tourism. The hydraulic fracking of natural gas has also led to air and water pollution and adverse health effects released from the chemicals and water used to shatter the rock formations (NIH, n.d.).

Nuclear energy comes from the mining and fission (splitting one atom into two) of radioactive elements, primarily uranium-235 (U-235). Although nuclear energy does not emit carbon like fossil fuels, the fission process results in radioisotopes that emit particles and radiation, collectively referred to as radioactive waste. Short- and long-term containment of radioactive wastes are complex, especially seeing that Plutonium-239 (one of the isotopes produced from

DOI: 10.4324/9781032668666-5

uranium fission) has a half-life (the amount of time for the radioactivity to fall to half of its original value) of 24,000 years.

Renewable energy is defined as energy sources that are replenished at higher rates than they are consumed and include solar, wind, geothermal, hydropower, and biomass (wood, municipal solid waste, landfill gas, biofuels). Some skeptics will argue nothing is truly renewable given cataclysmic events such as the burning out of our sun (supernovae), a meteor collision that implodes the planet, or a nuclear holocaust or global pandemic that ends all life as we know it. Although these very distant future scenarios are discussed later in this book, the aforementioned energy sources are still considered *mostly* renewable.

Utilizing renewable energy sources is a vital component in ecological design and in sustaining ecosystems. Solar energy from the sun is the primary source of sustenance for all living organisms on our planet. It takes about eight minutes for heat and light from the sun to reach the earth. The radiant energy reaching the earth's surface is in the form of ultraviolet, infrared, and visible light. Solar radiation is equal to 178,000 terawatts a year (or about 15 times the world's present energy supply). Of this, 30 percent is reflected in space; 50 percent is absorbed, converted to heat, and reradiated; and 20 percent creates the wind, powers the water cycle, and drives photosynthesis (SA, 1991). Due to the earth's rotation and axial tilt, the water and land absorb different amounts of heat energy. The temperature differentials create differences in air density, which thus create differences in air pressure. Wind results from air flowing from high pressure to low pressure. Both wind and ocean currents also move from east to west (clockwise in the northern hemisphere, counterclockwise in the southern hemisphere) due to the west-to-east rotation of the earth. Weather patterns, fronts, and precipitation are also driven by the interaction of different air masses.

Plants, algae, and bacteria also absorb sunlight and convert solar energy into chemical energy via photosynthesis. Photoautotrophs (plants, algae, and bacteria that perform photosynthesis), also known as "producers," absorb sunlight predominantly through chlorophyll and convert carbon dioxide and water to sugar and oxygen. Consumers (animals and humans) feed on organic material produced by the producers, and decomposers (bacteria and fungi) break down organic materials from dead plants, animals, and waste and convert them into nutrients in the soil, thus enabling the plants to grow and ensuring the consumers a continuous supply of food. As mentioned in Chapter 3, this balance is currently being disrupted as the intensity of human activities outpaces the natural regeneration cycle. The sun, therefore, is the critical piece of the puzzle. It can be tapped directly using passive or active technology or indirectly from the wind, water, and biomass it creates. As of 2023, renewable energy accounted for 8 percent of the total energy produced in the United States. Of the 8 percent, 61.2 percent was produced from biomass (biofuels, wood, and waste), 9.7 percent from hydro, 17.2 percent from wind, 10.4 percent from solar, and 1.4 percent from geothermal (EIA, 2023).

Solar energy

Solar energy can be captured by passive or active systems. A passive system uses ambient flows of energy and does not rely on external electrical or mechanical systems to convert energy. In the northern hemisphere, south-facing windows or collectors are often employed to maximize the absorption of sunlight in the winter. Collector boxes with glass tops can be mounted on the side of a structure and are insulated and painted black on the interior to minimize heat loss and maximize absorption. Cold air from the interior of the building is drawn into the collector from a vent in the bottom. The air is warmed in the solar collector and then flows back into the building by natural convection through a vent toward the top. In buildings that utilize south-facing windows, eaves/roof overhangs can be designed to block the sun in the warmer months but allow it to enter in the cooler months. Materials with high thermal storage capacity/high thermal lag (masonry, concrete, or water) are placed in proximity to the windows. These materials absorb the heat energy in the day without overheating and slowly release the heat when the sun goes down. Windows can be treated with shutters, curtains, smart glass, or glazing that traps the longer-wave infrared radiation that is absorbed on the interior of the building. Similar to passive solar air collectors, passive water systems also consist of a collecting box with a glass top and a black bottom with embedded water tubes. Cold water moves from the water source toward the collector box. The water is then heated in the box and flows into a storage tank. The collector must be lower than the tank to allow the heated water to rise by natural convection to the tank and the colder water from the tank to descend into the collector. Passive solar systems are discussed further in Chapter 11. Burrowing buildings into the earth (earth sheltering) is also a way to utilize the latent heat of the ground. This concept is discussed further in Chapter 9.

Active solar systems utilize similar components as passive collecting systems but use electrical and mechanical systems to pump the air or water. Active systems can also employ photovoltaic panels that produce electrical power. The panels are made up of small cells that consist of wafers composed of a semi-conductive material (usually silicon). The silicon absorbs photons and releases electrons. The electrons are then captured in the form of an electrical current along a contact. A group of solar cells is a module, and a group of modules is referred to as an array. Electricity is produced in the form of direct current (DC), so systems are then connected to an inverter to convert the DC to alternating current (AC). The AC is then connected to a fuse box, which is connected to a battery for storing the electrical power. Systems can be stand alone or connected to the grid, allowing a user to sell excess electricity to the power company. Photovoltaic arrays can be fixed or can rotate on one or more axes, thus allowing the array to track the sun across the sky and maximize absorption. Although the tracking units are more expensive, it is estimated they are roughly 10–25 percent more efficient than fixed units (Bushong, 2016). Photovoltaics can also be embedded in building materials, roofing, or ground cover.

Improvements in technology, lower costs, tax incentives, and an increase in design/installation companies have made solar the fastest-growing and most affordable source of renewable energy in most countries worldwide. As of 2022, China currently generates the most energy from solar with an annual capacity of 392.43 gigawatts (GW), accounting for 37.5 percent of the global market, followed by the United States at roughly 110.1 GW, Japan at 78.83 GW, Germany at 66.55 GW, and India at 62.80 GW (Bajpai, 2023). As of 2022, U.S. small-scale solar (less than 1-megawatt capacity) generated 39.5 gigawatts (GW), and as of January 2023, utility-scale solar generated 73.4 gigawatts, the latter accounting for roughly 6 percent of the total electrical power generation in the United States (EIA, 2022, 2023). California generated the most electricity from solar (14.2 gigawatts) in the United States, which can be a result of the 2020 policy requiring newly built single-family homes and multifamily buildings up to three stories high to have solar panels installed. The Office of Energy Efficiency and Renewable Energy (EERE, 2020) estimates that if solar panels covered ten million acres of land – which is approximately 0.4 percent of America's landmass – they could generate all the nation's electricity.

Generating energy from roads and sidewalks

Given the massive amount of surface area occupied by roads and walkways, there have been several installations in the past decade or so testing the viability of using these surfaces for solar power generation. In 2014, the Netherlands Organization for Applied Scientific Research (TNO) installed a bike path composed of solar panels known as SolaRoad in the town of Krommenie outside of Amsterdam. The pilot project consisted of a 230-foot (70-meter) section of path that contained rows of crystalline silicon solar cells embedded into the concrete and covered with a translucent layer of tempered glass (Renewable Energy World, 2014). Unfortunately, due to sustained damage and lower-than-expected energy generation, SolaRoad was removed in 2020. Other projects have had similar difficulties with durability, energy generation, and return on investment, including the Solar Roadways Project in Sandpoint, Idaho installed in 2017 that was intended to light up with LED lights and have heating elements capable of melting snow and ice, and the 0.5-mile (1 km) solar road constructed in Normandy, France in 2016 using the Wattway technology patented by Colas and the French National Institute for Solar Energy. Despite using silicon resin to protect the panels, the Wattway still could not withstand the wear and tear from traffic, and plans for expansion were halted. The Wattway technology was also installed in 2020 on 538 square feet (50 square meters) of Interstate 85 being dubbed "the Ray" in Peachtree Corners, Georgia. However, as of the writing of this book, the data is not currently available to determine if the Ray will endure the same damage as in Normandy. It seems more research and new materials are needed to make these installations viable. Until then, it seems trackable panels positioned along or down the medians of roads are more efficient.

Besides solar installations on roads and pathways, other researchers are examining the ability to generate energy from walking on surfaces. The company PaveGen has developed a technology that converts steps into electricity by harnessing the kinetic energy produced by the compression of recycled rubber triangular panels that are inserted within the sidewalks. As the triangles are compressed, copper coils and magnets transmit energy through electromagnetic induction, producing 2–4 joules that generate roughly five watts of energy that can be stored in batteries. The technology can also be integrated with smartphone applications via Bluetooth and can communicate with building management systems (Souza, 2019). The company has installations in over 50 locations across the world, including sidewalks, plazas, airports, tennis courts, and college campuses, to name a few.

As a result of the European Union's law for all cars to have zero carbon dioxide emissions by 2035, Sweden is developing the first electrified road on European Route E20 between Hallsberg and Örebro. The dynamic charging road allows vehicles to recharge as they drive, which enables the potential for traveling longer distances with smaller batteries. Although the charging system has not been selected yet, there are three types of systems that can be used: catenary systems that use overhead wires similar to the electrification of light rail systems and trolleys; conductive systems which transmit energy to the vehicles from contact with a rail on the ground; and electromagnetic inductive systems where energy flows from ground coils buried under the road to a receiver in the vehicle.

Roads, sidewalks, and rooftops are not the only sites where solar can be installed. In Arizona, the Gila River Indian Community and the U.S. Army Corps of Engineers have initiated a project that will install solar-photovoltaic shades over 1,000 feet (305 meters) of the 1–10 Level Top Canal. In addition to generating energy, the shade will decrease the evaporation of the water, which is an issue in the arid landscape. A similar project was completed in Gujarat, India, in

FIGURE 5.1 Top: roof-mounted solar panels; bottom: trackable solar panel in front of fixed, ground-mounted solar panels.

2015 called the Sardar Sarovar Canal Solar Power Project. Four projects have been installed over 8 miles (13 km) of the Vadodara Branch Canal that generate 35-megawatt (MW) capacity (SSNN, n.d.). Solar panels can also be installed on or along rail lines, rail terminals, or the rail cars themselves. In Switzerland, the firm Sun-Ways has developed technology that places solar panels in between the railroad ties. In the United Kingdom, Community Energy South and Network Rail are powering the lights and signals with 100 solar panels adjacent to a track near Aldershot in Hampshire. Indian Railways has installed panels on top of their rail coaches to power the lighting systems inside the train cars, and a two-mile (3.2 km) stretch of a rail tunnel between Paris and Amsterdam was lined with 16,000 solar panels – enough to power all the trains in Belgium for one day per year.

Wind energy

Similar to solar technology, lower construction costs, tax incentives, and renewable energy targets have increased wind installations in the past 20 years or so. The Energy Information Administration estimated the average construction costs for onshore wind generators in the United States to be $1,319 per kilowatt-hour in 2019, and as of January 2023, the U.S. wind capacity reached 141.3 GW (EIA, 2021, 2023).

Wind turbines usually consist of the foundation, the tower, the rotor that consists of the hub and the blades, and the nacelle that houses the gearbox, shafts, controllers, generators, yaw drives, pitch systems, and brakes. Turbines can be designed to face the wind (upwind turbines) or to face away from the wind (downwind turbines). Wind vanes communicate with the yaw drive on upwind turbines to rotate the nacelle to keep it facing the wind. Anemometers measure the wind speed and transmit data to the controller that can trigger the pitch systems to adjust the angle of blades to optimize flow, which may require reducing the rotation speed when winds are too strong.

Wind flows across the blades and creates lift and drag, ultimately converting aerodynamic force to rotation. The rotor spins and either directly connects to the generator or connects to a series of gears and shafts that can control the speed. The generator converts the force to electricity, producing AC electricity that is converted to direct current (DC) through a rectifier and back to AC through an inverter so that the electricity produced matches the frequency and phase when it is connected to the power grid. Larger rotor diameters allow the turbine to capture more wind and produce more electricity. The average rotor diameter (the length from tip to tip of the blades) is currently around 430 feet (131 meters), and the average hub height (distance from the hub in the center of the blades to the ground) for utility-scale, onshore turbines is 322 feet (98 meters); the average height for offshore turbines is currently around 330 feet (100.6 meters) but is projected to grow to 500 feet (152 meters) (Hartman, 2023).

Often, the biggest obstacle for wind farms is the visual impact, which can cause people who are in favor of wind energy in principle to oppose it because they do not want to see the turbines in the landscape. This is often dubbed a NIMBY (not in my backyard) scenario. In addition to aesthetics, turbine construction and operation can harm wildlife, especially along mountain ridges and/or along migratory bird or bat routes. According to the National Audubon Society, 140,000–328,000 birds are killed per year by wind turbines in North America alone (Bryce, 2016). These factors make wind power a double-edged sword (literally) in terms of environmental protection. Some technologies that are being tested to decrease bird fatalities include the use of GPS, radar, and/or cameras to detect approaching birds; using darker-colored blades

FIGURE 5.2 Top: construction of SolaRoad; bottom: SolaRoad in operation, running along rail corridor (rail with trail).

Source: Images courtesy of SolaRoad

that attract fewer insects than lighter-colored blades, thus reducing the number of birds that feed in the vicinity; the use of UV lighting as a bird/bat deterrent; and the production of blades that rotate vertically instead of the conventional "fan" style blade. According to the U.S. Fish and Wildlife Service, the best option is to site the turbines away from bird congregation spots at wetlands or migratory stopover points. However, as with everything these days, the data does vary. Between 1994 and 1998, a comprehensive study was conducted at the Searsburg Wind Facility in southern Vermont that examined the effects of turbine construction and operation on bird populations, behavior, migratory patterns, and fatalities and concluded there were virtually no adverse impacts. Researchers found no bird carcasses near the turbines and no evidence of scavenger activity (Walters, 2013).

Although bird deaths from wind turbines are disturbing, it is important to consider how these fatalities compare to deaths from other causes. Data from the U.S. Fish and Wildlife Service estimates there are roughly 1.8–4.8 billion total bird deaths a year, including 365–988 million from colliding with buildings, 6.6 million from collisions with communication towers, 8–57.3 million from collisions with transmission lines, 900–11.6 million from electrocutions, 89–340 million from collisions with cars, 1.4–3.7 billion from cats, 500,000–1,000,000 from oil pits, and up to 72 million from poison (USFWS, n.d.; Loss et al., 2014a; Longcore et al., 2012; Loss et al., 2014b; Loss et al., 2014c; Loss et al., 2013; Trail, 2006).

Constructing wind turbines offshore may be a solution that minimizes both the visual impacts and bird fatalities. Although the initial construction cost is higher, the more consistent wind speed that is present over water bodies yields increased power production. Offshore wind turbines can be constructed similarly to bridge construction over water by utilizing battered piles, cofferdams, or caissons. However, where the water is too deep – which is the case in the majority of offshore sites – turbines can be constructed to float on the water using spar buoys, tension lag platforms, semi-submersible platforms, or barges. Worldwide, there are 257 operating offshore wind sites with an annual capacity of over 50 GW, the largest being the 50 MW floating Kincardine Offshore Wind Farm in Scotland. With new targets and projects in planning, it is estimated the U.S. capacity from offshore wind will increase to about 20 GW and the worldwide capacity to about 260 GW by 2030 (EERE, 2022).

Biomass

Burning wood is probably the oldest way to produce energy, and it is still the most prominent form of renewable energy worldwide – especially in developing countries. Timber is sustainable if it is not harvested faster than it is regenerated. Sustainable forestry management practices encourage regeneration, biodiversity, and long-term sustainability of the forest. The most sustainable methods of harvesting timber consist of either a selective-thinning harvest, where cutting is confined to dead or diseased species or pruning the crowns of trees, or a single-tree harvest, which removes only trees of low value or those that compete with other trees. Both methods leave the forest intact and allow new seedlings to grow in place of the harvested trees. However, this selective harvesting may not yield enough timber to meet a community's energy needs and certainly would not be economically feasible for those selling the timber. In a shelterwood harvest, mature trees are left in place to "shelter" the growth of new trees. The mature trees are then removed over a period of 10–15 years once the new growth is well established. Other systems include seed tree harvests, which leave only a small percentage of mature trees to "reseed" the forest; group selection harvests, where trees are harvested in clusters over a

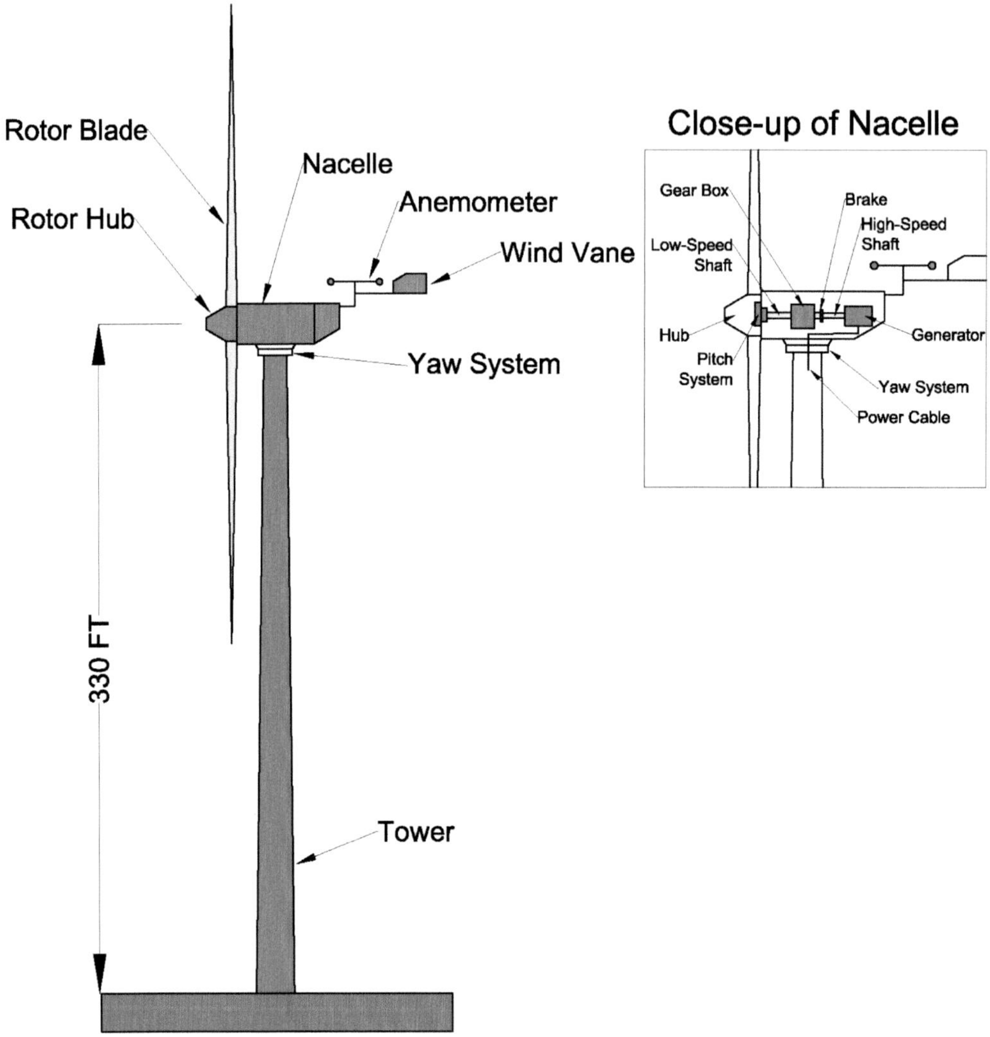

FIGURE 5.3 Wind turbine diagram.

Source: Diagram drawn by author, information derived from: Office of Energy Efficiency and Renewable Energy (EERE). (n.d.). *How a wind turbine works – text version.* www.energy.gov/eere/wind/how-wind-turbine-works-text-version

40–50-year span; and clear cutting, which removes all the trees in a specific area. Clear-cutting and any other large-scale group removal of trees are controversial due to the visual impact. However, in some cases, this can be an effective tool to regenerate a forest and create more biodiversity. Controlled burning to remove undergrowth on the forest floor can also be perceived negatively, but in fact it can be an effective way to prevent forest fires, foster new growth, and control disease. Once again, in any timber harvest, it is imperative to balance the removal with the regeneration to prevent deforestation.

In addition to the burning of wood or organic matter, biomass can be converted into liquid fuels such as ethanol or biodiesel to generate power. Ethanol can be made by fermenting sugar from plant material or by gasification that uses high temperatures to convert the biomass into

FIGURE 5.4 Top left: Maple Ridge Wind Farm in Lowville, New York; bottom left: wind turbine on the ridgeline of Bolton Valley Ski Resort; right: vertical-axis wind turbines at the Rochester Institute for Technology.

a synthetic blend of hydrogen and carbon monoxide that can be converted in turn into ethanol. The feedstocks used in fermentation can be starch- and sugar-based crops such as corn or sugar cane, or cellulosic, non-food-based feedstocks such as crop or wood residue or switchgrass. Corn-based ethanol is currently the most widely tapped energy crop in the United States; however, the biomass from the sugar cane crushing process (bagasse) is more efficient when examining the ratio between the amount of energy the crop yields and the amount it consumes in the manufacturing process. Bagasse provides about 17 percent of Brazil's total energy needs and has replaced more than 40 percent of the country's petrol needs (Farina, 2017).

Biodiesel is produced by combining lipids such as vegetable oil, animal fat, or cooking grease with alcohol. It can be used in pure form or blended with petroleum diesel at different concentrations. Both ethanol and biofuels emit fewer greenhouse gas emissions than conventional gasoline since the carbon dioxide released is offset by the carbon dioxide captured when crops used to make the ethanol are grown. Studies by Argonne National Laboratory found that corn-based ethanol released 40 percent fewer greenhouse gas emissions than gasoline and cellulosic-based ethanol released 88–108 percent fewer greenhouse gas emissions than gasoline depending on the feedstock type and the conversion process (Wang et al., 2020, 2012).

Hydro energy

Large-scale hydropower projects, which generate more than 30 megawatts per day, provide most of the power produced by water in the United States. Much of the power from large-scale facilities is purchased across state and country lines due to the limited locations of such projects. The amount of power produced by large-scale hydroelectric facilities is proportional to the height of the dam and the volume of water flowing through the turbogenerators. Although hydro power is clean in terms of emissions and the most common renewable source across the world, construction and operation can have ecological impacts both upstream and downstream from the dam. Fluctuating water levels due to increases/decreases in energy demand impact sediment and nutrient levels as well as fish migration and habitat.

Micro-hydro

There are alternatives to large-scale hydro projects, such as small-scale projects that generate .01–30 megawatts per day and micro-hydro power projects that generate around 100 kilowatts per day. Such projects can be constructed with minimal impact on habitat, stream morphology,

FIGURE 5.5 Robert Moses Hydroelectric Plant at Niagara Falls, New York. Water is channeled from the upper river (top) into a holding pond (middle) and then released into turbines in the gorge (bottom). The holding pond allows water to be increased or decreased on the basis of energy demand.

FIGURE 5.5 (Continued)

or water quality, especially at ski/snowboard resorts that already have the required infrastructure in place. Streams can be drained into existing snowmaking pipes and then channeled into an impulse turbine. The turbine can be co-located with existing pumps and generators used for snowmaking. A forebay is usually constructed near the intake to serve as a settling pond for debris before the water flows down the snowmaking pipe or "penstock." Additional filters are placed between the forebay and the penstock to prevent smaller sediment from clogging the system. Operations can be controlled remotely from a central maintenance facility, minimizing the need for intrusive roadways for service vehicles.

Similar to large-scale systems, the amount of power from a micro-hydro system is dependent on the product of flow volume (quantity of water) and head (the vertical distance between the intake and the turbine). These two factors determine the pipeline (penstock) length and diameter, the turbine type, the size of the generator, and the rotational speed. The greater the head, the less water is needed to produce a given amount of power. Although it is ideal to situate a micro-hydro system on a stream that flows consistently throughout the year, fluctuations in flow can be regulated with a holding pond located near the intake. The flow rate for a given project can be estimated by multiplying the net head (the vertical distance between the intake and the turbine after subtracting losses from friction) by flow (gallons per minute) and dividing the product by ten. This gives the system's output in watts.

There are three primary types of turbines used in hydroelectric systems: impulse turbines, submersible turbines, and reaction turbines. Impulse turbines, which include the Pelton Wheel and the Turgon Wheel, use the velocity of water to turn the runner. Impulse turbines can have multiple nozzle openings to increase power output. Submersible turbines, such as the Jack Rabbit, can be placed directly in streams and do not require a vertical drop (head). The Jack Rabbit generates about 100 watts, or 1.5 to 2.4 kWh per day (Cunningham & Woofenden, 2007). Reaction turbines, often used in large-scale hydro projects, produce energy by pressure from runners that maintain constant contact with the water. The turbine converts waterpower into rotational power at the shaft, which is converted into electrical power by the generator. Micro-hydro systems can operate off the grid with or without batteries or can connect directly to the grid via AC/DC generators.

In the late 1990s, Snowmass's snowmaking supervisor Jim Holton conceived the notion of utilizing the snowmaking distribution system for hydroelectric power at Snowmass Ski Resort in Colorado. Doug MacKenzie, Aspen Skiing Company's general manager at Snowmass, and Auden Schendler, the company's environmental affairs director, approved the idea and began planning for the project. The micro-hydro system at Snowmass functions like a typical large-scale hydroelectric power station. A storage pond located about 800 feet (244 meters) up the mountain collects water from a stream. The water is then drained from the storage pond via the snowmaking pipe to a small powerhouse at the base of the trail that contains a single-nozzle 115 kW Pelton Turbine, custom built by Canyon Hydro. Excess water is drained into a snowmaking retention pond located near the turbine and returns to the stream. Mountain Peak Controls in Brighton, Colorado, designed the control system that networks the snowmaking system and the turbine by regulating water flow, temperature, pressure, and flow direction. According to Schendler (2006), there were roughly 200,000 kilowatt hours of electricity generated in 2005 (enough to power 40 homes) and 400,000 pounds of carbon dioxide emissions were prevented. Currently, the system averages about 250,000 kilowatt hours per year. Without the snowmaking infrastructure in place, the cost of installing the pipe, the retention pond, and the turbine would be several million dollars. At Snowmass, the

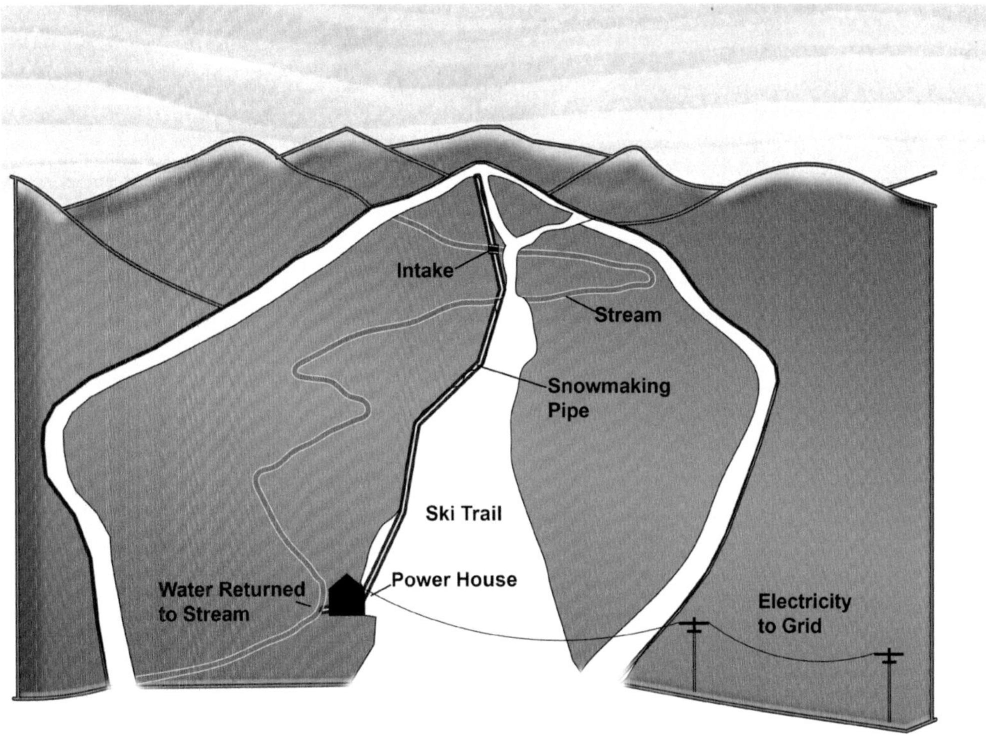

FIGURE 5.6 Diagram of micro-hydro project at a ski resort.

Source: Drawing by author. Information derived from: PDHengineer.com. (n.d.). *Design of microhydro electric systems*. Course No. M-1020.

entire system (including labor and permit fees) costs around $155,000 – an investment with a payback in as little as seven years.

In 2008, Whistler–Blackcomb Ski Area in British Columbia began construction of a 7.5-megawatt hydro system along the Fitzsimmons Creek known as the Fitzsimmons micro-hydro renewable energy project. It is estimated the project generates 32 gigawatt hours of electricity per year – enough to offset the resort's winter and summer operations, which include 38 lifts, 17 restaurants, 269 snowguns, and various service facilities. The creek provides an ample, year-round flow and is already networked with the resort's existing snowmaking system (Whistler-Blackcomb, n.d.).

Tidal and wave power

Another method of generating power through water is to capture energy from the rise and fall of ocean tides. Tides are cyclical movements of oceans created from the gravitational pull of the sun and moon on the earth. Energy from the tides can be harvested from the difference between the high and low tides by placing the turbines directly in the tidal streams, constructing a barrage or dam across the mouth of a bay, or by constructing artificial lagoons that capture and release water from the tides. Turbines can be rotational or vertical (turbine fences) and can capture both the incoming and outgoing flow of water. Depending on the system, turbines can be mounted along the barrages, float individually or in arrays, or be mounted directly to the sea floor. There are currently nine operational tidal power stations in the world and another dozen or so in planning. The largest is the Sihwa Lake Tidal Power Station in South Korea, which uses a seawall/barrage system and has a total power output capacity of 254 megawatts; the second largest (and oldest) is a barrage system in La Rance, France, with 240 MW capacity. Other locations include the 20 MW plant in Annapolis Royal in Nova Scotia, Canada, the 3.7 MW Jiangxia Tidal Power Station in China, the 1.7 MW tidal barrage in Kislaya Guba, Russia, and the 1 MW Uldolmok Tidal Power Station in South Korea (Power Technology, 2014). In the United Kingdom, the Mey Gen Tidal Energy project in Scotland is still in construction but is expected to produce up to 398 MW when all phases are complete. The United Kingdom is also still proposing a tidal system at Swansea Bay. The Swansea Bay Tidal Lagoon proposal called for a 5.9-mile (9.5 km), u-shaped seawall that forms a lagoon between the River Tawe and the River Neath. The seawall would house 16 reversible bulb turbines to capture water flowing in and out and is estimated to have a maximum capacity of 320 MW. However, progress on the project was halted due to the developer losing an appeal against a court ruling said claimed they failed to commence construction within the required five-year time frame. The project has now morphed into a new proposal called the Blue Eden project (TETHYS, n.d.). Although most of the existing systems utilize barrages/seawalls, the first commercial tidal-stream turbines were constructed as a demonstration project in 2008–2019 at the Strangford Lough in Northern Ireland. SeaGen, as it was known, comprised two 600 kilowatt turbines and produced 11.6 gigawatt hours of power during its lifecycle (Power Technology, 2020).

Some of the challenges with constructing tidal power stations include finding locations that contain a significant difference in the area between the high and low tide to produce a decent amount of electricity, the cost of developing an efficient tidal technology, and construction costs of the arrays and the infrastructure needed to connect them to the grid. There is also more research needed to understand the impact of the arrays on ambient flows, water quality, and impacts on marine habitats and species.

Researchers are also exploring technologies to generate power from waves by focusing them into channels, catch basins, or reservoirs where water flows to a turbine at a lower elevation. In addition, ocean thermal energy conversion (OTEC) is being explored to produce energy from temperature differences between the surface of the water and water at greater depths. Warm surface water is pumped into an evaporator where water vapor is produced that drives a turbine. The vaporized water is then piped into a condenser where it is cooled from water pumped from deeper locations in the ocean, turned back into a liquid, and returned to the ocean.

Geothermal

Geothermal energy uses steam produced from hot, underground reservoirs either to heat buildings directly or to drive turbines to produce electricity. Since the core of the earth consists of hot, molten rock, the temperature increases with depth. There are three types of geothermal power plants: dry steam, flash steam, and binary steam. Dry steam plants extract dry steam directly from underground and use it to drive turbines. Currently, the only dry steam plants in the United States are located at the geysers in the Mayacamas Mountains in northern California. There are currently 13 plants at the complex that draw from over 327 wells spread over 45 square miles (116.5 square km). The geysers are currently producing 725 megawatts of electricity, enough to power 725,000 homes (Calpine, n.d.). Since there are only two known sources of steam in the United States (the other being at Yellowstone National Park, which is protected from development), flash steam plants that draw from hot-water reservoirs beneath the surface are more commonly employed. Production wells are drilled into the reservoirs, allowing the hot water to ascend under its own pressure. As the water ascends, the pressure decreases and the hot water boils into steam, which is then extracted from the water and used to drive turbines. The water is then returned to the geothermal reservoir through a separate injection well. Binary steam plants use the extracted hot water to boil a secondary fluid with a boiling point less than water. The secondary fluid is then vaporized and used to drive the turbines.

Heat recovery

The U.S. Department of Energy (DOE, n.d.) estimates that between 20 and 50 percent of industrial energy input is lost as waste heat in the form of hot exhaust gases and cooling water and from hot equipment surfaces. The basic concept of a heat recovery system is to recover wasted thermal energy that would otherwise be exhausted to the outside and use it to generate electricity or to heat or cool incoming air or water. Thermal energy can be captured by heat exchangers such as thermal wheels/energy transfer wheels, fixed plate heat/recuperators, heat pipes, heat pumps, or closed-loop run-around coils. The exchangers are usually placed in the loft, attic, or on the roof with pipes feeding downward into the interior spaces. Thermal wheels consist of two wheels spinning in opposite directions, with one wheel drawing in the exhaust air that warms the air brought in from the outside by the other wheel. Thermal wheels are the most efficient of the heat exchangers but occupy more space than other heat exchangers. Fixed plate heat exchangers, on the other hand, are more compact and consist of a box of metal or plastic plates that mix the exhaust air with the outside incoming air. Heat pipes contain tubes with a working fluid or refrigerant that extract heat from exhaust air and transfer it to supply air. Vertical tubes can transfer heat via evaporation and condensation and horizontal pipes use capillary action through a porous wick. Closed-loop run-around coils are placed between the supply and exhaust air streams and can be used when the air streams are not close enough. Two or more tube coils are piped together and charged with a working fluid that is circulated by a pump. The fluid extracts heat from the exhaust air and transfers heat to the supply air. Heat pumps can transfer heat from the interior to the exterior during the cooling season similar to an air conditioning system. In the colder months, heat pumps work like an air conditioner in reverse. Even in cold temperatures, the refrigerant can still absorb heat on the outside and transfer it to the inside. Heat pumps can extract heat from air, water, or the ground and include ducted air-to-air systems, ductless systems, air-to-water systems, water-to-air systems, water-to-water systems, absorption heat pumps, and geothermal heat pumps that consist of horizontal arrays or vertical loops that source heat from the ground.

Carbon capture

Carbon capture and storage (CCS) and carbon capture, utilization, and storage (CCUS) involve the capture of carbon dioxide from point sources that use fossil fuels before they are emitted into the atmosphere. The carbon dioxide is compressed into a liquid and is transported, stored, or injected underground into deep geological formations such as depleted oil and gas reservoirs, coal beds, or saline aquifers. At the source, carbon dioxide can be captured by pre-combustion where fuel is gasified to produce a synthetic gas consisting of carbon monoxide and hydrogen; post combustion where chemical solvents separate carbon dioxide out of the fuel gas; or oxyfuel where the fuel is combusted in the presence of pure oxygen to ensure the products of combustion only contain carbon dioxide and water. Amine-based carbon dioxide capture is currently the most common method used to capture carbon. Carbon dioxide can also be captured away from the point source using biomass or by direct air capture (DAC). Since the carbon dioxide concentration is more dilute away from flue gas at a power station, it is less efficient and more expensive to capture.

The pressure from the carbon dioxide that is injected into underground wells can be used to extract oil from difficult spaces in the geological formations. Research is also being conducted to convert carbon emissions into biofuels and building materials or to grow plants, algae, and bacteria.

There are currently 41 operational commercial CCS/CCUS projects across the world with the capacity to store over 45 million metric tons of carbon dioxide, which is roughly 0.12 percent of the world's 37.12 billion metric tons of annual carbon dioxide emissions; there are another 26 facilities in construction and 325 in planning (IEA, n.d.; Global CCS, n.d.). It is estimated that CCUS can capture more than 90 percent of carbon dioxide from power plants and industrial facilities and can achieve a 14 percent decrease in greenhouse gas emissions by 2050 (C2ES, n.d.).

The biggest obstacle to CCS/CCUS deployment has been cost. According to the International Energy Agency (IEA), CCS costs about U.S.$15–25 per metric ton of carbon dioxide for industrial processes producing concentrated carbon dioxide streams for ethanol production of natural gas processing, and U.S.$40–120 per metric ton for dilute gas streams such as cement production and power generation. DAC remains the most expensive and is somewhere in the U.S.$134–332 per metric ton range (Stern & Berghout, 2021). Where storage is not feasible on-site, there are also costs associated with the transport of carbon via pipelines. However, given the market trends that have decreased the cost of other renewable energy sources, the IEA states there is potential to reduce costs in CCUS as the market grows, the technology develops, finance costs fall, economies of scale are reached, and experience of building and operating CCUS facilities accumulates (Stern & Berghout, 2021).

Food and agriculture

Sustaining life on our planet is directly connected to sustaining food production. Understanding where food comes from, what ingredients it contains, and how it is grown or produced has become commonplace in recent times. It is always shocking to see the amount of non-local products some grocery retailers sell. I recently went into a local grocery store in Vermont to purchase some honey and the only products they had available were from South America. Given Vermont has approximately 900 beekeepers with about 14,000 hives (ANR, n.d.), it is shocking to me that a store would not carry any local products. Likewise, given the state of Vermont has 124,000 cows on 662 dairy farms (NE Dairy, 2020), it is also shocking that some markets sell milk from outside the region. In addition to hurting local farmers, non-locally sourced food increases the distance food travels before reaching the consumer (food miles) and subsequently increases the carbon footprint. In the United States, recent estimates show processed food typically travels over 1,300 miles (2092 km) and fresh produce over 1,500 miles (2414 km) before it's consumed, and global food emissions account for nearly 20 percent of total food system emissions (Harris, 2022; Li et al., 2022). As food miles increase, there is also an increase in the number of times the food is handled and stored, ultimately increasing the chances of contamination. In addition to creating local jobs and offering transparency in how food is grown, locally sourced food can provide valuable farm-to-table connections and educational opportunities in communities.

Unlike the market referenced earlier that seems to blatantly refuse to sell local products, there are markets that are more proactive when it comes to sustainability. Wegmans, for example, has committed to sourcing products near stores, eliminating waste, reducing plastic packaging, and reducing its carbon footprint. They also have an organic farm that assists in sustainable growing and composting practices. Whole Foods also boasts environmental stewardship programs that aim to reduce food waste, reduce single-use plastic, support farmers that use regenerative agriculture, and a commitment to organics. Healthy Living in Vermont has pledged to provide all foods that are free from added hormones, antibiotics, artificial fats or trans fats, high fructose corn syrup, artificial colors, flavors, preservatives, sweeteners, and bleached or bromated flour. All their foods are also purchased directly from the producer or grown regionally, USDA-certified organic, certified eco-grown by the Integrated Pest Management Institute, and are non-genetically modified organisms (GMOs) (Wegmans, n.d.; Whole Foods, n.d.; Healthy Living, n.d.). Other markets have similar goals, but whether they are truly practiced or whether they are just "agro-green-washing" is something consumers will have to determine for themselves.

Community gardens and community-supported agriculture

In addition to markets that sell local products, community farmer's markets, community gardens, and community-supported agriculture (CSA) can foster a farm-to-table culture. Communities can facilitate community gardens by adopting zoning ordinances that permit agricultural uses in areas that would have been otherwise excluded. Codes can even go as far as to require a certain amount of community garden space for each planned development. Ordinances can also be more flexible to allow rooftop gardens, vertical gardens, balcony gardens, and/or small animal/species production, such as goats, chickens, rabbits, fish, and bees. Studies have shown that community gardens encourage healthy eating habits and provide equity by supplying nutritious food that some citizens may otherwise not have access to (Armstrong, 2000). In addition, community gardens can increase green space, lower carbon footprints, foster community engagement, and offer training and educational opportunities.

While community gardens are generally small-scale operations, CSA or crop sharing allows the consumer to subscribe to the harvest of a farm. The farmers and consumers essentially share the risks and the benefits of the harvest. Members usually purchase a share before the growing season that gives farmers upfront capital for their operations. Members then receive regular installments of food as it becomes available for the duration of the growing season.

There is also the concept of the not-for-profit farm. Having resided in Shelburne, Vermont, for close to 20 years, I was fortunate enough to be a short walk away from Shelburne Farms. The land that occupies Shelburne Farms was first settled by the Abenakis in the seventeenth century. As with much of the land in our country, the Abenakis were displaced by the European colonists in the eighteenth century. The land was eventually purchased by Eliza Osgood Webb (Lila) – the youngest daughter of William H. Vanderbilt – and her husband William Seward Webb. The Webbs purchased 33 farms across 4,000 acres along Lake Champlain on Shelburne Point and established a grand agricultural estate with a landscape designed by Frederick Law Olmsted and 30 estate buildings designed by Robert H. Robertson. The farm flourished until 1936, when financial constraints forced the Webbs to sell off some of the land. In 1972, fourth-generation Webb siblings turned the land into a non-profit and expanded the notion of the land as a showcase for agriculture into one that "repaired society's frayed relationship to agriculture and Nature." Today, the property functions as a working farm on 1,400 acres and contains a farm-dairy, cheesemaking, sugar-making, organic gardens, a solar orchard, woodlands, gardens, historic buildings, and walking trails. Sustainable agriculture and education remain a prime focus, and various pre-k–12 and professional development programs are offered through the Shelburne Farms Institute for Sustainable Schools (Shelburne Farms, n.d.).

Agroecology

The field of agroecology is the science of managing farm ecosystems and explores whole food systems, including the ecological, social, and economic impacts on the production, cultivation, transport, health, nutrition, resiliency, and supply of food products. The practices of the past that resulted in overcultivation, overgrazing, deforestation, and the use of chemical pesticides and fertilizers are gradually being replaced by new methods that minimize erosion and runoff, improve soil quality, minimize air and water pollution, conserve water, and foster natural and organic methods of growing crops and feeding animals.

The Union of Concerned Scientists (2022) has identified several best practices for sustainable agriculture, including: rotating crops and growing a mix of crops (intercropping); planting cover crops and perennials during the off-season to prevent soil erosion and keep soils healthy; eliminating or reducing soil till practices by inserting seeds directly into undisturbed soil; applying integrated pest management and reducing chemical pesticides; integrating livestock and crops to improve efficiency and profit; adopting agroforestry practices; and managing landscapes and vegetation to reduce erosion, absorb runoff, and support bees, pollinators, and biodiversity.

Waste reduction – rethink, reduce, reuse, recycle, recover

If we revisit the symbiotic relationship between producers, consumers, and decomposers in the natural world, we can begin to see the contrast between the ways humans and nature handle waste. In nature, waste is recycled as food; in the human or "built" environment, waste is merely an end product that stockpiles into pollution, which ultimately disrupts the health and nutrient cycle in our communities. In the early 1990s, the three-R concept – reduce, reuse, and recycle – was the conventional triage in dealing with waste. However, over the years, this seems to have expanded to five or more terms, including refuse, reduce, reuse, repurpose, recycle, recover, rethink, redesign, repair, refurbish, reclaim, regift, etc. – and the list goes on. I'm not sure if adding more terms makes understanding the circular economy of consumption easier or more difficult. There is certainly an overlap on all the terms, but without going into a full dissertation of how each is defined, this book will focus on five: rethink, reduce, reuse, recycle, and recover.

Rethink

Minimizing the amount of waste generated often begins with rethinking our consumption and waste practices to transform from a linear economy to a circular economy. In a traditional, linear economy, profit is prioritized over sustainability. Raw materials are collected, transformed into products, used by consumers, and thrown away. In a circular economy, materials are collected, transformed into products, used by consumers, and then recycled, reused, repurposed, or recovered. Rethinking often begins with questioning what we buy or consume, how we design products and packaging, and what will happen to the product when it is no longer needed or when it no longer works. Rethinking practices also might reveal ways to retain, reuse, or repurpose items. Before simply throwing something in the trash, it is always important to consider donating the item to a charity, shelter, or thrift store. Some items that you think have no value may also be of interest to museums or movie production companies for props.

Rethinking practices also require gaining an understanding of the ecological, social, and economic impacts of waste, including emissions to soil and water, emissions to air, the consumption of land for landfills, the impacts on biodiversity and human health, the costs of disposal and/or shipping waste, and an understanding of how such activities have disproportionately affected lower-income communities, indigenous communities, and people of color. A 2017 report from the Clean Air Task Force (CATF) and the National Association for the Advancement of Colored People (NAACP) showed that African Americans are 75 percent more likely than other Americans to live in areas situated near facilities that produce hazardous waste (Fleischman & Franklin, 2017). The Toxic Waste and Race at Twenty reports also reviewed data over a 20-year time frame and found that more than half of the people living within 1.86 miles (3 km) of toxic facilities were people of color (Bullard et al., 2008). As stated by Denise Patel – the U.S. Program Director at the Global Alliance for Incinerator Alternatives (GAIA):

> Research shows that 79 percent of municipal solid waste incinerators are located in environmental justice (EJ) communities . . . the end result is that low-income and Black, Indigenous, and people of color (BIPOC) communities are most impacted by waste policies and are often targeted by the waste industry and others as potential sites for building incinerators and landfills. These industries know that these communities lack the ability and capacity to fight back to protect themselves. Policies like redlining have further concentrated polluting facilities, including waste facilities, in EJ and BIPOC communities.
>
> *(Yang, 2021)*

Reduce

According to the World Bank (Kaza et al., 2018), the world generates about 2.01 billion tons of municipal solid waste (MSW) per year with projections showing an increase to 3.4 billion tons by 2050. Worldwide, about 37 percent of waste is disposed in some type of landfill, 33 percent is openly dumped, 19 percent is recycled/composted/recovered, and 11 percent is incinerated. Food and green waste make up the largest waste category, accounting for 44 percent of the total, while dry recyclables – including plastic, paper, cardboard, metal, and glass – amount to 38 percent of total waste. Daily, the global average is 0.74 kg/per capita/per day, with the North American region generating the highest average at 2.21 kilograms per capita per day. In North America, dry recyclables account for more than 55 percent and green and food waste account for less than 30 percent. The United States generates 12 percent of the world's trash and averages 2.2 kilograms (4.9 pounds) per capita per day (Environment America, 2021).

Many of us have probably heard stories of our great (or great-great) grandparents who lived during and/or after the Great Depression saving jars, cans, containers, fabric scraps, flour bags, and food scraps and living by the motto, "use it up, wear it out, make do or do without." My great-grandmother used to sheer the sheep, spin the wool, make her own clothes, and then use the scraps (and any other textile scraps) to weave rag rugs. All food scraps were put into a meat grinder at the end of the week to make goulash. In the late twentieth century, culture changed, and now it is more common to hear expressions like, "they don't make them like they used to." The "throw-away society" is a term that has been coined to describe a society engaged in over-consumption of short-lived items. Much of this pattern is a result of consumerism and the planned obsolescence manufacturing strategies in which products are designed to last for only short periods of time so that consumers are forced to replace them. Several years ago, when I upgraded one of my smartphones, the salesperson at the store was utterly shocked that I had kept the original phone working for six years and that I was able to tolerate the slowness of its operating system due to more memory-intensive apps. In the auto industry, if you can get 50,000 miles out of a car without any major repairs, you have essentially beaten the system. My grandmother's refrigerator that she bought in the early 1970s lasted until the day she died in the early 1990s; I've gone through three in the past ten years. Likewise, the tube-style RCA television set I took to college with me was still going strong when flat-screen technology came out, but none of the donation centers wanted it and no one took it when I put it out by the curb with a "free" sign on it. It is also common to hear that it costs the same (or even more) to repair something as it does to buy a new one. This is in part due to the outsourcing of manufacturing to countries where labor costs are cheaper than where we bring items for repair in the United States. Items that are assembled in the United States often have components that are imported, also adding to the cost, complexity, and difficulty in obtaining the parts to complete the repair. And in the digital age, this consumerism is worse than ever. With the pace of technological change, items such as computers, cameras, phones, operating systems, software, apps, and hardware all become quickly outdated, and like my smartphone I kept for six years, simply slow down or stop functioning altogether. The paradox is that capitalism and production are needed for a healthy or "sustainable" economy, but the sheer amount of production and subsequent waste degrades the physical environment and requires management, which in turn ultimately hinders the economy.

Indeed, there is a delicate balance. If everything lasted forever, how would companies make any money unless they charged a fortune for said infinite product? But, at the same time, if the product doesn't last long enough, will consumers keep buying the same brand? Believe it or not, people do have the power to forge a change in the way products are manufactured. For example, consider the product reviews that are now ubiquitous on online marketplaces. Prior to the internet, such reviews were confined to a select group of magazine and newspaper product testers. But we are all reviewers now. Granted, this leaves room for bogus or biased comments, but, overall, it has led to a level of transparency in the marketplace that simply did not exist in the past. If a product is bad, that fact quickly becomes known, and this forces competition for better-quality products. Higher quality equals a longer lifespan, which equals less frequent waste. Although more durable, longer-lasting items may be more expensive initially, the consumer has the choice to weigh this investment against the longer lifespan.

In addition to reducing their carbon footprint and amount of municipal solid waste, companies can save money by reducing packaging. Although much of this is a "less is more" design challenge, another option is to use biodegradable packaging derived from renewable raw materials that will completely decompose in a composting area. Such bioplastics are often made from polylactic acid (PLA), derived from corn starch, tapioca roots, or sugar cane. Public policy can also have an impact on reducing packaging. In 1991, Germany passed a packaging ordinance known as VerpakV that obliges manufacturers and distributors to take back used sales packaging from consumers and send it to be recycled.

Another effective way to reduce waste is to charge households, industries, and businesses for the quantity of waste they dump. Such programs are often referred to as unit-pricing or pay-as-you-throw (PAYT) programs, which differ from the more conventional fixed-fee waste disposal costs. Pricing systems include proportional pricing where a customer is charged by can, bag, or weight; variable rate pricing, where customers are billed for a set container size and pay additional fees if they exceed said container size; and multi-tiered pricing, where customers pay a base level of service and pay a second-tier fee on the amount of waste they discard. Of the three, the proportional system seems to have the most direct relationship between waste produced and cost and therefore might provide the best incentive to reduce waste. All systems pose the risk of unethical trash dumping whereby a customer tries to put more trash in the recycling container, in nearby trash receptacles, on the ground/in the woods, and so on to avoid having to pay additional costs. However, PAYT programs have been shown to work. Data collected by Pay-T-Now shows a 17 percent diversion of the tonnage disposed of in trash cans – with one-third going to recycling, one-third to composting, and one-third to source reduction – and an over 50 percent increase in recycling tons (Skumatz et al., 2015).

Reuse/repurpose

Growing up in the 1980s, unless an older sibling gave you a hip rock tee shirt, most hand-me-downs were often a source of ridicule. However, in the new millennium, this paradigm is changing. Retro and vintage are now considered "cool." In the past 20 years, there has been a resurgence in antique stores, thrift shops, and used-clothing stores. Online trading/selling sites and forums such as eBay, Craigslist, Facebook Marketplace, Reverb, and Front Porch Forum have made it easier for people to resell or give away items they no longer want. Television shows such as *Pawn Stars*, *American Pickers*, and *Antiques Road Show* have piqued people's interest in hunting for treasures. Indeed, one person's trash is another person's treasure. In addition to reusing something as intended, items can be repurposed with an entirely different use. In several locations throughout the world, people have converted old cars into rolling grills and started mobile catering companies. Green Mountain Ski Furniture in Stowe, Vermont, uses old skis to construct its tables, chairs, benches, and clothing racks. Indeed, the possibilities are endless: old tires can be converted into playgrounds or garden planters; ladders can be made into bookshelves; old television sets can be turned into aquariums; doors can be used as tables; old card catalogs can be used for hardware storage; CD cases can be coasters; old bottles can be light fittings; crates can become shelves; bathtubs can become couches; and, as discussed in Chapter 10, building materials can be repurposed and buildings themselves can be adapted for new functions. Any sort of reuse is better than an item ending up in a landfill.

Recycle

Recycling is similar to reuse, but the material is collected at the curbside or at drop-off locations and sent to a treatment center where it is sorted, cleaned, melted down, and remolded into a new item. Out of the total 292.4 million tons of MSW generated in the United States in 2018, 94 million tons of MSW were recycled and composted, which resulted in 23.6 and 8.5 percent of the total, respectively. Landfilling accounted for the biggest portion at 50 percent; combustion with energy recovery accounted for 11.8 percent and other food management pathways accounted for 6.1 percent. Of the 23.6 percent that was recycled, 23.05 percent was paper or

FIGURE 5.7 Repurposing items. Christmas tree decorations made from old funnels, by Phyllis A. Richards.

paper board, 21.59 percent food, 12.20 percent plastics, 12.11 percent yard trimmings, 8.76 percent metals, 6.19 percent wood, 5.83 percent textiles, 4.19 percent glass, 3.13 percent rubber and leather, 1.56 percent other materials, and 1.39 percent miscellaneous organic waste (EPA, 2020).

Why these numbers are not higher is a bit upsetting, to say the least. One of the issues is that the recycling infrastructure in the United States has not kept up with the demand. Up until 2018, over half of the world's discarded materials were sent to China to be recycled. In 2016, the United States exported 16 million tons of plastic, paper, and metal to China, and up to 30 percent of this was not able to be recycled because of contamination by the mix of non-recycled material. As a result, roughly 1.3 million to 1.5 million tons of plastic ended up in the ocean off China's coast (Cho, 2020). As more countries are banning the import of plastic, U.S. recycling facilities are either incinerating or burying the recycled materials in landfills. Although single-streaming recycling where all materials can be placed in the same bin has made it easier for customers to recycle, data shows about 25 percent is contaminated and cannot be recycled (Cho, 2020). In addition, the technology at many facilities is unable to effectively sort the materials.

There are three types of plastic recycling methods: mechanical, where plastic is ground down and reused in similar products; incineration that can be used to produce heat and electricity; and chemical conversion of plastic into fuels. Mechanical remains the most common, incineration results in the release of toxins, and chemical currently requires extremely high temperatures. However, researchers are studying ways to make chemical extraction more efficient. For example, by using ruthenium metal and carbon as catalysts, 90 percent of plastic waste can be converted into fuel at a lower temperature of 220 degrees Celsius compared to current standards of 300 degrees Celsius (Jia, 2021). Another interesting option is to convert plastic into the filament used in three-dimensional printers. In 2011, Tyler McNaney founded the Vermont-based company Filabot. The Filabot technology grinds, melts, and extrudes various types of plastic materials, including milk jugs, detergent bottles, soda bottles, product packaging, and discarded print plastic, into custom-sized material compatible with three-dimensional printers utilizing a desktop unit that is no larger than a typical paper printer. What started as a kick-starter has turned into a 26-person operation with customers including NASA, the Pentagon, LEGO, Xerox, Dow, and Intel (Filabot, n.d.).

Since China's ban on accepting plastic, there has been funding allocated to improve the sorting and recovery of materials at U.S. recycling facilities. The Infrastructure Investment and Jobs Act has allocated $275 million for Solid Waste Infrastructure for Recycling grants (EPA, 2023). In the 2019 National Framework for Advancing the U.S. Recycling Program, the National Recycling Strategy was established to increase the recycling rate by 50 percent by 2030. Five strategic principles were identified, including improving markets for recycling commodities; increasing collection and improving materials management infrastructure; reducing contamination in the recycled materials stream; enhancing policies to support recycling; and standardizing measurement and increasing data collection (EPA, 2021).

Other strategies to improve recycling include requirements for composting food, banning single-use plastic bags, providing water filling stations to limit plastic bottle use, increasing container deposit criteria, incorporating more recycled material into products and building materials to create demand, and providing adequately sized and/or different containers for specific materials. It is shocking to see that in 2024, many waste companies provide extra-large containers for trash but only one small bin for recycling. Many customers are forced to buy their own containers if they are recycling savvy, but others will simply just throw whatever doesn't fit into the small bin into the trash. Combining incentives or penalties with some of the PAYT programs discussed earlier can also be effective. Customers can be rewarded with tax incentives or discounts on their PAYT subscriptions for more material that is recycled; likewise, more fees can be charged for material that is disposed of in the trash and/or for food scraps not composted. In South Korea, for example, customers are required to separate their trash, recycling, and organic waste in dedicated containers equipped with Radio Frequency Identification (RFID). Recycling is free, but charges are incurred by the weight of the trash and compost. Customers are encouraged to remove the moisture content from the compost, thereby making the compost container lighter. As a result, the country recycles 54 percent of its materials and 95 percent of its food waste (Cho, 2020).

Battery recycling/recovery

Small, portable electronic devices, power tools, cars, and renewable energy storage needs have all led to an increase in the use of batteries which raises the issue of what to do with the batteries when their lifespan runs out. Batteries can be single use or rechargeable, with the former being alkaline/zinc carbon, button cell or coin, or lithium single use and the latter consisting of nickel-cadmium, lithium-ion, nickel-metal halide, nickel-zinc, or small sealed lead acid. Automotive batteries can be lead acid or lithium-ion. Seeing batteries contain metals such as mercury, lead, cadmium, nickel, and silver along with critical minerals like cobalt, lithium, and graphite, improper disposal can cause harm to human health and to the environment. In addition, many of the elements in batteries are mined, which poses issues with meeting demand and adverse conditions for the people working to extract the materials.

However, with advancements in technology, batteries can be efficiently recycled and materials can be recovered. Alkaline and zinc carbon batteries are mechanically separated into zinc and manganese concentrate, steel, paper, and plastic. Nickel-cadmium and nickel-metal halide are melted using high-temperature metal reclamation (HTMR) that separates out high-melt and low-melt metals. Lithium-ion batteries are first shredded to separate out the copper, aluminum, plastics, canisters, electrolytes, and black mass from cathodes and anodes. Black mass is then sent through a heat-based smelting process (pyrometallurgy) or a liquid-based leaching process (hydrometallurgy) to recover cobalt, nickel, and lithium. There are also several emerging technologies that make the process cleaner and more efficient. For example, researchers at Sweden's Chalmers University of Technology have developed a recycling method that allows recovery of 100 percent of the aluminum and 98 percent of the lithium in electric car batteries by using oxalic acid – an organic ingredient found in plants (Chalmers, 2023).

Recover

In some cases, waste can be converted into energy. One of the simplest ways to convert waste into energy is through composting organic materials, such as plants, leaves, grass clippings, fruit and vegetable scraps, wood ash, coffee grounds, and shredded paper. Materials have varying degrees of carbon and nitrogen – to allow for enough oxygen, a ratio of two-thirds carbon to one-third nitrogen is recommended. Compost can be put into piles or into containers or "tumblers." In both cases, it is essential to aerate the material, either by tilling the pile or by rotating the container. Layering dry and wet material also allows oxygen to penetrate and nourish the pile. In addition to oxygen, microbes in the soil and detritus feeders such as worms and grubs decompose the organic material and reduce the waste volume into a humus-like material that can fertilize and regenerate the soil. As of the writing of this book, nine states – including California, Connecticut, Maryland, Massachusetts, New Jersey, New York, Rhode Island, Vermont, and Washington – have enacted laws mandating composting of organic materials. However, without proper fencing and/or containers, composting is bound to attract wildlife. Since the law was passed in Vermont, Vermont Fish and Wildlife has reported an uptick in bears roaming on people's properties. The state does provide drop-off locations, but it seems providing dedicated containers would improve participation and minimize human-animal interactions, rodent surges, and damage to property. The state also recommends using three parts brown material for one-part green material, frequent mixing of material, and electric fencing (VT ANR, 2020).

Waste can also be converted into fuel through the process of anaerobic digestion. Animal manure is put into a sealed chamber (often referred to as a methane digester), diluted with water, and sits at 101 degrees Fahrenheit for 21 days. The absence of oxygen allows anaerobic bacteria to break down the organic material into methane, carbon dioxide, and water. The methane is then extracted and combusted to generate electricity or refined to produce biomethane, which is comparable to natural gas. The solid waste that is left over is used as bedding for animals or to fertilize soil. The United States currently has 343 farms with operational biogas recovery systems. The U.S. Environmental Protection Agency (EPA, n.d.) estimates there are 8,000 farms across the country that could potentially generate nearly 16 million megawatt-hours of energy per year and displace about 2,010 megawatts of fossil-fuel-fired generation.

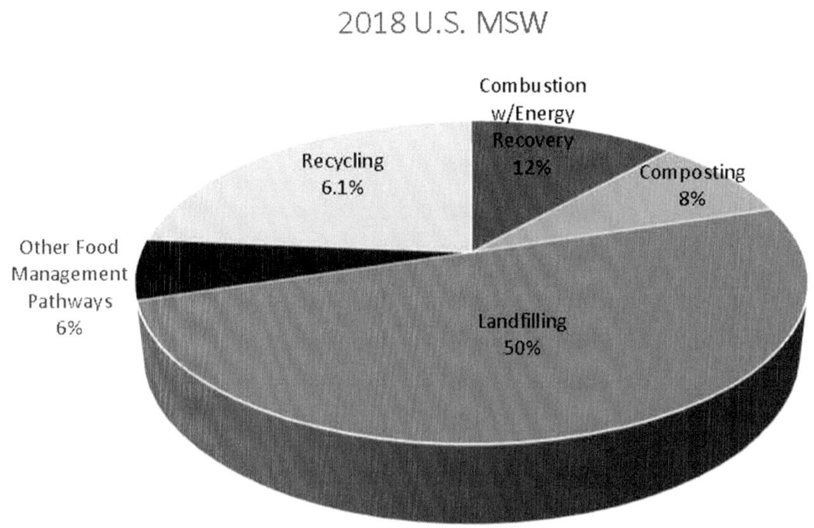

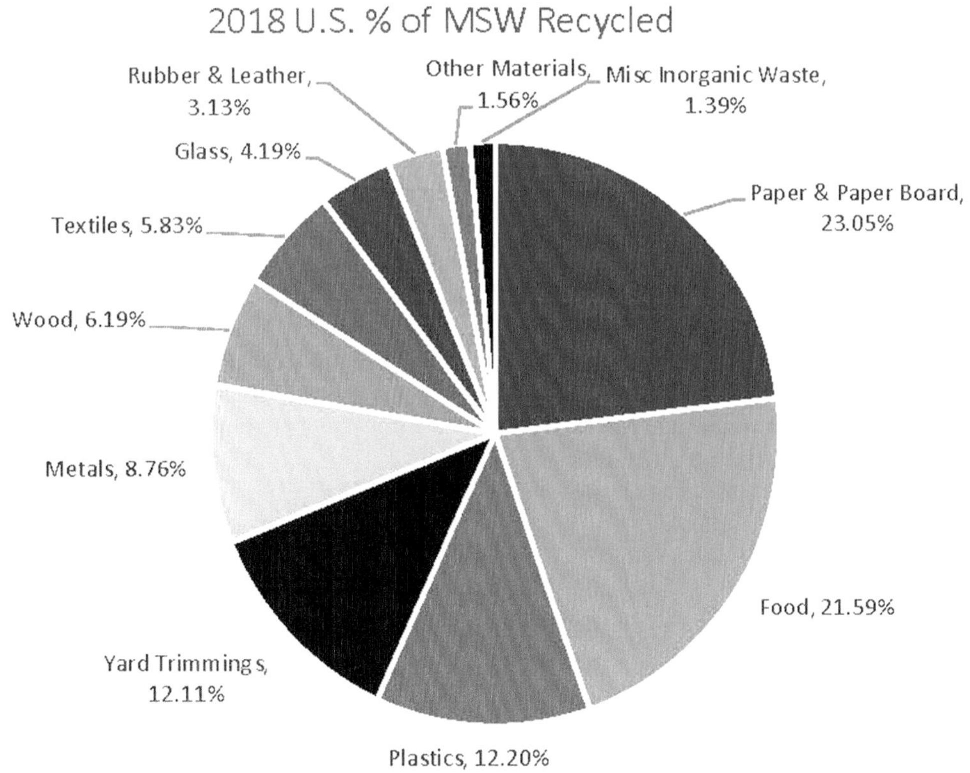

FIGURE 5.8 Top: 2018 U.S. MSW; bottom: 2018 U.S. percentage of MSW recycled.

Source: U.S. Environmental Protection Agency (EPA) (n.d.). *National overview: Facts and figures on materials, wastes, and recycling.* https://www.epa.gov/facts-and-figures-about-materials-waste-and-recycling/national-overview-facts-and-figures-materials

In regions where there are no connections to municipal wastewater treatment plants, natural, biological processes can also be used to break down human waste on-site. A conventional on-site wastewater treatment system usually consists of an underground septic tank to decompose the waste and a leach field with gravel-lined trenches to distribute the wastewater. Engineered mounds can also be used in soils that are too permeable or too impermeable. Such systems usually have a dosing chamber between the septic tank and the mound that distributes the effluent to the mound at a metered rate. The septic tank collects waste from the building and uses anaerobic bacteria to decompose organic solid material known as "scum" that floats to the top of the tank. The by-product of the scum digestion – known as sludge – sinks to the bottom of the tank. The gray water between the scum and the sludge is then released into the leach field, where soil microbes further break down bacteria, viruses, and nutrients. The sludge needs to be periodically removed from the tank or else it will overflow and either flood the leach field and/or back up into the building. Leach fields also can clog with age due to reduced soil percolation rates from the buildup of bacteria under the drainage trenches. Where soil conditions and/or lot size are unable to accommodate the leach field, aerobic septic systems can be used to produce a high-quality secondary effluent that can be spread across surfaces. Although designs can vary, aerobic systems usually consist of a pretreatment tank to remove solid materials that bacteria cannot break down, an aerobic unit that contains an aeration chamber that mixes air with wastewater, a settling chamber for sludge, and a disinfection chamber/unit that uses chlorine or ultraviolet radiation to produce a sterile effluent which may be required if the effluent is distributed on the surface rather than underground in a drain field.

Constructed wetlands, biofilters, and plants can also be used in conjunction with conventional on-site wastewater treatment systems. Omega's Eco Machine and John Todd's Living Machine both clean water by mimicking biological processes and create closed-loop hydrological cycles. At the Omega Center for Sustainable Living, wastewater flows into solid settlement tanks, settles as sludge, and is injected with microorganisms to accelerate decomposition; the remaining wastewater flows into equalization tanks that equalize the flow of water; water is then released to anoxic tanks where naturally occurring microbial organisms digest ammonia, phosphorus, nitrogen, potassium, and other substances in anaerobic conditions; water then enters four constructed wetlands, then into two aerated lagoons, then to a recirculating sand filter, and then into dispersal fields where reclaimed water is released back into the groundwater (Omega, n.d.).

Dealing with waste is something most of us usually do not think about. We throw things in the garbage, flush the toilet, and put items out on the curb, and then they are magically out of sight, out of mind. Stormwater is another element that most people do not pay much attention to – that is, until it causes flooding, beach closures, water pollution, or even floating sewage in the case of combined sewage overflows (CSOs). On-site composting and wastewater treatment force people to see and deal with their waste first-hand, which fosters circular ecological connections. When people see the amount of waste they generate, they often become more inclined to reduce it.

References

Armstrong, D. (2000). A survey of community gardens in Upstate New York: Implications for health promotion and community development. *Health & Place*, *6*, 319–320.

Bajpai, P. (2023). *Top five nations in solar energy generation.* NASDAQ. www.nasdaq.com/articles/top-five-nations-in-solar-energy-generation

Bryce, E. (2016). *Will wind turbines ever be safe for birds?* National Audubon Society. www.audubon.org/news/will-wind-turbines-ever-be-safe-birds

Bullard, R. D., Mohai, P., Saha, R., & Wright, B. (2008). Toxic wastes and race at twenty: Why race still matters after all these years. *Environmental Law*, *38*(2), 371–411. www.jstor.org/stable/43267204

Bushong, S. (2016). *Advantages and disadvantages of a solar tracker system*. Solar Power World. www.solarpowerworldonline.com/2016/05/advantages-disadvantages-solar-tracker-system/

Calpine. (n.d.). *The geysers*. https://geysers.com/geothermal

Center for Climate and Energy Solutions. (C2ES). (n.d.). *Carbon capture*. www.c2es.org/content/carbon-capture/

Chalmers University of Technology. (2023, October 17). New recipe for efficient, environmentally friendly battery recycling. *ScienceDaily*. www.sciencedaily.com/releases/2023/10/231017123501.htm

Cho, R. (2020). *Recycling in the U.S. is broken. How do we fix it?* Columbia Climate School, State of the Planet. https://news.climate.columbia.edu/2020/03/13/fix-recycling-america/

Cunningham, P., & Woofenden, I. (2007, February/March). Micro-hydroelectric systems simplified. *Home Power*, 117. www.homepower.com/articles/microhydro-power/equipment-products/microhydro-electric-systems-simplified

Environment America. (2021). *Trash in America: Moving from destructive consumption towards a zero-waste system*. https://environmentamerica.org/center/resources/trash-in-america-2/#:~:text=Americans%20throw%20out%204.9%20pounds,landfills%20or%20burned%20in%20incinerators

Farina, E. (2017). *How Brazil got sugar smart to meet its energy challenge*. Friends of Europe. www.friendsofeurope.org/insights/how-brazil-got-sugar-smart-to-meet-its-energy-challenge/

Filabot. (n.d.). www.filabot.com/

Fleischman, L., & Franklin, M. (2017). *Fumes across the fence-line: The health impacts of air pollution from oil and gas facilities on African-American communities*. NAACP/CATF. https://naacp.org/resources/fumes-across-fence-line-health-impacts-air-pollution-oil-gas-facilities-african-american

Global CCS Institute (Global CCS). (n.d.). *198 new facilities*. https://status23.globalccsinstitute.com/new-facilities/

Harris, L. (2022). *What are food miles?* Connect 4 Climate. www.connect4climate.org/initiative/what-are-food-miles

Hartman, L. (2023). *Wind turbines: The bigger the better*. Office of Energy Efficiency and Renewable Energy. www.energy.gov/eere/articles/wind-turbines-bigger-better

Healthy Living. (n.d.). *What we stand for*. https://healthylivingmarket.com/what-we-stand-for/

International Energy Agency (IEA). (n.d.). *Carbon capture, utilization, and storage*. www.iea.org/energy-system/carbon-capture-utilisation-and-storage

Jia, C. (2021). *Converting plastic waste into fuel*. Harvard University, Science in the News. https://sitn.hms.harvard.edu/flash/2021/converting-plastic-waste-into-fuel/

Kaza, S., Yao, L. C., Bhada-Tata, P., & Woerden, V. (2018). *What a waste 2.0: A global snapshot of solid waste management to 2050*. Urban Development. World Bank. http://hdl.handle.net/10986/30317 License: CC BY 3.0 IGO.

Li, M., Jia, N., Lenzen, M., Malik, A., Wei, L., Jin, Y., & Raubenheimer, D. (2022). Global food-miles account for nearly 20% of total food-systems emissions. *Natural Food*, *3*, 445–453. https://doi.org/10.1038/s43016-022-00531-w

Longcore, T. R. C., Mineau, P., MacDonald, B., Bert, D. G., Sullivan, L. M., Mutrie, E., Gauthreaux Jr., S. A., Avery, M. L., Crawford, R. L., Manville, A. M., Travis, E. R., & Drake, D. (2012). An estimate of mortality at communication towers in the United States and Canada. *PLoS ONE*. https://doi.org/10.1371/journal.pone.0034025

Loss, S. R., Will, T., Loss, S. S., & Marra, P. P. (2014a). Bird-building collisions in the United States: Estimates of annual mortality and species vulnerability. *Condor*, *16*, 8–23.

Loss, S. R., Will, T., & Marra, P. P. (2013). The impact of free-ranging domestic cats on wildlife of the United States. *Nature Communications*, *4*. https://doi.org/10.1038/ncomms2380

Loss, S. R., Will, T., & Marra, P. P. (2014b). Refining estimates of bird collision and electrocution mortality at power lines in the United States. *PLoS ONE*, *9*(7). https://doi.org/10.1371/journal.pone.0101565

Loss, S. R., Will, T., & Marra, P. P. (2014c). Estimation of annual bird mortality from vehicle collisions on roads in the United States. *Wildlife Management*, *78*, 763–771.

National Institute of Environmental Health Services (NIH). (n.d.). *Hydraulic fracturing and health*. www.niehs.nih.gov/health/topics/agents/fracking/index.cfm

New England Dairy (NE Dairy). (2020). *Fact sheet*. www.newenglanddairy.com/wp-content/uploads/Moo-IQ_VT_Fact-Sheet-2020.pdf
Office of Energy Efficiency and Renewable Energy (EERE). (2020). *Top reasons to love solar energy*. www.energy.gov/eere/articles/top-reasons-solar-energy
Office of Energy Efficiency and Renewable Energy (EERE). (2022). *Offshore wind market report: 2022*. U.S. Department of Energy. www.energy.gov/sites/default/files/2022-09/offshore-wind-market-report-2022-v2.pdf
Office of Energy Efficiency and Renewable Energy (EERE). (n.d.). *How a wind turbine works – text version*. www.energy.gov/eere/wind/how-wind-turbine-works-text-version
Omega Center for Sustainable Living (Omega). (n.d.). *The Eco Machine*. www.eomega.org/center-sustainable-living/eco-machine
Power Technology. (2014). *Tidal giants – the world's five biggest tidal power plants*. www.power-technology.com/features/featuretidal-giants-the-worlds-five-biggest-tidal-power-plants-4211218/?cf-view
Power Technology. (2020). *SeaGen Turbine, Northern Ireland, UK*. www.power-technology.com/projects/strangford-lough/?cf-view
Renewable Energy World. (2014). *Netherlands installs first solar bike path*. www.renewableenergyworld.com/solar/netherlands-installs-worlds-first-solar-bike-path/
Sardar Sarovar Narmada Nigam (SSNN). (n.d.). *Solar power project*. www.sardarsarovardam.org/-Solar-Power-Project.aspx
Schendler, A. (2006, February/March). *Sustainable skiing: Snowmass ski area gets hydro*. Home Power. www.homepower.com/articles/microhydro-power/project-profiles/sustainable-skiing
Scientific American. (1991). Energy for planet earth (p. 2). W.H. Freeman & Company.
Shelburne Farms. (n.d.). *History of the property*. https://shelburnefarms.org/about/history-property
Skumatz, L., D'Souza, D., & BeMent, D. (2015). *Pay-as-you-throw/variable rates for trash collection: 2014 update, Region 9 grant report volume 1 of 2, Executive Summary and Key Results*. Econservation Institute. www.paytnow.org/PAYT_EI_R9_v25_Vol1
Souza, E. (2019). *Sidewalks that generate energy through the steps*. ArchDaily. www.archdaily.com/911965/sidewalks-that-generate-energy-through-the-steps
Stern, A., & Berghout, N. (2021). *Is carbon capture too expensive*? IEA. www.iea.org/commentaries/is-carbon-capture-too-expensive
TETHYS. (n.d.). *Swansea Bay Tidal Lagoon*. https://tethys.pnnl.gov/project-sites/swansea-bay-tidal-lagoon-sbtl#project-progress
Trail, P. (2006). Avian mortality at oil pits in the United States: A review of the problem and efforts for its solution. *Environmental Management*, *38*, 532–544. https://doi.org/10.1007/s00267-005-0201-7
Union of Concerned Scientists. (2022). *What is sustainable agriculture*? www.ucsusa.org/resources/what-sustainable-agriculture
U.S. Department of Energy (DOE). (n.d.). *Waste heat recovery basics*. www.energy.gov/eere/iedo/waste-heat-recovery-basics
U.S. Energy Information Administration (EIA). (2021). *Average U.S. construction cost for onshore wind generation decreased by 27% since 2013*. www.eia.gov/todayinenergy/detail.php?id=49176
U.S. Energy Information Administration (EIA). (2022). *Today in energy*. www.eia.gov/todayinenergy/detail.php?id=60341
U.S. Energy Information Administration (EIA). (2023). *Total energy*. https://www.eia.gov/totalenergy/data/monthly/index.php#renewable; https://www.eia.gov/totalenergy/data/monthly/pdf/sec10_3.pdf
U.S. Environmental Protection Agency (EPA). (2020). *Advancing sustainable materials management: 2018 fact sheet*. www.epa.gov/sites/default/files/2021-01/documents/2018_ff_fact_sheet_dec_2020_fnl_508.pdf
U.S. Environmental Protection Agency (EPA). (2021). *National recycling strategy*. www.epa.gov/circulareconomy/national-recycling-strategy#:~:text=The%20National%20Recycling%20Strategy%20recognizes,communities%20with%20environmental%20justice%20concerns
U.S. Environmental Protection Agency (EPA). (2023). *Solid waste infrastructure for recycling grant program*. www.epa.gov/infrastructure/solid-waste-infrastructure-recycling-grant-program#:~:text=The%20Infrastructure%20Investment%20and%20Jobs%20Act%2C%20also%20referred%20to%20as,to%20remain%20available%20until%20expended
U.S. Environmental Protection Agency (EPA). (n.d.). *AgSTAR data and trends*. www.epa.gov/agstar/agstar-data-and-trends

U.S. Fish & Wildlife Service (USFWS). (n.d.). *Threats to birds*. www.fws.gov/library/collections/threats-birds

Vermont Agency of Natural Resources, Fish and Wildlife Department (ANR). (2020). *Composting with bears in mind*. https://vtfishandwildlife.com/newsroom/composting-with-bears-in-mind#:~:text=The%20Vermont%20Fish%20%26%20Wildlife%20Department,can%20avoid%20attracting%20hungry%20bears

Vermont Agency of Natural Resources, Fish and Wildlife Department (ANR). (n.d.). *Bee industry overview*. https://agriculture.vermont.gov/public-health-agricultural-resource-management-division/apiary-program/bee-industry-overview

Walters, J. (2013). *Vermont's long, careful path to renewable energy*. Georgia Mountain Community Wind. http://georgiamountainwind.com/vermonts-long-careful-path-to-renewable-energy-pt-2-the-regulatory-record/

Wang, M., Han, J., Dunn, J., Cai, H., & Elgowainy, A. (2012). Well-to-wheels energy use and greenhouse gas emissions of ethanol from corn, sugarcane and cellulosic biomass for U.S. use. *Environmental Research Letters*, *7*. https://doi.org/10.1088/1748-9326/7/4/045905

Wang, M., Le, U., Kwon, H., & Xu, H. (2020). *Life-cycle greenhouse gas emission reductions of ethanol with the GREET model*. Presentation at the 2021 National Ethanol Conference. Argonne National Laboratory. https://afdc.energy.gov/files/u/publication/ethanol-ghg-reduction-with-greet.pdf

Wegmans. (n.d.). *Helping creating a healthier, better planet*. www.wegmans.com/values-in-action/sustainability-at-wegmans/#:~:text=Sourcing%20from%20Near%20Our%20Stores,sourced%20on%20the%20East%20Coast

Whistler-Blackcomb. (n.d.). *Environment energy*. www.whistlerblackcomb.com/explore-the-resort/about-the-resort/environment/energy.aspx

Whole Foods. (n.d.). *Environmental stewardship*. www.wholefoodsmarket.com/mission-in-action/environmental-stewardship

Yang, C. (2021). *Addressing the environmental justice implications of waste*. Environmental and Energy Study Institute (EEST). www.eesi.org/articles/view/qa-addressing-the-environmental-justice-implications-of-waste

6

TRANSPORTATION

The adage "if you build it, they will come" is applicable when it comes to roadways and automobiles: if you plan for more roads, cars, and traffic, you end up with more roads, cars, and traffic. The historical trend has been to widen, expand, and/or add lanes to existing roads in response to increasing congestion. Two lanes are expanded to four lanes, four to six, six to eight, and eventually the once natural landscape is transformed from "green to gray." The theory that road expansion saves fuel, reduces emissions, and decreases travel times has been proven false in several target sites. Initially, an expanded road may alleviate some congestion, but the subsequent increased travel speeds lead to more frequent trips and higher fuel consumption. Over time, the addition of lanes also often induces more traffic, and congestion and bottlenecking issues reemerge (Newman & Kenworthy, 1999, p. 59).

However, many people, including this author, still love the freedom of personal automobiles. Indeed, cars allow us to schedule trips on our own schedule, haul all sorts of groceries, goods, and equipment, and afford us a level of privacy that is forfeited when riding on public transit (the exception being personal rapid transit). Throw in a global pandemic and/or a pre-existing germ phobia, and public transit may be the last thing on anyone's mind. Indeed, data from the American Public Transportation Association showed significant declines in all modes of public transit during 2019–2020 (APTA, 2023). For these reasons, it may not be feasible to completely eliminate automobiles in the near future.

Before the wide availability of hybrid cars, it was common to see climate change advocacy groups placing flyers (not on recycled paper, mind you) on the windshields of SUVs in parking lots telling the drivers that they were killing the planet. While their hearts may have been in the right place, some of these groups failed to realize that a majority of the middle class could not afford a hybrid vehicle at that time and/or that others were commuting from remote areas with no access to the bus line that required four-wheel drive to navigate the terrain, especially in the winter. When Toyota launched the Prius in 1997 (which was the first mass-produced hybrid passenger vehicle), many anecdotal reports claimed that they did not perform well in the snow. Now that technology and vehicle options have improved for electric and hybrid fuel/electric vehicles (including hybrid SUVs), this dialogue certainly has changed. However, until all vehicles are converted to be more fuel efficient, people will most likely still choose the most appropriate

DOI: 10.4324/9781032668666-6

vehicle for their specific needs. But it won't be long until gas-powered options are not available. As of the writing of this book, nine states in the United States have announced they will ban the sales of gas-powered vehicles after 2035, including California, Connecticut, Maryland, Rhode Island, Massachusetts, New Jersey, New York, Oregon, and Washington. Worldwide, at least 19 countries have adopted bans with similar timelines, with several of the Scandinavian countries announcing bans before 2030.

Although zero-emission vehicles, including battery-electric (BEV), plug-in hybrid-electric (PHEV), and hydrogen fuel cell (FCV), are indeed a mandatory step to reduce our carbon emissions, the technologies still do not address vehicular congestion and the subsequent fragmentation of communities by roadway and parking infrastructure. Although flying cars are on the distant horizon, they may not be affordable for the masses for some time to come, and if/when they are, it is unclear if they will alleviate congestion or just expand it in another dimension. For these reasons, it is also imperative to adopt more auto-alternative forms of transportation that can decrease vehicle miles traveled and congestion on roads and avert the need to widen roads or add more parking. To address the complex dispersal of habitation across the landscape, a variety of modes of public transportation need to be improved, established, and networked, including fixed rail (heavy rail, light rail, streetcars, aerial cable transit, automated guideway transit, monorail, magnetic levitation, pneumatic tubes), bicycle, bus/bus rapid transit, micromobility (e-bikes, e-scooters, hover boards, electric skateboards, Segways, etc.), ferries/water taxies, demand response transit (DRT, including taxicabs, Uber, Lyft, paratransit, van pools), and air.

Data compiled by the American Public Transportation Association (APTA, n.d.) shows positive correlations between public transportation and sustainable economic, environmental, and social factors. Economically, public transit creates and sustains employment, saves individuals money, connects workers to jobs in rural areas, and increases real estate and development. APTA estimates that every \$10 million invested in public transportation yields \$30 million in increased business sales and \$1 billion invested creates about 50,000 jobs. Environmentally, public transit decreases congestion, pollution, and energy consumption and reduces dependence on oil. Overall, public transportation consumes 95 percent less carbon monoxide, 90 percent less volatile organic compounds, and roughly half the carbon dioxide and nitrogen oxide per passenger mile compared to vehicles. APTA estimates public transit saves six billion gallons of gasoline and reduces 63 million metric tons of carbon being emitted into the atmosphere annually. Socially, public transport provides freedom, opportunity, access, and mobility for all ages, abilities, and socio-economic groups and is estimated to be ten times safer than travel by personal automobile.

Fixed-guideway systems

Although all modes of public transit are essential to provide solutions to public mobility, unless dedicated lanes are constructed with roads or above or below roads, modes of transit that operate with vehicular traffic (buses, taxis, van pools, streetcars) are susceptible to the same conditions as automobiles, including traffic lights, road construction, weather conditions, congestion, and accidents. These factors – coupled with the number of stops at stations along the route – can inevitably make travel time significantly longer for public transit modes shared with vehicles than by car or by a fixed rail/guideway system, given the same route. As confirmed by Eriksson et al. (2008), shorter travel time is one of the key factors that make all public transit more effective. A study conducted in New York City by Redman et al. (2013) found a 15-minute shorter commuting time corresponded to about 25 percent higher ridership for rail service. However, depending on the route, the number of stops, the travel distance, the walking distance to and from destinations/stations, and the population density of the area, comparing any mode of public transit to the personal automobile can yield varied results. For example, riding a subway in a dense city with car congestion would most likely be faster than traveling by car, but could take longer than car travel in a less dense region with less auto congestion. Understanding the threshold at which people will change their behaviors and ditch their cars is critical to the success of any planned public transit system. Traffic analysis and qualitative surveys conducted during the planning process can reveal travel patterns, preferences, and projected ridership that can inform the mode of transit, route alignments, stations, capacities, headways, and frequency of trains.

In addition to more competitive travel times, the permanence of fixed guideways can influence urban form by fostering transit-oriented development (TOD) that can cluster a mix of commercial, recreational, and residential uses around compact, walkable nodes where the stations reside. Such sustainable patterns of urban design can outweigh the higher initial capital investments for fixed-guideway systems compared to conventional road expansion. As stated by Warren (1998, p. 80),

> A primary value of fixed-guideway transit is that it creates the pre-conditions for reversing our current trends toward far-flung dispersal of development; investments in it should be regarded as a downpayment on a more resource conserving future.

Fixed-guideway systems can operate on either fully exclusive rights of way (Category A) or semi-exclusive rights of way (Category B) and can compete with cars since transit speeds equal or exceed those of automobiles. This is one of the reasons why train travel is popular in European and Asian countries (others include networking, frequency, technology, high funding, and marketing), where trains travel an average of 4–6 mph (6.4–9.7 km/h) faster than cars (Newman & Kenworthy, 1999, p. 59). In addition to providing speeds that are competitive with (or better than) those of the personal automobile, the exclusive rights of way provided by fixed guideways – whether at-grade, elevated, or underground – can provide dedicated routes for pedestrians and bicycle lanes, which themselves can be crucial components in improving circulation and revitalizing communities.

Various criteria are used to classify transit systems, including right-of-way category, system technology, propulsion, speed, size, weight, capacity, and scale of service. For the purposes of this book, fixed-guideway transit includes transit methods that utilize exclusive rights of way, including, but not limited to heavy rail transit (HRT; or rapid rail transit – RRT), light rail transit (LRT), cogwheel rail, monorail, magnetic levitation, pneumatic tubes, and automated guideway transit (AGT), which can include aerial cable transit (gondolas, trams) and surface cable transit (funiculars, moving sidewalks, people movers).

Heavy rail transit

Heavy rail systems include underground, rapid rail transit (RRT or metros) found in dense, urban areas, commuter rail systems that link outlying areas to the periphery of the city, and regional rail lines or high-speed rail that span between regions. Within cities, heavy rail systems are often referred to as "subways" or "undergrounds" since they are usually constructed below grade. RRT/metro systems usually have up to eight vehicles per train that house 800–1,600 passengers; they can transport roughly 22,000 passengers per hour per line and have maximum speeds between 50 and 60 mph (80 97 km/h) (Vuchic, 2007). As Cervero (1998, pp. 17–22) points out, "The relationship of heavy rail transit to dense, urban areas is symbiotic: the cities could not survive without the rail and the rail could not be sustained without very high population densities." In comparison, commuter rail lines are usually 25–50 miles (40–80 km) long, have up to 12 vehicles per train that can house 2,000 passengers and transport up to 48,000 passengers per hour per line, and have maximum speeds of 50–80 mph (80–129 km/h) (Vuchic, 2007; Cervero, 1998). RRT is usually powered by an electrified third rail, thus the underground alignments are ideal since it is easier to keep pedestrians off the tracks. When third-rail systems run at-grade, fencing must be installed. Commuter and inter-city rail, on the other hand, is usually powered by diesel or hybrid diesel–electric locomotives.

Light rail transit

Light rail transit (LRT) is essentially an upgraded version of the streetcar with increased speed and capacity and the ability to operate at-grade, underground, or elevated. Although LRT can run on streets, it is usually constructed on exclusive or semi-exclusive rights of way (categories A and B) and is usually powered by overhead wiring rather than the third-rail systems favored by heavy rail transit. Although costlier, underground and elevated systems permit higher speeds of up to 75 mph (121 km/h) compared to at-grade systems mixed with auto traffic, which average only 12–30 mph (19–48 km/h) (Vuchic, 2007). The lightweight vehicles permit higher rates of acceleration and more efficient deceleration from high speeds with less power consumption and

FIGURE 6.1 Buffalo Metro light rail transit, powered by overhead wiring, operates at- and below grade.

degradation compared with heavy rail transit. LRT is generally composed of two- to four-axle, hinged vehicles that operate on steel wheel/steel rail or rubber tire/concrete rail technologies. Vehicles can accommodate 260–520 passengers, resulting in capacities of up to 11,000 passengers per hour per direction for four-car trains running as closely as three minutes apart, or up to 20,000 passengers per hour per direction with headways of 90 seconds or less (Cervero, 1998). Although at-grade systems are the most cost-effective form of LRT, slow speeds and interference with other modes of transport can lead to increased congestion and longer headways. If constructed at-grade along streets, tracks can be placed on medians and preferential signal timing can be granted at intersections. Compared to AGT and monorail systems, elevated LRT requires larger infrastructure and produces more noise. Light rail transit is generally less expensive than heavy rail transit and is often applicable in small- to medium-sized metropolitan areas.

Bus rapid transit

Bus rapid transit (BRT) – also referred to as busways or transitways – are bus systems that provide higher capacity, reliability, and speed/frequency than a typical bus with a traffic system. To accomplish this, BRT typically uses dedicated lanes and/or priority signaling at intersections to provide continuous travel. The lanes can be placed along the side of a road, down the center of the road, or can be elevated or tunneled like light and heavy rail systems. Alignments at-grade are usually down the center of the road to keep buses away from the curb where cars park and turn. Often, cars are limited to where they can cross bus lanes or even prohibited altogether from turning across bus lanes to prevent delays in the BRT system. Some notable BRT systems have been implemented in Yichang, China; Belo Horizonte, Brazil; Bogata, Colombia; Buenos Aires, Argentina; Mexico City, Mexico; Johannesburg, South Africa; Pimpri-Chinchwad, India; Jakarta, Indonesia; Hartford, Connecticut; the Lincoln Tunnel, New Jersey; and Pittsburgh, Pennsylvania, to name a few.

BRT is often pitted against LRT. There are ample studies comparing the advantages and disadvantages of both as well as subjective spins and opinions arguing for one over the other. A study conducted at the University of Texas (Zhang, 2009) found on average, BRT costs less than half of LRT to construct per mile; however, this includes BRT systems that can take advantage of existing at-grade lanes already in place. When comparing the construction of dedicated guideways, BRT had higher land acquisition costs per mile than LRT. The study also found BRT to have lower average line capacities per hour than LRT or HRT. Lambas et al. (2017) also compared advantages and disadvantages, citing more track/station flexibility in BRT, but LRT consumes less space than BRT. There is also the question of fuel consumption and emissions, which until all buses are retrofitted with electric/hybrid-electric systems, LRT would have an advantage – especially in underground systems. But setting the systems as rivals against each other is not productive, for every site/region will have unique characteristics, budgets, environmental impacts, population densities, and economic development opportunities that can guide the selection of a particular system. Often, both systems could prove viable and/or can be networked with each other to provide the most efficient outcome.

Automated guideway transit

The American Society of Civil Engineers (ASCE, 2013) defines automated people movers (APMs) – which are also often referred to as automated guideway transit (AGT) – as a guided transit mode that is fully automated and features vehicles that operate on guideways with exclusive right of way. The Boston-based organization Trans-21 further delineates APMs based on their specific use and scale:

> People movers of architectural scale are those built by private entities within one property line, essentially as horizontal elevators to serve airports, amusement parks, hotel complexes, recreational areas, casinos, museums, science parks, and medical centers, generally with no fees. People movers of metropolitan scale are regional metro systems which are both owned and operated by public transit authorities, and for which fares are invariably charged. A third, intermediate category, people movers of institutional scale, are owned and operated by universities or corporations in controlled environments, and, similar to metropolitan systems, generally charge fares.
>
> *(Warren, 1998, p. 129)*

There is a variety of APM antecedents dating back to the mid-nineteenth century, and in general they can be classified as discontinuous, semi-continuous, or continuous systems. Within these categories, APMs may be further delineated by guideway/vehicle relationship, including vehicles that are suspended, bottom supported, or side supported; track system (monorail, standard/international gauge); suspension systems, including steel wheel on rail, polyurethane, rubber, or pneumatic tires, air cushions, and magnetic levitation; type of propulsion, including self-propelled vehicles with passive guideways and mechanically, electrically, or pneumatically active guideways with passive vehicles; and volume of passengers, including personal rapid transit (PRT), shuttle/loop transit (SLT), and group rapid transit (GRT).

Continuous APMs usually consist of perpetually moving walkways, guideways, or cable lines propelled by cables, belts, chains, or rotating wheels. The "moving pavement" was one of the first forms of continuous APM and can be dated back to 1874, when Alfred Speer proposed an elevated loop in New York City. The system consisted of a series of articulated platforms running on wheels moving at 15 mph (24 km/h) boarded from several 6-seat trolleys running parallel with the platforms, which by means of a friction brake could be used for transfer (Richards, 1976, p. 58). In 1871, Wilhelm and Heinrich Retting improved Speer's concept for a moving plane to transport people with the stepped platform railway. In section, there are three stepped platforms with a 1–2 mph (1.6–3 km/h) speed differential. Interchange zones contained circular rotating sections in the center of the two opposite-moving tracks that facilitated passage from one plane to another. The concept of using parallel platforms moving at different speeds eventually came to fruition at a number of sites, including a loop at the 1893 Columbian Exhibition in Chicago, a two-speed linear system in Berlin, a 2-mile (3.2 km) system at the Paris Exhibition in 1900, and a test loop built below 42nd Street in New York City in 1923 that consisted of three parallel moving platforms with an estimated capacity of 32,000 seated passengers per hour in each direction (Richards, 1976, p. 58). Unfortunately, the 42nd Street proposal was eventually halted in favor of the subway, so the "futuristic" image of a massive network of moving pavements as seen in the cartoon series *The Jetsons* would have to wait. In modern times, the moving pavement concept is most common in airports to connect terminals and in department stores that use escalators to connect multiple levels.

Semi-continuous APMs reduce headways down to 30 seconds or less by slowing the speed of the vehicles at stations to roughly 1 mph (1.6 km/h) or less to allow passengers to enter or exit. Guideways may be solely or a combination of active or passive technologies. The most common type of semi-continuous APMs runs along an active guideway and disengages or detaches from the main line at stations to allow for passenger exchange. Examples of such systems are gondolas and detachable chairlifts at ski/snowboard resorts, the Tomorrowland Transit Authority people mover at Walt Disney World, Orlando, the Metromover, Miami, both in Florida, the MIA people mover between Miami International Airport and Miami Intermodal Center, the SK system at Charles De Gaulle Airport, Paris, France, and the Montmartre and Laon funiculars, also in France.

Discontinuous APMs constitute most existing people movers and are characterized by self-propelled, active vehicles that run along passive guideways and come to complete stops at stations. Vehicles may be summoned on demand with the use of a push button or operate at scheduled intervals ranging from 30 to 90 seconds. In 1972, a system known as Vehicule Automatique Leger (VAL) was developed in Lille, France in response to a design competition launched by Matra Transport International (now Siemens). The VAL technology boasted a completely automatic system with an unobtrusive and exclusive right of way that could offer frequent service at times of high traffic and still provide regular service during times of low traffic. The initial line was opened in 1983 and has since been extended from Lille north to the Belgium border, amounting to roughly 28 miles (45 km) with over 60 stations. The stations have adopted urban and architectural design guidelines that promote compact development, respect local heritage, and integrate with adjacent neighborhoods and communities. Other forms of mass transport were reorganized around the metro to ensure smooth interchange and maximum usage. In addition to the Lille lines, VAL people movers are on several lines throughout France, including the Paris Orly Airport (Orlyval), the Toulouse Metro, the Rennes Metro, the Charles de Gaulle Airport (CDGVAL), and the Lyon Metro. In 2015, the city of Toulouse added passenger trains that could hold up to 320 passengers, increasing the capacity of Line A to 400,000 passengers per day (Siemens, n.d.). The technology has also been implemented abroad at the Turin Metro in Italy, the Taipei Metro in Taiwan, and the U Line in Uijeongbu, South Korea.

Aerial cable transit

Although aerial cable transit (ACT) systems are a form of automated guideway transit, their variability warrants extended discussion. ACT systems are commonly used in mountainous terrain, at ski/snowboard resorts, or in areas that require large spans where construction at-grade is not possible. ACTs include mono-cable, bi-cable, tri-cable, tramway, pulse systems, and jig-back/dual line systems, with mono-cable detachable gondola (MDG) being the most common due to their low construction and operation costs. MDG can carry between 1,000 and 4,000 passengers per hour/per direction, 8–15 passengers per car, can travel at speeds up to 13.6 mph (22 km/h), and generally requires 40–45 feet (12–14 meters) of right of way in the air. Tower spacing is dependent on tower height and typical MDG systems are in the 100–800 feet (30–244 meters) range (MDMPO, 2016). However, tramway, bi-cable, and tri-cable systems can have much higher towers and spans. For example, Whistler's Peak 2 Peak tri-cable gondola is the longest of its kind, spanning 1.88 miles (3 km), and has the record for the highest lift of any kind at 1,427 feet (435 m) above the valley floor. The Seilbahn Zugspitze aerial tramway in Germany holds the record for the longest free span at 1.99 miles (3.2 km) and the highest support tower at 417 feet (127 m) (Whistler-Blackcomb, n.d.; Cable Car Zugspitze, n.d.).

FIGURE 6.2 Top: the Montmartre funicular (Paris, France) was originally constructed in 1900 to transport people to the Sacré Coeur Basilica at the summit of the outlier of Montmartre. It was reconstructed in 1931 and 1991 and currently consists of two independent shuttles that run approximately 350 feet at a 35 percent incline. The frequency and speed of the funicular are fully automated and adjust relative to the passenger demand. Bottom: the Tomorrowland Transit Authority people mover at Walt Disney World, Orlando, Florida, formerly known as the WEDway, is an open-air, semi-continuous system using a linear synchronous motor.

Source: Images courtesy of Kelly L. Richards (top) and Dr. Kimberly A. Richards (bottom)

FIGURE 6.3 Aerial cable transit is commonly utilized at ski/snowboard resorts and comes in a variety of forms. Top left: continuous, fixed-grip single chair; top right: continuous fixed-grip double chair; bottom left: semi-continuous, mono-cable detachable gondola (MDG); bottom right: semi-continuous, detachable quad chairlift.

If roads are not wide enough to accommodate the support columns down the median and/or if there are not adequate setbacks, an arch support can be designed that straddles the road. Designer Jared Ficklin at Argodesign devised a suspended gondola system in the Wire One proposal for the city of Austin. The proposal spans eight miles (12.9 km) and includes 19 stations from the University of Texas to Slaughter Lane over South First Street. The system contains ten-person, climate-controlled cars with a capacity of 6,000 riders per hour, which is the equivalent of running 100 full buses an hour (Argodesign, n.d.). Although the proposal did not move forward – mostly due to the estimated $290–$550 million cost estimate – the feasibility study provided a boost to the concept (The Associated Press, 2017).

While there are approximately 1,500 operational cable transit systems in the world, a majority have been utilized at theme parks and mountain resorts. However, cable transit is becoming more common for urban transport, accounting for about 50 systems worldwide. Some notable systems include the Metrocable Gondola in Medellin, Colombia, which has three integrated lines that connect the city center with outlying areas in the hillsides and has been credited with reducing poverty and crime by providing more equitable transit access; the 2.2-mile (3.5 km) Teleferico do Alemao in Rio de Janeiro that contains six stations and connects Complexo do Alemao to a suburban rail line; the Portland Aerial Tram that connects the city's streetcar system in the South Waterfront District to the top of Marquam Hill where the Oregon Health and Science University resides; the Roosevelt Island Tram in New York City that connects Roosevelt Island to Manhattan; and the Emirates Airway Gondola that crosses the River Thames in London. There are also various proposed systems for Brooklyn, Washington, D.C., Chicago, San Diego, Seattle, Cleveland, Cincinnati, Buffalo, Baton Rouge, Austin, Tampa Bay, Miami, and Utah (Barber, 2017; MDMPO, 2016; The Gondola Project, n.d.; Ski Industry News, 2020).

FIGURE 6.4 Wire One Proposal.

Source: Image courtesy of Jared Ficklin and Argodesign

FIGURE 6.5 Urban gondola alignments down the center of the road, straddling over the road and running alongside the road.

FIGURE 6.5 (Continued)

In addition to their use for transporting skiers/snowboarders to the summits of mountains and/or linking different peaks, many mountain towns have utilized ACT between the resort and adjacent towns to curb auto congestion. For example, the Zlatibor Gold Gondola in Serbia connects Zlatibor to the ski center of Tornik, and at 5.6 miles (9 km), it is currently the longest gondola in the world. At Telluride Resort in Colorado, the Telluride Mountain Village Gondola was opened in 1996 and spans 3 miles (4.8 km) between the towns of Telluride and Mountain Village. There are three primary stations, including Station Telluride at Oak Street, Station St. Sophia at a 1,750 feet (533 meters) mid-mountain elevation, and Station Mountain Village. A shorter span also runs between Station Mountain Village and a free parking garage (Station Village Parking) and Market Plaza. The cabins travel at 11 mph (17.7 km/h), thus making the travel time 13 minutes between the town of Telluride and Mountain Village and another three minutes between Station Mountain Village and Market Plaza. The gondola operates roughly 287 days per year with a capacity of 1,070 passengers per hour, providing nearly three million passenger rides annually – the equivalent of 21 passenger buses. The cost to build the gondola was $16 million, and the annual operational cost is $3.5 million. However, the gondola is free to use due to funding contributions from the Telluride Mountain Village Owners Association, the town of Telluride, and from 1 percent of the lift ticket sales at Telluride Resort (Town of Mountain Village, n.d., Telluride Mountain Village Gondola, n.d.). Since the gondola uses approximately 2.3 million kilowatt hours per year, the Green Gondola Campaign was established to offset the gondola's energy consumption. Initially, the campaign relied on purchasing renewable energy credits, but, as of 2011, Mountain Village took over the campaign with a goal of replacing 20 percent of the gondola's energy use with renewable energy sources by 2020. In 2013, Mountain Village installed a 10-kilowatt solar system at the Station Mountain Village that offset about

FIGURE 6.6 Telluride Mountain Village Gondola. Top: Map of the gondola, courtesy of Tor Anderson/True North Designworks; bottom: Image of the gondola, courtesy of the town of Mountain Village.

1.5 percent of the gondola's electrical usage. The town needs to develop about 270 kW worth of solar resources to meet its 20 percent goal and 1.3 MW worth of solar resources on-site to fully power the gondola. Since large plots of flat land are limited in the area, the town is exploring rooftop solar as an alternative (Movellan, 2014).

In 2006, Breckenridge Resort in Colorado opened the Breck Connect – a 7,592 feet (2,314 meters) gondola by Leitner-Poma connecting downtown Breckenridge and the South Gondola

Garage and North Gondola Lot to Shock Hill and the bases of Peak 7 and Peak 8. The gondola is free, can transport 3,000 people per hour, and takes about 10 minutes. Since it opened, the Breck Connect has drastically reduced bus transit to and from the mountain. To further reduce congestion, the town of Breckenridge hired SE Group to conduct a feasibility study of creating a north–south gondola. Three options were presented ranging from a 7,530 feet (2,295 meter) extension from the Breck Connect Station to the Satellite North Lot, a 10,395 feet (3,168 meters) line that includes the north extension as well as an extension south of the Breck Connect Station to Lot F, and a 12,630 feet (3,850 meters) extension that includes the north extension and a southeast extension to the ice rink. The projected costs of the three proposals were $34,500,000, $51,750,000, and 52,800,000, respectively. Operations and maintenance costs ranged from $4.5 million to $7 million a year, but these costs could be offset by eliminating bus service which currently costs about $1 million a year and taxes from lift tickets which are roughly $3.5 million per year (Landsman, 2017). The plan has not received much support from the town council, mostly due to cost. However, some members also felt that there was a much larger need for a comprehensive transit plan that drastically changes the way people circulate in the region (Pace, 2017).

To reduce traffic congestion in Little Cottonwood Canyon, the Utah Department of Transportation (UDOT, 2022) has identified a proposed gondola (preliminarily known as "Gondola Alternative B") as the preferred alternative mode of transportation. Gondola B would start at a proposed 2,500-space parking structure at the entrance of Little Cottonwood Canyon in La Caille and travel east approximately eight miles (12.8 km) with two angling stations and two loading stations at Snowbird and Alta resorts. Each gondola cabin would hold 35 people and arrive every two minutes, ultimately carrying 1,050 passengers per hour and taking approximately 55 minutes for the entire route. The gondola would run independently of State Route 210, allowing it to avoid traffic delays and weather-related travel difficulties. Although this alternative had the highest visual impact, it had the lowest impacts on watersheds, wildlife movement, and topography alterations compared to the other alternatives, which included enhanced bus service, a dedicated bus lane, and a cog-rail option. The total cost of the Gondola B Alternative is $550 million, which includes tolling infrastructure, transit parking, adding snow sheds, addressing trailhead parking, and widening Wasatch Boulevard. Operations and maintenance costs are estimated to be $4 million for winter use only and an additional $3 million if used in the summer months. Although the initial construction cost was the third highest of the alternatives, the overall 30-year life cycle cost was the lowest (UDOT, 2022). As of the writing of this book, the Utah Department of Transportation has selected Gondola Alternative B with phased implementation of Enhanced Bus Service Alternative components. Planning and design is expected to start in summer 2023, and the project will include three phases: Phase 1 will include improved and increased bus service, constructing resort bus stops and a mobility hub, tolling, and winter roadside parking restrictions; Phase 2 improvements (contingent on funding) will include widening and other improvements to Wasatch Boulevard, constructing snow sheds, and implementing trailhead parking improvements; and Phase 3 (also contingent on funding) will be the construction of the gondola and a base station with 2,500 parking spaces near the mouth of Little Cottonwood Canyon (UDOT, 2023).

Monorail

Monorail transit is characterized by rubber-tired vehicles that straddle or suspend from single, narrow guideways. Although monorails can be constructed underground and at-grade, elevated systems are most frequently employed due to low installation costs, nominal operational and maintenance demand, and negligible impact on the environment due to ease of construction, minimal infrastructure, and low levels of noise. In addition to providing a dedicated right of way separated from roadway traffic, monorails require smaller footprints than light rail systems, bus lanes, and elevated busways and can be positioned on or along existing roadways, thus eliminating the need to obtain easements from property owners and/or disturb undeveloped areas and/or natural habitats. Compared to heavy and light rail systems, monorails are generally easier to construct since they essentially employ a "post and beam" system with a moderate footprint that can be assembled with minimal disruption to street traffic, businesses, and residences. Monorails are also safer than heavy and light rail systems because they usually operate on dedicated and elevated rails that have minimal chance of colliding with pedestrians or traffic; the straddle and suspended designs also make the likelihood of derailment extremely small. Monorails also offer more flexibility in design aesthetics. Rails and beams can be customized to match a particular design theme, and the quiet, sleek operation enables them to negotiate curves and weave through buildings and tight spaces. The track generally requires a 6 feet by 6 feet (2 meters by 2 meters) concrete base every 80–120 feet (24–37 meters) (The High Road Foundation, n.d.) and requires roughly 20–30 feet (6–9 meters) of right of way depending on the design (The Monorail Society, n.d.a). Monorails can be fully automated or manually operated and their capacity, speed (typically between 50 and 65 mph/80 and 105 km/h), and headways are comparable to those of rapid rail transit systems. For example, the Walt Disney World monorail in Orlando, Florida, has an average speed of 30 mph/48 km/h (with a maximum of around 40 mph/64 km/h) and carries over 150,000 people per day (WDW, n.d.).

There are approximately 50 monorails in operation worldwide with another 30 or so planned (including magnetic levitation monorails), with Japan taking the lead in the municipal sector with the 11-mile (17.7 km) Tokyo–Haneda Monorail, the 9.9-mile (15.9 km) Tama Monorail, the 8-mile (12.8 km) Okinawa Monorail, the 5.5-mile (8.9 km) Kitakyushu Monorail, and the 17.4-mile (28 km) Osaka Monorail, which is used to connect various rail lines that fan out from the center of the city and will eventually encircle Osaka with over 30 miles (48 km) of track. The system currently transports more than 100,000 passengers per day (The Monorail Society, n.d.b). Japan also has two suspended monorail systems, including the Chiba City Townliner, which is currently the longest suspended monorail in the world, at 9.6 miles (15.4 km), and the Shonan Monorail between Ofuna and Enoshima. The largest and highest-capacity monorail is the Chongqing Metro in China, which is roughly 50 miles (75 km) in length and transports over 400,000 passengers per day (Railway Technology, 2012). In Germany, the H-Bahn at the TU Dortmund University and the Skytrain in Dusseldorf utilize automated suspended monorail technology developed by SAFEGE (Société Anonyme Française d'Etude de Gestion et d'Entreprises) and later adopted by Siemens Transportation Systems and Aerorail. The H-Bahn (also known as the Siemens People Mover or SIPEM) contains two primary lines and approximately 1.9 miles (3 km) of track. The cabin holds 45 people, travels up to 30 mph (48 km/h), completes 36–60 runs per hour, and can carry approximately 15,000 passengers per hour per direction (H-Bahn21, n.d.; The Monorail Society, n.d.c). The H-Bahn switching is achieved through swiveling blades on the rail that allow the car to transfer to the desired side of the bypass loop in the track. This design allows cars to pass each other while maintaining a single-line track, thus minimizing visual disruption to the landscape. The suspension system also navigates curves better than bottom-supported monorails, and the track enclosure provides protection from ice and snow. The CRRC Corporation in Beijing has also developed a suspended monorail based on the SAFEGE technology that can travel 35–50 mph (56–80 km/h), climb grades up to 10.4 percent, and has a minimum curve negotiation radius of 100 feet (30 meters) (CRRC, n.d.).

Although monorails can climb steeper grades than heavy and light rail systems, existing systems usually max out in the 8–12 percent grade range (The Monorail Society, n.d.c). The Colorado Intermountain Fixed-Guideway Authority is currently studying the option to construct a 163-mile (262 km) high-speed magnetic levitation monorail between Denver and Vail utilizing linear induction motors being developed by Sandia National Laboratories in New Mexico. The system can handle grades up to 8 percent (TRB, n.d.). Although they are limited in numbers, there are also cable-car monorails that use monorail guideways but are propelled by cable-driven power units. The Dresden Suspension Railway in Dresden, Germany climbs at a 39.2 percent grade at 2.5 meters per second, and the Skyrail Midorizaka in Hiroshima, Japan, climbs at a 27 percent grade (The Monorail Society, n.d.d). Intamin Transportation has recently developed the People Mover P6 that can climb grades up to 20 percent and transport 3,500 passengers per hour and the Mountain People Mover P8 that can climb grades up to 50 percent and transport 3,000 people per hour (Intamin, n.d.). Although these systems have slower travel speeds than larger capacity systems (~11 mph/17.7 km/h), their ability to navigate mountainous and curvilinear spaces makes them competitive with aerial cable systems. The maximum 8–12 percent slope grades can be managed by fluctuating column heights, excavation, or tunneling.

FIGURE 6.7 Top: straddle-style monorail at Walt Disney World when the park opened in 1973; middle: suspended H-Bahn monorail at the TU Dortmund University in Germany, image courtesy of Magdalena Gruber; bottom: Shanghai Mag-Lev, courtesy of Thyssen Krupp Transrapid/Stoiber Productions.

FIGURE 6.7 (Continued)

Funiculars and cog rails

Another option for steep slopes and curved terrain is to use funicular or cogwheel (or rack railway) rail systems, the former utilizing a bottom-supported cable and suited for distances up to three miles (4.8 km) or so (and is technically considered an AGT as discussed earlier), the latter utilizing self-propelled rail cars fitted with cogwheels that mesh with a toothed third-rail line and better suited for distances greater than three miles. The Parsenn Resort in Davos, Switzerland, utilizes two Doppelmayr funicular systems: a 6,142-feet (1,872-meter) section that climbs 2,172 feet (662 meters) of vertical and carries 2,200 people per hour; and another section that travels 7,264 feet (2,214 meters) up 1,453 feet (443 meters) of vertical and carries 970 people per hour (Skiresort.info, n.d.). The use of cog rail is prominent throughout the Alps as a means of linking villages and mountain resorts. The Intamin Mountain People Mover discussed earlier uses a combination of technology used in roller coasters and cogwheel rails. Two installations of this system currently exist at the Tschuggen Grand Hotel and at Arosa Lenzerheide in Switzerland. The Mount Washington Cog Railway in New Hampshire is the world's oldest cog rail and the Broadmoor Manitou and Pikes Peak cog rail in Colorado is the world's longest and highest cog railway, spanning 9 miles (14.5 km) to the top of Pikes Peak at 14,115 feet (4,302 meters) (Broadmoor, n.d.). The Pilatus Railway in Switzerland is the world's steepest cog rail climbing grades up to 48 percent. Cog rails can be powered by steam (Mt. Washington), biodiesel (Broadmoor Manitou and Pikes Peak), a third rail (Intamin Mountain People Mover), or an overhead cable (Pilatus).

FIGURE 6.8 Top: Davos Funicular, image courtesy of Doppelmayr/Garaventa; bottom: Tschuggen Express in Switzerland using the Intamin Mountain People Mover.

Source: Image courtesy of Coastersandmore.de

FIGURE 6.8 (Continued)

Magnetic levitation and pneumatic tubes

Developed in Lathen, Germany by Transrapid, magnetic levitation (mag-lev) technology uses the repulsive – attractive principles of magnetism to float trains above guideways with minimal noise and virtually no friction, which enables extremely high speeds suitable for transcontinental travel. In addition, maintenance and energy costs are low compared with normal rail and wheel, and systems can be designed with sharper turns and steeper inclines.

There are currently six operational mag-lev systems in the world: the Shanghai Mag-Lev in China, which is the longest mag-mag-lev, spanning 18.6 miles (30 km) between Pudong International Airport and the Longyang Road metro station and running up to 267 mph (430 km/h); the 5.6-mile (9 km) Tobu–Kyuryo Line (the Limino Line) in Aichi, Japan; the 0.6-mile (1 km) Korean Rotem mag-lev in the city of Taejeon; the Incheon Airport Mag-lev in Daejeon, South Korea; the 11.5-mile (18.5 km) Changsha mag-lev in China; and the 6-mile (9.7 km) S1 line of the Beijing subway system in China. In April 2015, a test track in Yamanashi, Japan set the world record for a mag-lev system with a speed of 374 mph (602 km/h). Japan Railways has plans for a 178-mile (286 km) route between Tokyo and Nagoya scheduled to begin service in 2027 along with further expansion into Osaka by 2037, which would extend the system to 319 miles (514 km). At these speeds, the system could cut the commute between Tokyo and Nagoya to just 40 minutes and Tokyo to Osaka to 67 minutes, which would otherwise take five and up to eight hours by car, respectively. Several systems have been proposed in the United States, including a link between Los Angeles and Nevada, a 47-mile (75.6 km) link from the city center of Pittsburgh to its surrounding suburbs, a 14.9-mile (24 km) line between Orlando and Orange County, and the 40-mile (64 km) Northeast Mag-lev between

Washington, D.C., and New York, with the first section between Baltimore and Washington, D.C., slated to open by 2035.

Pneumatics

Although utilizing pneumatics can be commonly seen at drive-through banks and in medical facilities to transport blood samples, using the technology for human transport has also been explored, dating back to the London–Croydon Railway in 1845, the Crystal Palace atmospheric railway in 1864 (also in London), and Alfred Beach's pneumatic subway train in New York in 1967. There have also been more recent proposals that utilize magnetic levitation inside vacuum tubes, thereby eliminating air resistance and allowing even faster speeds. Such systems are often referred to as vactrains or hyperloops. ET3 Inc. envisions an evacuated tube transport system in which capsules will reach speeds of up to 4,000 mph (6,437 km/h), which would cut the journey time from New York to Beijing to two hours (Chan, 2015). Hyperloop Transportation Technologies (Hyperloop TT) has a test track in France and is currently developing a feasibility study for a Great Lakes hyperloop corridor that would connect Chicago, Cleveland, and Pittsburgh and is also developing a pneumatic cargo transport system called Hyper Port. In 2020, Hyperloop One (formerly Virgin Hyperloop) became the first company to test passenger travel on its 1,640-feet (500-meter) DevLoop test track in Nevada. Although the two-person test was touted as a success, as of 2022, the company has abandoned passenger travel development and is focusing on freight. The passenger test also did not run the pod at the theoretical maximum speed of 760 mph (1,223 km/h), so much is still unknown about safety and human comfort. In addition, there are questions about the turning radius of the track given the high sustained speeds, pressurization over long distances, and, of course costs, with some preliminary estimates showing a significantly higher cost per mile to construct than high-speed rail.

Solar and hybrid rail

Although electrified rail has no direct carbon emissions, how the electricity is generated may produce emissions if powered by burning fossil fuels. According to the International Energy Agency (IEA, n.d.), on a wheel-to-well basis, electrified rail still emits one-fifth less carbon emissions per passenger kilometer than air travel and only 1 percent of all transport emissions. Converting diesel locomotives to electric and using renewable sources of energy to power rail can reduce emissions even further. Aside from transforming the entire energy grid to fully use renewable sources of energy, another strategy is to generate power from the rail infrastructure itself. In Belgium, a 2-mile (3.2 km) rail tunnel on the Paris–Amsterdam rail line was topped with 16,000 solar panels by solar developer Enfinity. The panels will produce 3.3 GWh of electricity a year, which is equivalent to that needed to power 950 households. The system will provide energy for 4,000 trains per year and reduce carbon dioxide emissions by 2,400 tons (UIC, 2011).

There are also several designs for hybrid-electric rail systems that, similar to hybrid cars, capture and store heat energy from braking and/or surplus energy from the primary power source that can be used later for acceleration. The regenerative braking system on the Las Vegas monorail reduces power consumption by up to 20 percent; the entire system reduces emissions by more than 27 tons annually and prevents two million vehicle miles annually (Las Vegas Monorail, n.d.). In 2007, General Electric unveiled its 4,400 horsepower Evolution hybrid diesel–electric locomotive that uses 10 percent less fuel and produces 10 percent fewer carbon dioxide emissions than conventional diesel systems (GE, 2007). In 2021, Amtrak awarded Siemens a contract to build 73 multi-power push-pull trains for use on the Northeast Corridor and on state-sponsored routes across the United States. The locomotives can be powered by diesel or biodiesel or electrically powered and are estimated to produce a 95 percent reduction in particulate matter and an 89 percent reduction in emissions compared to existing Tier 0 locomotives (Clinnick, 2021). Frazer-Nash Research has also developed Metrail, a self-powered, hybrid-electric monorail that uses roof-mounted solar panels to supplement the on-board hybrid-electric powertrain.

Micromobility

Micromobility modes of transit include small, low-speed, electric-powered transportation devices, such as e-bikes/pedal-assist bikes, electric scooters, hover boards, electric skateboards, and Segways. E-bikes have rear motors in the 1,000-watt range, can have multiple pedal-assist speeds that can be equipped with full throttle modes, and use lithium-ion batteries that can last up to roughly 40–120 miles (64–193 km) per charge, depending on battery size and the level of pedal-assist used by the rider. Some companies, universities, and municipalities have introduced bicycle ride-share programs, such as Lyft, Veo, Lime, Citi Bike, and Bird. There are two major types of micromobility share systems: station-based bike share systems where there is a defined set of locations where a trip can be started and ended; and free-floating bike share systems, which are also referred to as dockless systems where people can unlock and pay for the bikes with an app, ride the bikes to their destination, and leave them where they want. The locations of the bikes are tracked with GPS. According to the National Association of City Transportation Officials, there were 112 million micromobility trips in 2021, including 47 million trips on station-based bike share systems, 62.5 million trips on dockless e-scooters, and 2.5 million trips on dockless bikes. The data also showed that several cities, including New York, Chicago, Philadelphia, Washington, D.C., Portland, and Los Angeles, strengthened commitments to equitable access by expanding services in previously underserved neighborhoods (NACTO, 2021).

Water taxis/ferries

Given that 71 percent of the earth's surface is water, it is no surprise we have utilized our oceans, lakes, rivers, bays, channels, and canals for transport. In the United States, there are 29,000 nautical miles (53,708 km) of navigable waterways with 190 operators, 500 ports, and 700 active vessels in 37 states and three U.S. territories (DeSimone & Sullivan, 2014). Most of the services are concentrated in coastal cities, with New York having the most of any state at 75. In Venice, Italy, the public transportation service Azienda del Consorzio Transporti Veneziano (ACTV) and the Azienda Veneziana della Mobilita operate 160 water buses/water taxis (also known as vaporetto) and 150 floating pontoons that transport roughly 190 million passengers per year (ACTV, 2010).

Transit-oriented development

The Transit Oriented Development Institute (TODI, n.d.) defines transit-oriented development (TOD) as "the creation of compact, walkable, pedestrian-oriented, mixed-use communities centered around high-quality train systems." In general, TOD is most successful when it is planned and implemented before development is widespread and/or at the onset of regional growth. A pre-planned transit system can influence land use patterns by clustering mixed-use development around transit stations and conserving land between the stations. In Singapore, for example, a centralized planning concept known as the "Ring Plan" was conceived in the 1970s in response to rapid population growth. The plan called for the configuration of high-density housing, industrial sites, and urban centers in a ring around the urban core, linked together by a high-capacity and efficient transportation network. Each of the nodal, satellite towns is made up of five to seven interlocking residential neighborhoods that orbit a higher-order town center. Each neighborhood is roughly 100 acres (40 hectares) in size, contains some 4,000–6,000 dwelling units, and has a small neighborhood center of retail shops, schools, and recreational facilities within a five-minute walking radius; a system of pedestrian walkways links precincts with neighborhood centers and is in turn separated from busy roads and tied to the town center (Cervero, 1998, p. 160). The mass rapid transit (MRT) system currently has over 140 stations, spans roughly 93 miles (150 km), and carries roughly 2.75 million passengers per day. The system runs underground in the dense, urban core and is elevated toward the outskirts. It is also complemented by several light rail and bus systems that provide feeder runs to the MRT stations. The LRT system has 41 stations and carries over 180,000 passengers per day (Statista, 2023). With planned expansions of trains, the Singapore government estimates that eight out of ten people will live within a ten-minute walk to a train station (Gov.sg, 2020).

Much of Singapore's MRT success was due to the semi-autocratic government and centralized control over urban/regional planning and development. For example, many government initiatives were put in place to limit automobile usage and ownership, such as automobile surtaxes, road-use surcharges, the Vehicle Quota System, the Certificate of Entitlement program, and a series of use-related charges, including the Electronic Road Pricing system and high fuel taxes. All motor vehicle tax proceeds go into a consolidated fund and are recirculated into public development. The result is that, despite having one of the highest GDPs per capita in Asia, fewer than 12 percent of Singaporean households own two cars and less than 3 percent own three or more cars (Tjoe, 2022). Although this "socialist" form of governance may not be applicable to other regions with different governing structures, many of Singapore's centrally planned land use concepts can certainly be precedents for regional planning commissions and/or in comprehensive transportation plans.

In contrast to Singapore, many regions failed to plan their rail transit before or during development, and post-implementation can be difficult since people have settled in dispersed patterns throughout the landscape. Therefore, the central question becomes: can new or expanded fixed-rail transit systems reverse existing patterns of sprawled development, or is the damage already done and too difficult to reverse? There are case studies that both support and oppose this question, but ultimately, the success or failure of any given transit system and its subsequent impact on transit-oriented development is unique to each site and often dependent on many variables, including population density, climate, the socioeconomics of the region, changing culture, politics, budgets, specific development patterns, topography, and the design characteristics of the transit system, the stations, and the areas surrounding them. Zoning can certainly play a part in re-shaping communities, particularly when nodes around transit stations are up-zoned to allow denser development and land in between stations is down-zoned to allow less dense development. Obtaining land and convincing businesses to relocate to a transit node seems to be the bigger challenge. As discussed in Chapter 4, tax incentives, tax increment financing, and transfer of development rights (TDRs) are some strategies that can be used to re-centralize development around a transit station.

Some systems, such as the Champlain Valley Flyer in Vermont, which is discussed in the following text, were implemented but failed to attract enough ridership to cover the operating costs. Other systems, such as Buffalo's light rail Metro, have decent ridership but have failed to revive the city center or reverse sprawled patterns of development. However, many regions have had success. For example, successful TOD communities have been established in Concord, Massachusetts, surrounding the commuter rail station, in the village of Arlington Heights, Illinois, along Metra's Union Pacific Northwest Line, and in Emeryville, California, at the transfer point of Amtrak's California Zephyr service. In Concord, 20 rental apartments, as well as retail, office, and restaurant space, were developed along with pedestrian pathway improvements (NHHS, n.d.). In the village of Arlington Heights, moving the Metra station closer to the downtown core has resulted in over 1,500 residential units, 157,000 square feet of retail space, a performing arts facility, parks, and public art (Bernick & Cervero, 1997, p. 341). In Emeryville, the reconstruction of Emery Station Plaza into a three-building, 550,000-square-foot mixed-use residential, retail, and business complex has breathed new life into 20 acres of former brownfield that surrounds the site (NHHS, n.d.).

Case study: regional rail for Vermont

The Champlain Valley Flyer was a 12.9-mile (20.7 km) commuter train that ran between Charlotte and Union Station in Burlington, Vermont, as a pilot program in response to a rail feasibility study conducted by the Vermont Agency of Transportation in 1993. Ridership was originally projected at 214,562 passenger trips annually and was expected to generate more than $160,000 in revenue. Instead, the train managed to attract only about 83,000 riders annually and it generated only about $53,000 in its last year of operation, thus preventing any further studies to extend the line (Crabtree, 2003). The route is currently used for excursion trains and for Amtrak's Ethan Allen Express that runs from Burlington to New York City. Amtrak also runs the Vermonter in the central part of the state from St. Albans to Washington, D.C.

The lessons learned from the failure of the Champlain Valley Flyer can be used to inform the success of a new, regional transit network that provides adequate regional connections to cultural and recreational attractions, public services, and other modes of public transit; establishes efficient and safe pedestrian connections with and along the corridor within a quarter-mile radius and effective short-haul connections greater than a quarter of a mile; provides appropriate frequency and headways; establishes mixed-use development at transit stations; and implements effective marketing campaigns to promote and/or incentivize ridership and activity along the corridor.

Regional connections

One of the major shortcomings of the Champlain Valley Flyer was the lack of connections to schools, colleges/universities, hospitals, ski/snowboard resorts, village/town centers, the Burlington International Airport, and the Amtrak station in Essex Junction, to name a few. Given the imminent issue of traffic congestion at mountain resorts in recent times, connecting regional transit lines to the resorts is crucial to preserve the scale and aesthetics of the small villages that host the resorts. The Vermont Ski Areas Association (VSAA) reported 3,762,047 skier/rider

FIGURE 6.9 Vermont mountain resort congestion.

visits in 2021–2022, up from 3,532,186 in 2020–2021 and 3,664,326 in 2019–2020. Vermont reported numbers as high as 4,670,903 in 2014–2015, but the VSAA acknowledges this difference could be due to RFID scanning systems yielding more accurate counts in recent times (Lynn, 2022, June 9). Regardless, three million to four million annual skier/rider visits in a state with a population of only 624,340 is significant on any measure, especially seeing the roadway, parking, water, and sewer infrastructure in many cases is practically the same as it was when the resort towns were originally chartered. Collective ski/snowboarding passes such as the Ikon, Epic, and Indy have also made skiing/riding more affordable, which has further exacerbated the congestion. On weekends, holidays, and powder days, some resorts are experiencing parking shortages and one- to two-hour traffic delays on the roads used to access the resorts (Murray, 2017). Increased tourism is a double-edged sword: on the one hand, it creates jobs and ignites the economy, but, on the other hand, it creates a need for more development to accommodate the flux of visitors; and if development exceeds a threshold that degrades the landscape, it detracts from the very things people seek at the mountain resorts in the first place: nature, open space, and tranquility. However, not adapting infrastructure to population fluxes also results in traffic congestion, long wait times at the chair lifts, overcrowding, and animosity, ultimately degrading the experience. An informal poll was recently conducted on Unofficial Networks asking people what their least favorite thing is about skiing/riding. As one avid skier reports, "Resorts are literally turning into the cities you are trying to escape from" (Haas, 2022).

Short-term solutions for traffic congestion

Some of the more common short-term strategies to control mountain resort congestion include capping ticket sales, utilizing reservation systems, charging for parking and/or toll roads to encourage car-pooling, and offering passes with black-out days during holidays and/or weekends. Although these strategies might modestly decrease congestion at the resorts at peak visit times, they do not significantly impact the traffic congestion on arterial and access roads leading to the resorts – especially given that most of the roads are two lanes with low shoulders.

In addition, not all visitors to mountain towns are necessarily skiing or snowboarding at a lift-serviced resort. There are other activities at or near resorts in the winter months, including Nordic skiing, backcountry skiing, snowmobiling, ice skating, snowshoeing, spa visits, and hiking, not to mention locals commuting to and from their place of employment. Many resorts and villages also see surges in the summer months for weddings, festivals, and conferences and in the fall months for leaf peepers coming to see the foliage. And short of coordinating a regional shift in the dogmatic Monday–Friday, 9:00 AM–5:00 PM work schedule, it is difficult to evenly distribute population fluctuations.

Another short-term fix that is often utilized is to expand bus transit. While bus transit offers route flexibility that can be difficult to engineer with fixed-guideway transit (especially in mountainous regions), unless the infrastructure allows for dedicated bus lanes, buses still succumb to traffic delays and road conditions that cause delays and unreliability that disincentivize ridership. Many resorts also believe that increasing parking, constructing parking structures, and/or adding road lanes will solve their congestion woes. There are several issues with this approach, the first and foremost being that the mountainous terrain often doesn't allow expansion without major feats of engineering and/or detrimental impacts on natural systems. Of the 22 active mountain resorts in Vermont, seven currently utilize portions of state forest or state park land for ski/snowboard area development under a long-term lease from the Vermont Department of Forests, Parks, and Recreation, three reside on and have agreements with Green Mountain National Forest, and two resorts have agreements or are part of town parks/forests. Between these agreements and the state of Vermont's stringent Act 250 development permitting process that was designed to protect the state's natural capital, justifying clear cutting trees and/or bulldozing chunks of mountains to expand roads would be difficult. More importantly, expanding lanes and parking/parking structures also usually leads to induced traffic demand, ultimately making skier/rider congestion on trails and traffic congestion on access roads as bad or worse than they were before the roads and parking were expanded, not to mention the endangerment of the historic character, scale, and aesthetic of the villages near the resorts.

Long-term solution: ski/ride Vermont by rail

Solving congestion issues merely at the resorts will not be effective without addressing congestion issues and transportation needs at the regional scale. Given the topographical, ecological, and aesthetic constraints mentioned earlier, a regional transit plan that augments existing rail infrastructure with new forms of fixed-guideway transit would provide a more holistic, long-term solution for the entire state and would be easier to construct, more reliable, and more ecologically sensitive than constructing new and/or expanding existing at-grade bus, rail, roadway, and parking infrastructure.

Figure 6.10 shows a conceptual proposal that links 17 of the 22 mountain resorts in Vermont using a network of primary routes utilizing existing fixed rail lines, secondary routes between the primary routes and the villages and/or resorts using elevated monorails and aerial cable transit, and tertiary routes between the secondary systems and the resorts using a combination of suspended monorails, aerial cable transit, people movers, cogwheel rail, and funiculars. This concept utilizes existing rights of way and developed areas for the transit systems to minimize destruction and/or disturbance to natural systems, minimize visual impacts to communities, and minimize car–pedestrian–transit intersections. Site characteristics – including the number of crossings, distance, the number of headways needed, passenger volume, right-of-way dimensions, the number of segments, set back distances along rights of way, slope, speed, the number of stations, structures (bridges, tunnels, excavation), and utilities – informed the selection of the transit system for each route. Due to their speed, ease of construction, and ability to navigate curves, elevated monorails were the most frequently used secondary system, while aerial cable transit, funiculars, and cogwheel rails were utilized for the tertiary systems due to their ability to climb steep grades that are prevalent between the villages and the base of the resorts. A suspended monorail was selected for the Sugarbush-Waterbury-Stowe route due to its ability to navigate tighter curves than the straddle-style monorail. There were also a few exceptions where aerial cable transit was used as a secondary system given the terrain and proximity to the existing primary fixed rail route.

A skier/rider can board the commuter rail that utilizes the existing Amtrak route and transfer to secondary and tertiary systems to get to Jay Peak, Cochran's, Bolton Valley, Stowe, Smugglers' Notch, Mad River Glen, Sugarbush, Middlebury Snow Bowl, Killington, Pico, Okemo, Saskadena Six, Mount Snow, Magic Mountain, Bromley, Stratton, and Brattleboro Ski Hill and enjoy the après ski/snowboarding night life without the worry or hassle of having to drive home. In addition to providing connections to mountain resorts, the transit system links several cities, towns, universities/colleges, hospitals, train/bus stations, airports, and tourist attractions throughout the northeast region, ultimately providing more equity, opportunities, and efficiency for personal, business, and recreational travel. In addition, given the proposed network's juxtaposition through unique features in Vermont's landscape, merely riding the rail can be an attraction itself.

Mixed-use and transit-oriented development

Transit systems that have failed are also often caused by a lack of mixed-use development around the transit stations. The stations can be designed as activity nodes that combine residential, commercial, and recreational uses. To engage the attractions of a community, the station needs to be either positioned in the center of the community or mixed-use development needs to be planned directly around the station. This reprograms the station from being a mere platform for loading the train to an attraction itself. Planning for mixed-use development requires input from the community and local businesses. Planning at the county scale, for example, might reveal specific types of development that would make each transit node unique. Uses in each node do not necessarily have to be the same, since one of the main functions of a transit line is to connect disparate activities. There are currently several special event excursion trains offered on the Champlain Valley Flyer, including a Mother's Day express brunch, a train for the fireworks on the Fourth of July, and several dinner/entertainment trains on weekends in the summer months. This special event model is a good precedent for increasing ridership on a year-round basis. If each node were programmed with unique activities, the demand to travel to different sites would increase. Working in Burlington, an evening concert in Shelburne, and dinner and drinks in Middlebury could all be part of a day's adventure – without the worry or hassle of driving.

Pedestrian and short-haul connections

Although the proposed rail network provides transit options at three scales, there will still be a significant amount of people who do not live in proximity to a station – even in a relatively

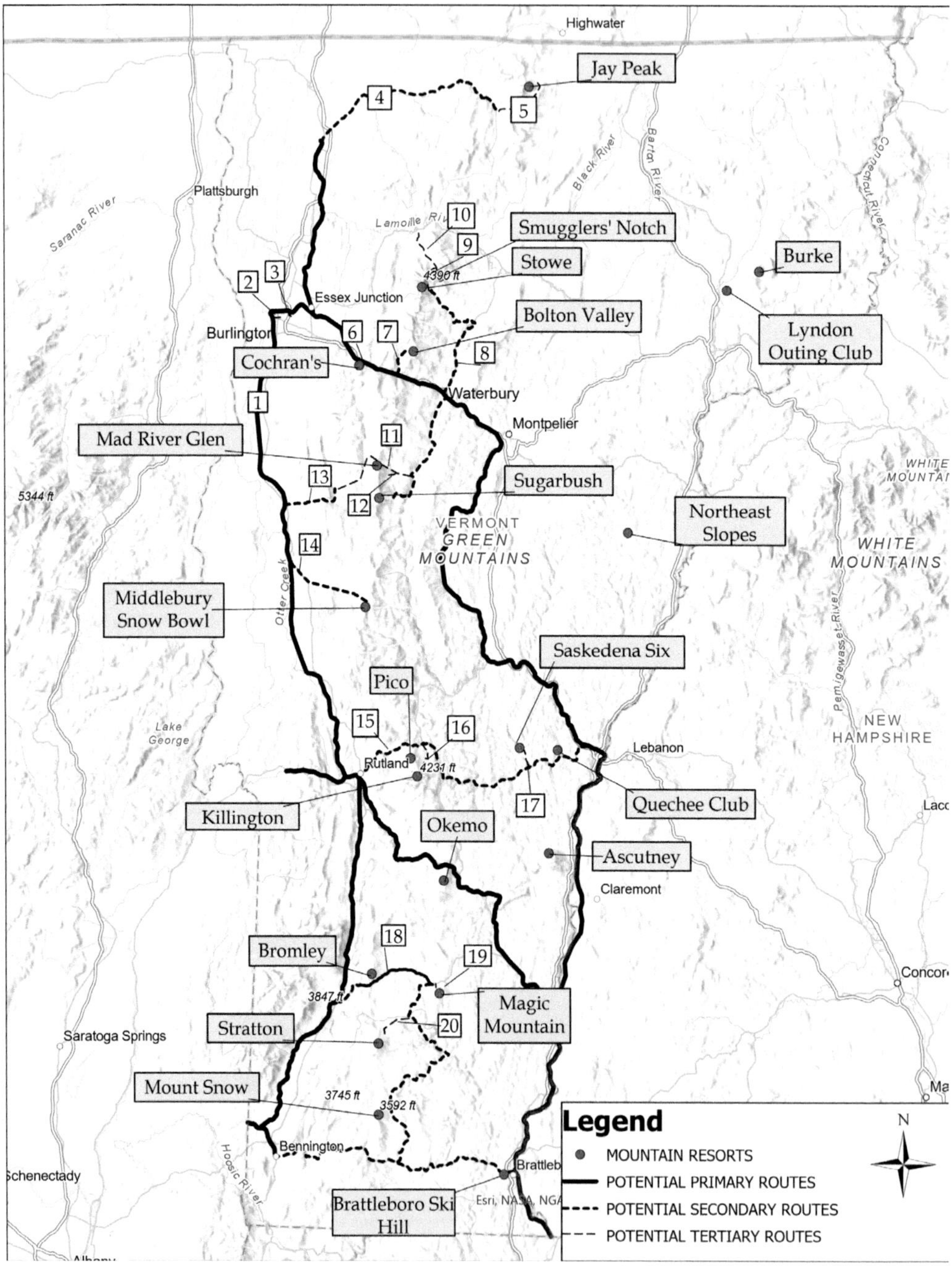

FIGURE 6.10 Proposed regional rail system for Vermont. World Topo Map courtesy of Esri, DeLorme, HERE, TomTom, Intermap, increment P Corp., GEBCO, USGS, FAO, NPS, NRCAN, GeoBase, IGN, Kadaster NL, Ordnance Survey, Esri Japan, METI, Esri China (Hong Kong), swisstopo, MapmyIndia, and the GIS User Community.

TABLE 6.1 Proposed routes

Map #	*Proposed route*	*Route*	*System*
1	Amtrak/Vermont Commuter Rail	Primary	Heavy rail/light rail
2	Main Street–UVM/UVM Med Ctr	Tertiary	Mono-cable gondola
3	Airport-St Michael's College	Tertiary	APM
4	St Albans-Montgomery Ctr	Secondary	Mono-cable gondola
5	Jay Peak Mountain Road	Tertiary	Mono-cable gondola
6	Richmond–Cochran	Tertiary	Mono-cable gondola
7	Bolton Access Road	Tertiary	Tri-cable gondola
8	Sugarbush-Waterbury-Stowe	Secondary	Suspended monorail
9	Notch Road	Tertiary	Funicular
10	Jeffersonville–Smugglers' Notch	Tertiary	Suspended monorail
11	Appalachian Gap	Tertiary	Tri-cable gondola
12	Mt Ellen Gondola	Tertiary	Mono-cable gondola
13	New Haven–Bristol	Secondary	Monorail
14	Middlebury Snow Bowl	Secondary	Tri-cable gondola
15	Rutland–White River Junction	Secondary	Monorail
16	Killington Rd	Tertiary	Mono-cable gondola
17	Woodstock–Saskadena Six	Tertiary	Mono-cable gondola
18	Manchester–Brattleboro–Bennington	Secondary	Monorail
19	Thompsonburg–Magic Mountain	Tertiary	Mono-cable gondola
20	Rawsonville–Stratton	Tertiary	Mono-cable gondola

FIGURE 6.11 Proposed suspended monorail in Stowe. Bottom image of suspended monorail courtesy of Magdalena Gruber.

FIGURE 6.11 (Continued)

FIGURE 6.12 Proposed suspended monorail to Smugglers' Notch and Sugarbush. Bottom image of suspended monorail courtesy of Magdalena Gruber.

FIGURE 6.12 (Continued)

FIGURE 6.13 Top: proposed suspended monorail connecting to Waterbury Amtrak Station; bottom: proposed Notch Road Funicular.

Source: Image of funicular courtesy of Doppelmayr/Garaventa

FIGURE 6.13 (Continued)

FIGURE 6.14 Tri-cable gondolas ascending Bolton Valley Access Road, the Appalachian Gap (Route 17), and traveling above the existing Amtrak line in Middlebury, Vermont.

Source: Images of tri-cable gondolas courtesy of Doppelmayr/Garaventa

FIGURE 6.14 (Continued)

FIGURE 6.15 Top: proposed Killington Road gondola; bottom: proposed Route 4 monorail connecting to Killington Gondola Base Lodge.

dense area like Chittenden County. Since it is more difficult to relocate residences and businesses to be closer to a rail station, the other option is to connect remote communities more efficiently to the rail line using pedestrian-scaled street design within the five-minute, quarter-mile radius, and implementing various short-haul transit systems for neighborhoods greater than a quarter of a mile from a train station. Some of the issues facing pedestrian walking networks include weather, dangerous street crossings, air and noise pollution from cars and trucks, and accessibility issues, which themselves can be exacerbated by different physical ability levels and challenging terrain. Pedestrian-centric street design, which is discussed in more detail in Chapter 7, should provide dedicated lanes for walking and bicycling as well as safe crossings over and/or under roadways. The walking and bicycling lanes should be connected to the transit station and can be co-located with the rails to foster further multi-modal transportation. As evident in the ski/ride Vermont concept, secondary and tertiary transit systems can be used as short-haul connectors outside of the quarter-mile/five-minute transit station radius to link the primary line to other points in the community.

Marketing

In addition to the activity-based marketing in the ski/ride Vermont by rail concept, there is a need for incentive marketing to keep ridership at consistent levels. For many, the environmental benefit of decreasing auto congestion and fossil fuel consumption is certainly a sufficient motive for riding the train; unfortunately, not all people share these ideals. One incentive-based model that is a successful precedent is a frequent-flyer scheme, such as those offered by airlines and/or credit card companies. For every mile traveled, points can be redeemed for further travel and/or goods and services. Amtrak's Guest Rewards program is similar, offering two points for every dollar spent on Amtrak travel. A similar model could be implemented on a commuter rail system and could be coordinated with Vermont businesses to stimulate the local economy. Given the connection to the mountain resorts, discounts on the price of a ski/snowboard ticket/pass could be offered for every train trip.

The state could also work with local employers to provide tax incentives for promoting train ridership. For example, the Cook County Regional Transportation Authority in Illinois implemented a Transit Ridership Incentive Program (TRIP) that provides up to $1,700 to Cook County businesses that enroll employees in a pre-tax transit benefit program where commuters can purchase their transit fares before they are taxed, thereby lowering their taxable income. This lowers the cost of commuting by 30–40 percent for commuters, while employers receive $10 for each employee enrolled (up to a total of 50 employees) as well as $200 for an initial startup with five or more employees enrolled (CRTA, 2011). Chapel Hill, North Carolina implemented a "fare-free" system in 2002 for bus transit. This program is not technically free but rather "pre-paid" from property taxes and vehicle registration fees. The University of North Carolina also purchases passes for all its students and employees. The system is credited with increasing ridership from three million passengers a year to seven million (Jaffe, 2013).

Employers can also initiate internal "challenge" programs to promote ridership. In 2012, the Transportation Research Center at the University of Vermont sponsored a program that hooked up meters on employees' cars to measure fuel consumption. Simply documenting the miles per gallon persuaded many to start car-pooling or riding the bus. Other companies have taken this monitoring a step further and initiated "green challenges," whereby employees with the lowest miles per gallon win prizes. One company in Vermont, NRG Systems, offers cash incentives to employees to purchase hybrid vehicles and/or a bicycle. They also cover all costs for buses, trains, and subways – even away from work.

Reducing cars and providing auto-alternative forms of transit are both imperative to reduce the roadway and parking infrastructure discussed in the following chapters. As evident throughout this book, achieving sustainability often relies on combining many strategies, policies, and cultural changes.

References

American Public Transportation Association (APTA). (2023). *2022 public transportation fact book* (73rd ed.). www.apta.com/wp-content/uploads/APTA-2022-Public-Transportation-Fact-Book.pdf

American Public Transportation Association (APTA). (n.d.). *Public transportation facts*. www.apta.com/news-publications/public-transportation-facts/

American Society of Civil Engineers (ASCE). (2013). *Automated people movers standards, ANSI/ASCE/T&DI 21–13*. https://doi.org/10.1061/9780784412985

Argodesign. (n.d.). *Wire One urban cable transit systems*. Argodesign. www.argodesign.com/work/wire-one.html

The Associated Press (AP). (2017). Proposal for Austin gondola lift grounded by officials. *The Seattle Times*. www.seattletimes.com/nation-world/proposal-for-austin-gondola-lift-grounded-by-officials/

Azienda del Consorzio Transporti Veneziano (ACTV). (2010). *ACTV home page*. https://actv.avmspa.it/en/content/home-actv-eng

Barber, M. (2017, September 21). *11 gondolas changing the way people move*. Curbed. https://archive.curbed.com/2017/9/21/16340394/urban-gondolas-cable-cars-cities

Bernick, M., & Cervero, R. (1997). *Transit villages in the 21st century* (p. 341). New McGraw Hill.

Broadmoor and Manitou Pikes Peak Cog Rail (Broadmoor). (n.d.). www.cograilway.com/

Cable Car Zugspitze. (n.d.). https://zugspitze.de/en/Zugspitze/Summer/Cable-car-Zugspitze

Cervero, R. (1998). *The transit metropolis: A global inquiry* (pp. 17–22, 160). Island Press.

Chan, W. (2015, April 21). *Japan's maglev train sets world record: 603 kph*. CNN. www.cnn.com/2015/04/21/asia/japan-maglev-train-world-record/index.html

Chicago Regional Transportation Authority (CRTA). (2011, December 8). Pre-tax transit benefit program to save Cook County employers, employees money. *Mass Transit Magazine*. www.masstransitmag.com/press_release/10476721/pre-tax-transit-benefit-program-to-save-cook-county-employers-employees-money

Clinnick, R. (2021). Siemens to build 73 trains for Amtrak including first battery-hybrid. *International Railway Journal*. www.railjournal.com/fleet/siemens-to-build-73-trains-for-amtrak-including-first-battery-hybrid%E2%80%AF%E2%80%AF/

Crabtree, P. (2003, February 28). Governor discusses transportation, energy. *Rutland Herald*.

CRRC. (n.d.). *Suspended monorail vehicle*. www.crrcgc.cc/en/g7389/s12949/t282574.aspx

DeSimone, A., & Sullivan, J. (2014). *Where the water meets the road*. U.S. DOT, FHA. https://highways.dot.gov/public-roads/marchapril-2014/where-water-meets-road#:~:text=There

Eriksson, L., Friman, M., & Gärling, T. (2008). Stated reasons for reducing work-commute by car. *Transportation Research Part F: Traffic Psychology and Behaviour*, *11*, 427–433.

General Electric (GE). (2007). *GE unveils first hybrid road locomotive*. www.ge.com/news/press-releases/ge-unveils-first-hybrid-road-locomotive

The Gondola Project. (n.d.). *5 Cities that are using ski lifts as public transit*. www.gondolaproject.com/5-cities-that-are-using-ski-lifts-as-public-transit/

Hass, C. (2022, November 22). *Answer to poll on Unofficial Networks*. Retrieved November, 2022, from www.facebook.com/UnofficialNetworks/posts/pfbid06X3F8AhtEk2Mk4yRS3jrjzRSF6wdv8MSqWcvXhaocA2PBHYcDWws4KeJkJiv7YC5l?comment_id=517726180072423&reply_comment_id=454682603480908&notif_id=1669151887368735&notif_t=comment_mention&ref=notif

H-Bahn21. (n.d.). *Technology*. https://h-bahn.info/en/technology/vehicles-h-bahn/

The High Road Foundation. (n.d.). *Advantages: Narrow footprint*. www.thehighroadfoundation.org/faq/small-footprint

Intamin Transportation. (n.d.). *Technical data mountain people mover P8*. www.intamintransportation.com/product/mountain-people-mover/#1607004981828-31a04905-50879107-2b03

International Energy Agency (IEA). (n.d.). *Rail*. www.iea.org/energy-system/transport/rail
International Union of Railways (UIC). (2011). *First green train in Belgium thanks to solar tunnel*. www.uic.org/com/enews/nr/242/article/first-green-train-in-belgium
Jaffe, E. (2013, March 6). *How free transit works in the United States*. Atlantic Cities Place Matters. www.theatlanticcities.com/commute/2013/03/how-free-transit-works-united-states/4887/
Lambas, M. E. L., Giuffrida, N., Ignaccolo, M., & Inturri, G. (2017). Comparison between bus rapid transit and light rail transit systems: A multi-criteria decision analysis approach. *WIT Transactions in the Built Environment*, *176*, 143–154. http://doi.org/10.2495/UT170131
Landsman, P. (2017, July 18). *Town of Breckenridge considers gondola transit*. Liftblog. https://liftblog.com/2017/07/18/town-of-breckenridge-considers-gondola-transit/
Las Vegas Monorail (n.d.). *Operational excellence*. www.lvmonorail.com/operations/
Lynn, L. (2022, June 9). *Vermont, nation see growth in skier/rider visits*. VT Ski + Ride. https://vtskiandride.com/vermont-nation-see-growth-in-skier-rider-visits/
Miami-Dade Metropolitan Planning Organization (MDMPO). (2016). *Aerial cable transit feasibility study*. http://miamidadetpo.org/library/studies/aerial-cable-transit-feasbility-study-final-report-2016-02.pdf
The Monorail Society. (n.d.a). *Technical pages*. www.monorails.org/tMspages/TPindex.html
The Monorail Society. (n.d.b). *Monorails of Japan*. www.monorails.org/tMspages/Osaka.html
The Monorail Society. (n.d.c). *Siemens H-Bahn*. www.monorails.org/tMspages/TPSiem.html
The Monorail Society. (n.d.d). *Monorails of the world*. www.monorails.org/tMspages/Where.html
Movellan, J. (2014, July 18). *Solar powered gondola saving clean mountain air in Telluride, Colorado*. Renewable Energy World. www.renewableenergyworld.com/solar/solar-powered-gondola-saving-clean-mountain-air-in-colorado/#gref
Murray, E. (2017, January 1). Stowe sees unprecedented traffic congestion. *Burlington Free Press*. www.burlingtonfreepress.com/story/news/local/2017/01/01/stowe-sees-unprecedented-traffic-congestion/96065722/
National Association of City Transportation Officials (NACTO). (2021). *Half a billion trips by micromobility since 2010*. https://nacto.org/wp-content/uploads/2022/12/2020-2021_shared_micro_snapshot_Dec7_2022.pdf
New Haven – Hartford – Springfield Rail Program (NHHS). (n.d.). *Transit oriented development success stories*. www.nhhsrail.com/pdfs/TODcasestudydraft_100311.pdf
Newman, P., & Kenworthy, J. (1999). *Sustainability and cities: Overcoming automobile dependence* (p. 59). Island Press.
Pace, E. (2017, July 16). North-south town gondola gets little support at Breckenridge council meeting. *Summit Daily*. www.summitdaily.com/news/north-south-town-gondola-gets-little-support-at-breckenridge-council-meeting/
Railway Technology. (2012). *Chongqing Metro, China*. www.railway-technology.com/projects/chongqing-metro/.
Redman, L., Friman, M., Gärling, T., & Hartig, T. (2013). Quality attributes of public transport that attract car users: A research review. *Transport Policy*, *25*, 119–127.
Richards, B. (1976). *Moving in cities* (p. 58). Westview Press.
Siemens. (n.d.). *VAL – Automated people mover*. www.mobility.siemens.com/global/en/portfolio/rail/rolling-stock/val-systems.html
Singapore Government (Gov.sg). (2020). *Keeping Singapore moving*. www.gov.sg/article/keeping-singapore-moving
Ski Industry News. (2020, May 22). *Zlatibor Gold Gondola, Serbia: Longest gondola in the world now in construction phase*. www.snowindustrynews.com/articles/2020/may/zlatibor-gold-gondola-serbia-longest-gondola-in-the-world-now-in-the-construction-phase/
Skiresort.info. (n.d.). *Funicular lifts Davos Klosters*. www.skiresort.info/ski-lifts/lift-types/davos-klosters/lift-type/funicular/
Statista. (2023). *Passenger transport in Singapore – statistics and facts*. www.statista.com/topics/5880/passenger-transport-in-singapore/#topicOverview
Telluride Mountain Village Gondola. (n.d.). *Gondola system*. https://tmvoa.org/gondola-landing-page/
Tjoe, L. N. (2022). 12% of car-owning households have 2 cars, less than 3% own 3 cars or more. *The Straits Times*. www.straitstimes.com/singapore/politics/12-per-cent-of-car-owning-households-in-singapore-have-2-cars-less-than-3-per-cent-own-3-or-more
Town of Mountain Village. (n.d.). *Gondola-Chondola*. https://townofmountainvillage.com/explore/getting-around/gondola-chondola/

Transit Oriented Development Institute (TODI). (n.d.). *Transit oriented development*. http://www.tod.org/
Transportation Research Board (TRB). (n.d.). *Seraphim urban magnetic levitation program*. https://rip.trb.org/view/1227693
Utah Department of Transportation (UDOT). (2022). *UDOT identifies Alternative B as the preferred alternative in Little Cottonwood Canyon*. https://udot.utah.gov/connect/2022/08/31/udot-identifies-gondola-b-as-the-preferred-alternative-in-little-cottonwood-canyon/
Utah Department of Transportation (UDOT). (2023). *Current phase*. https://littlecottonwoodeis.udot.utah.gov/
Vuchic, V. R. (2007). *Urban transit: Systems and technology* (pp. 70–76). John Wiley & Sons.
Walt Disney World Magic (WDW). (n.d.). *Walt Disney monorail system*. www.wdwmagic.com/transportation/monorail.htm
Warren, R. (1998). *The urban oasis* (pp. 80, 129). McGraw Hill.
Whistler-Blackcomb. (n.d.). *Peak 2 Peak gondola*. www.whistlerblackcomb.com/explore-the-resort/activities-and-events/summer-activities/peak-2-peak-360-experience/peak-2-peak-gondola.aspx
Zhang, M. (2009). Bus versus rail: Meta-analysis of cost characteristics, carrying capacities, and land use impacts. *Transportation Research Record*, *2110*(1), 87–95. https://doi.org/10.3141/2110-11

7

ROADWAYS AND PARKING

Although designs for transport vehicles can be dated back to Leonardo da Vinci in the fifteenth century, other notable designs were produced by Nicolas-Joseph Cugnot in 1769, Robert Anderson around 1832, Karl Friedrich Benz's three-wheeled, internal combustion engine in 1885, and Gottlieb Wilhelm Daimler and Wilhelm Maybach's four-wheeled, gas-powered vehicle in 1886. In the United States, George Baldwin Selden patented a design for an internal combustion engine around 1876, and, in 1893, Frank and Charles Duryea introduced the first successful gas-powered, "motorized carriage" in Springfield, Massachusetts (LOC, n.d.). By the early twentieth century, there were several players in the auto market. With the development of assembly-line construction by Henry Ford in the 1920s, costs could be lowered to allow more Americans to be able to purchase a vehicle. Since then, the automobile has grown exponentially. In 1920, there was one passenger car registered for every 13 citizens in the United States; in 1930, roughly one per every five; by the late 1970s, one car per two people; in 2020, with the total number of registered vehicles in the United States at 281,312,446 and a total population of 329.5 billion, the United States is quickly approaching one vehicle per person with 0.85 vehicles per person (NGS, 1975; WHO, 2020). Worldwide, the patterns are similar. According to the World Health Organization, there were 2,022,200,847 registered vehicles in the world as of 2020, with China, the United States, and India taking the top three positions, respectively. With a world population of 7.821 billion in 2020, that is roughly one vehicle for every four people. Given these exponential increases, the world could approach one car per person or even exceed that in the very near future. Although fuel efficiency and alternative fuel vehicles (biodiesel, electric, hybrid, ethanol, hydrogen, propane, natural gas) are increasing, such technological improvements still do not address the vehicle miles traveled, congestion, and the fragmentation of communities caused by roadways and parking surfaces.

The automobile gave citizens the freedom to flee the congested and polluted urban centers to the countryside for a peaceful retreat, and it was not long before the desire to live in such serenity became proletarian. After World War II in the United States, Baby Boomers and war veterans flocked to the suburbs in search of this "American Dream." Demand for housing escalated, and the once cruising roadways that traversed the natural contours of the land were no longer adequate for the flow of commuters to and from the suburbs. As cars increased, so did the roads and parking. In

DOI: 10.4324/9781032668666-7

1921, there were only 387,000 miles of surfaced roads – this number would triple over the next 20 years (Warren, 1998, p. 18). According to the Bureau of Transportation Statistics (U.S. DOT, 2020), there were 4.16 million miles of public roads in the United States, which is an increase of 615,000 miles since 1960.

Much of the highway development in the United States in the mid-twentieth century was largely due to the Interstate Highway Act of 1956. The highways were originally designed to link metropolitan regions and reduce commuting times between rural areas and urban centers. Ironically, as more highways were built, the development of open land in remote locations also increased. In turn, the same highways that were intended to alleviate congestion eventually fostered more car journeys from distant locations. As history has demonstrated, this vicious cycle perpetuates: congested roads = highway expansion = increased development = more cars = more congestion = the need for more and wider roads. Transportation planners often refer to this as induced traffic demand, which is derived from economists' concept of induced demand whereby increasing the supply of something makes people want it even more. Although the concept has been studied for more than a century, a 2009 study analyzed data for all metropolitan areas in the United States between 1980 and 2000 and found an almost perfect correlation between lane mile construction and vehicle miles driven (Blumgart, 2022, February 28).

As the number of cars and roads increases, the demand for parking also increases. Given that the area of typical parking space is about 150 square feet (14 square meters) and that an average aisle width is around 20–25 feet (6–7.7 meters), one car requires anywhere between 250 and 350 square feet (23–33 square meters) of space. Multiply this by minimum parking space requirements (around 5 spaces per 1,000 square feet for retail, 2–4 spaces per 1,000 square feet for offices), and the amount of asphalt starts to increase, especially since a typical large box-retail store is around 200,000 square feet. As stated by Warren (1998),

> when we build, we provide not one, but two or three parking places per car – one at home, one at work or school, one at the shopping center, entertainment or vacation spot, and the like. We are dedicating some 700–1,050 square feet (65–97.5 square meters) to each private car for the purposes of storage alone, even before streets and highways are counted.

The Federal Highway Administration defines roads by purpose and function, including arterials (interstates, highways, freeways), collector roads (avenues, boulevards, and parkways), local roads/streets, alleys, and parking lots. Since their inception, the arterials exacerbated sprawl from the city centers into the suburbs (FHWA, n.d.a). Over time, the suburbs have ironically become plagued with the same congestion that people were trying to get away from in the cities, just with a more auto-centric aesthetic. In contrast to higher densities, mixed-uses, buildings addressing the street fronts, and walkable and continuous street networks inherent in urban environments, suburban environments are typically characterized by lower density, single-use zoning, and cul-de-sac street networks that do not provide contiguity, walkability, or connection to the commercial districts. The commercial districts in suburbs generally consist of multi-lane collector roads that are flanked by box retail, strip malls, and shopping centers that have large setback distances to the street/sidewalk and are surrounded by large surface parking lots. The sprawl of cities into suburbs and suburbs into the countryside has created contiguous, concrete-line megalopolises that have fragmented communities, green space, and natural ecosystems. As pointed out by Warren (1998, p. 18),

> as woods and grasslands have been sliced up into housing developments, malls, industrial parks, and highways, the whole complex fabric of the natural world has been gradually unraveling, leaving only isolated islands of nature preserves, and rendering untold numbers of plant and animal species vulnerable to extinction.

In the twenty-first century, there is a growing movement to rethink the way highways and roads are designed and juxtaposed in relation to cities, neighborhoods, and natural features. As introduced in Chapter 3, highway expansion through and around cities often purposefully fragmented neighborhoods and businesses and disproportionately impacted minorities and people of color. This, coupled with the government's practice of eminent domain to obtain the land for the highways, further isolated minority communities from the urban core and displaced many citizens to fringe locations with poor environmental conditions and others to homelessness. Fortunately, many of these highways are starting to reach the end of their lifespan and there is a movement by planners and designers to remove, tunnel, cap, retrofit, or convert the highways into boulevards, parkways, green spaces, and pedestrian-centric public places. In 2024, the U.S. Department of Transportation announced $3.33 billion in grant awards for 132 projects through the *Reconnecting Communities Pilot and Neighborhood Access and Equitygrant programs*, which is an initiative to reconnect communities that are cut off from opportunity and burdened by past transportation infrastructure decisions (U.S. DOT, 2024). More specially, funding is being allocated for "transformative community-led solutions, including capping interstates with parks, filling in sunken highways to reclaim the land for housing, converting inhospitable transportation facilities to tree-lined streets, and creating new crossings through public transportation, bridges, tunnels and trails" (U.S. DOT, 2023).

Roadway and parking reduction and retrofitting

Roadway reduction and retrofitting strategies can be applied to all the Federal Highway Administration (FHWA) road types both within village/urban centers and between urban centers and peripheral communities. Many of the strategies discussed in the next subsections are consistent with the Complete Streets policy initiative developed by the National Complete Streets Coalition of Smart Growth America. As defined by Smart Growth America (n.d.),

> Complete Streets advocate for street designs that integrate people and place in the planning, design, construction, operation, and maintenance of our transportation networks. This helps to ensure streets put safety over speed, balance the needs of different modes, and support local land uses, economies, cultures, and natural environments.

Traffic calming strategies

The goal of traffic calming is to make cities and villages safer, quieter, and more pedestrian friendly. Calming strategies generally include reducing the width of lanes; providing curb extensions (bulb-outs) to decrease the distance pedestrians have to navigate to cross the road; decreasing speed limits; incorporating speed bumps/cushions/tables; utilizing traffic circles, roundabouts, diverters, chicanes, and raised medians; using different materials, textures, or colors for roads, sidewalks, and crossings; and providing on-street parking, which reduces the need for large surface parking, provides a buffer between pedestrians and traffic, shortens

FIGURE 7.1 Top: Auto congestion on Routes 7 and 2 in Vermont; bottom left: South Burlington, VT in 1937 (left) and 1995 (right). Although Vermont has some of the most stringent land use regulations in the country, it is not immune to auto-centric design and urban sprawl.

Orthophotography courtesy of the University of Vermont, Planning, Design, and Construction.

crossing widths due to the curb extensions that define the parking spaces, and creates more driver awareness due to vehicles backing out of the spaces. Frontage roads that run parallel to through roads can divert traffic to businesses, retail, and residences, and if strategically planned, can decrease the number of intersections and traffic lights on the main roads. Frontage roads can also be designed with diagonal or parallel on-street parking in lieu of large surface parking. Using rumble striping or materials such as brick or granite block paving on roads in urban and village centers has also been shown to reduce vehicular speeds and could therefore create safer pedestrian walking conditions. Nogueria and Mennis (2019) found that brick reduced speeds by approximately 3 mph (~5 km/h) and granite block reduced speeds by approximately 7 mph (~11 km/h) compared to asphalt-paved city streets. This can be attributed to both noise and roughness that cause drivers to slow down. Changing one-way streets to two-way streets also has been shown to decrease driving distances, provide better access, and decrease confusion since drivers do not have to drive out of the way several blocks to get to a desired destination (PPS, 2008).

Pedestrian circulation

Anyone who has ever attempted to navigate a multi-lane roadway – particularly on a suburban commercial collector road – by any means other than a personal automobile has most likely felt frustrated, uncomfortable, and unsafe. The vintage arcade game *Frogger* comes to mind with the frog desperately trying to cross the road without getting hit by a car or truck. Sidewalks are often scarce, and where they do exist, they are often so close to the road that the passing of a large truck nearly deafens you or knocks you completely to the ground. Sidewalks also sometimes mysteriously end, forcing you to walk through high grass/swales or cross the street where sidewalks may be totally absent. When you do arrive at your destination, you take your life into your own hands navigating parking lots that often have no designated pedestrian routes.

Street and parking design should provide dedicated lanes for walking and bicycling. Ideally, the lanes should be physically separated from the street with a curb and/or vegetative strips, bioswales, or street trees to provide adequate safety from car and truck traffic. The distance between the street and the sidewalk should be at least the width of a car length to allow cars to turn without blocking the sidewalk. Many communities have implemented bike lanes along the sides (shoulders) of roads. While this is better than no bike lane at all, it is not an adequate design strategy given the recent increase in auto-veering/swaying because of distracted drivers using smartphones and/or digital handheld devices. In cases where a vegetative buffer is not possible, a guard rail can be placed between the road and the pedestrian path. Some designs provide separate paths for walking and "wheeled" conveyance (bicycling, rollerblading, ski-skating, wheelchairs, etc.). Often, different materials are used for each function, such as gravel, brick pavers, or concrete for walking and pervious pavement or asphalt for wheeled conveyance. Striping can also distinguish bicycling and walking lanes where paths are constructed from one path material. Curb bulb-outs and lane reductions (discussed earlier) make crossing intersections easier along with pedestrian tunnels and overpasses (discussed in the following text).

Reducing the setback distance between the building and the street is also integral to fostering walkable and engaging streetscapes. In the context of urban design, setback is defined as the required distance a structure is offset from each of its property lines. The setback is also a function of the size, height, and shape of the building and the site. Zoning ordinances usually govern these ratios but are often lackluster in promoting a decent aesthetic. Many ordinances are concerned only that there is adequate distance between the structure and the curb but fail to regulate whether there is too much distance between these two entities. As explained by Calthorpe (1993, p. 110), "An active pedestrian environment is stimulated by buildings at the sidewalk with numerous entries; surface parking lots are 'dead' spaces for pedestrians and drain the life of the street."

Sidewalks should also be of adequate width to accommodate people, wheelchairs, bicycles, benches/sitting areas, lighting, street art, street trees, and bike parking/bike share systems. The

FIGURE 7.2 Top left: pedestrian path with bike lanes separated from the road by median and from the walking sidewalk; top right: guardrails used on the shoulder to protect bicycle lane from vehicle lane; bottom left: recreation path under bridge to provide a continuous pathway separated from the intersection; bottom right: ample distance between road and sidewalk allows cars to turn without blocking pedestrians on sidewalks.

National Association of City Transportation Officials (NACTO, n.d.) defines four zones for sidewalk design: the frontage zone that is the extension of the building that contains entry ways and street cafes; the pedestrian through zone where people circulate, which should be 5–7 feet (1.5–2.1 meters) in residential zones and 8–12 feet (2.4–3.7 meters) in downtown or commercial areas; the street furniture/curb zone, which is the section between the through zone and the buffer zone that contains street furniture (benches, lighting, trash cans, plantings/GSI, and street trees); and the enhancement/buffer zone, which is the area between the street curb and the furniture zone that can include curb extensions, parklets, stormwater management features, parking, bike racks, bike share stations, or bike lanes.

Overpasses and underpasses for pedestrians and wildlife

Overpasses and tunnels provide a safer means of navigating across roads than at-grade crossings. Rather than merely providing a connection between two points, the overpass/underpass infrastructure can also provide programmable space. The Vancouver Land Bridge in

Washington State, for example, bridges Route 14 and reconnects historic Fort Vancouver with the Columbia River. The bridge is 40 feet (12.2 meters) wide and lined with a pedestrian path, indigenous vegetation, and Native American art that retells the story of the site as a crossroads for Native Americans and European settlers (The Confluence Project, n.d.). Utilizing rooftops of buildings or structures also provides a means of bridging roadways. Where access rights can be obtained, this strategy minimizes the need for extensive ramping, stairway, or elevator infrastructure that is needed to elevate the bridges and make them accessible for people with different abilities.

Tunneling under roadways can also be implemented where there is limited space at-grade to construct an overpass. Tunnels can be gradually ramped under roads and/or accessed by elevators or stairs with stair lifts and bike rails. Natural light can be brought in via light tubes, but tunnels should also be augmented with ample artificial lighting to provide a safe environment. Tunnels do not need to be dark, dingy spaces with the sole purpose of connecting two points. They can be designed to include benches, artwork, murals, sitting areas, and sound systems.

FIGURE 7.3 Top left: Confluence Land Bridge courtesy of Washington State Department of Transportation; top right: conceptual land bridge over United States 2 in Vermont connecting student center to proposed green roof on water reservoir; lower left: conceptual pedestrian bridge over U.S. Route 2 in Vermont bridging UVM's Main Campus with the Athletic Campus; lower right: conceptual strategy to tunnel U.S. Route 2 under an at-grade land bridge.

FIGURE 7.4 Pedestrian bridges using ramping with accessible slopes.

Overpasses and underpasses can also be designed specifically for wildlife. With the proliferation of development and rapid loss of forests and open space, the need for dedicated circulation paths for wildlife has increased. A study conducted by the Cornell University Cooperative Extension in 2005 reported that there were more than 1.5 million deer–vehicle collisions in the United States in 2002, causing at least $1.1 billion in damage and killing 150 humans and at least 1.5 million deer (Curtis & Hedlund, 2005). Several sites have been proactive in addressing this. At Banff National Park in Canada, various sizes of wildlife crossings were constructed

FIGURE 7.5 Top left: pedestrian tunnel under U.S. 2 at UVM in Burlington, Vermont; top right: pedestrian tunnel options using stairs/stair lift; bottom left: pedestrian tunnel under Route 20 in Geneva, New York, connecting Lakefront Park to the downtown; bottom right: pedestrian tunnel under I-90 on the Auburn Trail in Fishers, New York.

both above and under the Trans-Canada Highway. The crossings consist of two major 164 feet (50-meter) wide overpasses that allow animals to safely cross the highway between the town of Banff and Castle Junction along with 21 less visible underpasses that vary from small culverts suitable for reptiles and rodents to open-span underpasses and creek bridges (Benedict & McMahon, 2006). The Netherlands has over 600 underpasses and ecoducts, including the 164 feet-wide by 2,625 feet long (50 by 800 meters) Natuurbrug Zanderij Crailo (Sand Quarry Natural Bridge) that crosses the N524 roadway, a railway, a business park, a river, and a sports complex. In the state of Washington, the I-90 Wildlife Bridges Coalition grew out of the Cascades Conservation Partnership and is currently working to provide wildlife crossings over portions of I-90 east of Snoqualmie Pass, a stretch of land that is an active and vital crossing area in the north–south migration of wildlife in the Cascade Range. In July 2015, ground was broken for the construction of a wildlife and pedestrian overpass at Price Creek. The National Wildlife Federation's Northeast Regional Center has also started a project known as Critical Paths. Through computer modeling and field observation, the project has identified 11 priority zones out of the 38 wildlife crossings that were surveyed along the spine of the Green Mountains in

Vermont (NWF, n.d.). In addition to overpasses and underpasses, other mitigation measures include ensuring the guard rails have efficient gaps for wildlife to pass through, using less salt in the priority zones to decrease "salt licks" (animals staying on roads to lick the salt), decreasing speed limits, increasing signage, conserving adjacent land, and implementing wildlife overlay districts in the zoning regulations to ensure development is minimized.

FIGURE 7.6 I-90 wildlife overpass in Washington State.

Source: Images courtesy of Washington State Department of Transportation

GSI for streetscapes and parking

Roadway and parking design can also incorporate the green stormwater (GSI) infrastructure concepts introduced in Chapter 2 to reduce impervious surfaces, capture stormwater runoff, cool and filter the air, and provide wildlife habitat. Some of the designs include infiltration trenches, vegetated swales, pervious pavement, planters, drainage wells, bulb-outs, rain gardens/bioretention areas, and street trees and/or vegetative buffers along medians and between car lanes, parking, and pedestrian paths.

In 2001, the city of Seattle employed bioretention cells as part of the Street Edge Alternatives Project (SEA Streets) in a residential neighborhood in the city. The original street was redesigned in a curvilinear fashion to mimic natural drainage patterns. Vegetated bioretention cells and more than 100 evergreen trees and 1,100 shrubs were constructed and planted along the edge of the road. The result was an 11 percent reduction in impervious surfaces compared to a traditional street, which has subsequently yielded a 99 percent reduction in the total volume of stormwater leaving the street (Seattle Public Utilities, n.d.). The street, which was also lined with pedestrian walkways, has become an attraction by providing residents and tourists with a pleasant, natural oasis in an otherwise hard-edged environment.

In Portland, Oregon, the Green Streets Program started in 2003 with several pilot demonstration projects that provided information before the program was implemented citywide. In 2007, the city council approved the Green Street Resolution and the One Percent for Green Fund which allocates 1 percent of construction funds from any project requiring a street opening permit toward green infrastructure projects. The projects thus far have reduced peak flows by 84–90 percent; reduced suspended solids, organic pollutants, oils, and heavy metals by 90 percent; and reduced costs by 40 percent cost compared to traditional upgrades for pipe systems (City Parks Alliance, n.d.).

In 2016, Geneva, New York, received a $10 million Downtown Revitalization Initiative (DRI) grant from New York State to improve the vitality of its downtown center. The DRI covered 100 acres (40.5 hectares) from Seneca Lake to the Main Street Corridor and focused on connectivity and transportation, civic amenities and greenspace, arts and culture, rehabilitation of key buildings, infill development, entrepreneurial endeavors, poverty eradication, and social justice (City of Geneva, 2017). Many of the completed projects to improve the streetscape incorporated GSI infrastructure strategies – including curb extensions with bioretention, bioswales, pervious pavement/pavers, storm drain marking, landscaped medians, and street trees – along with sidewalk improvements (lighting, sitting areas, increasing width), dedicated bike lanes, reduced width of car lanes, crosswalk improvements, mixed-use development, pedestrian tunnel enhancements, marina development, construction of a solar-powered residential village, and improved signage.

GSI can also be incorporated adjacent to the parking lots and/or on vegetated medians that bisect rows of parking. In addition to softening the parking lot, vegetated islands and medians can incorporate pedestrian pathways that make it easier to navigate the lot. Pervious pavement, porous asphalt, and turf-block systems can be used in lieu of typical asphalt paving to increase water infiltration. Figure 7.12 shows a typical commercial layout with box stores surrounded by large surface parking. The retrofitted design includes some of the GSI and Smart Growth strategies discussed earlier, including moving the buildings toward the street front, placing diagonal parking along the street and behind the buildings, and placing vegetative medians and trees down the center of the primary street and between the diagonal

FIGURE 7.7 Top: Street Edge Alternatives Project on 2nd Avenue NW, Seattle, Washington; bottom: close-up of vegetated bioretention area and curb-less roads.

Source: Images courtesy of Seattle Public Utilities

FIGURE 7.8 Streetscape improvements in Geneva, New York. Top left: vegetated bioretention swales in medians; top right: bioretention, seating, street trees, street parking, and curb extension; bottom left: curvilinear porous pavement used for bike lanes; bottom right: close-up of curb extension with bioretention that frames in on-street, diagonal parking; porous asphalt in the parking area.

parking and the pedestrian paths. The existing site inside the boundary is about 28.5 acres, of which 26.6 are impervious surfaces – the site is thus 93 percent impervious. The reconfigured site has slightly fewer parking spaces but is only 49 percent impervious – almost a 50 percent reduction in impervious surfaces.

Retrofitting the commercial strip

When I was in graduate school in the early 2000s, many of my architecture/planning colleagues often pitted the suburbs and the cities against each other, where a gain to one would be a loss to the other. Some theorists adopted a "death to the suburbs" attitude and advocated for a complete retraction of development in the suburbs to re-concentrate businesses and residences toward the city centers. Although regreening large swaths of land in the suburbs is desirable, whether we like it or not, suburbs have become their own entities and are here to stay. Apart from different urban forms, the city and the suburbs function more as collective megalopolises. As described by Dunham-Jones and Williamson (2011, p. xxvii), "It is no

FIGURE 7.9 Top: bioretention garden capturing runoff from building and parking at Gas Works in Middlebury, Vermont; bottom left: bioretention garden directly off parking area at Redstone Lofts in Burlington, Vermont.

FIGURE 7.10 Top: vegetated bioretention median breaking up a parking lot; bottom: bulb-out with bioretention at the end of the median framing parking area.

FIGURE 7.11 Left: bioswale between porous paved parking and sidewalk at Cornell Botanical Gardens; right: vegetated swale showing drain (top) and curb inlet (bottom).

longer useful to talk about city centers versus suburbs . . . the suburbs are behaving more like city centers and metropolises embrace both as they become more polycentric." The authors go on to describe the concept of suburban urbanization, where more people are centralizing around suburban centers and suburbs are subsequently becoming more diverse in age, income, race, and ethnicity. These demographical shifts, along with the increase of people being able to work remotely, are creating new demands for different types of housing, pedestrian circulation, transit, and public gathering spaces – all of which have been historically scarce in the suburbs.

Transforming commercial raceway collector roads into more pedestrian-friendly, multiway boulevards can utilize many of the traffic calming and GSI infrastructure strategies discussed earlier. Successful examples have been developed on Palm Canyon Drive in Cathedral City, California; Ocean Parkway in Brooklyn, New York; Octavia Boulevard in San Francisco, California; and Avenue in Berkeley, California, to name a few. On Palm Canyon Drive, a 45-mph, six-lane arterial was converted to a four-lane, 35 mph multiway boulevard with local access lanes that contain angled parking. Landscaped and tree-lined medians were placed down the center median, between the main road and local access road, and between the angled parking

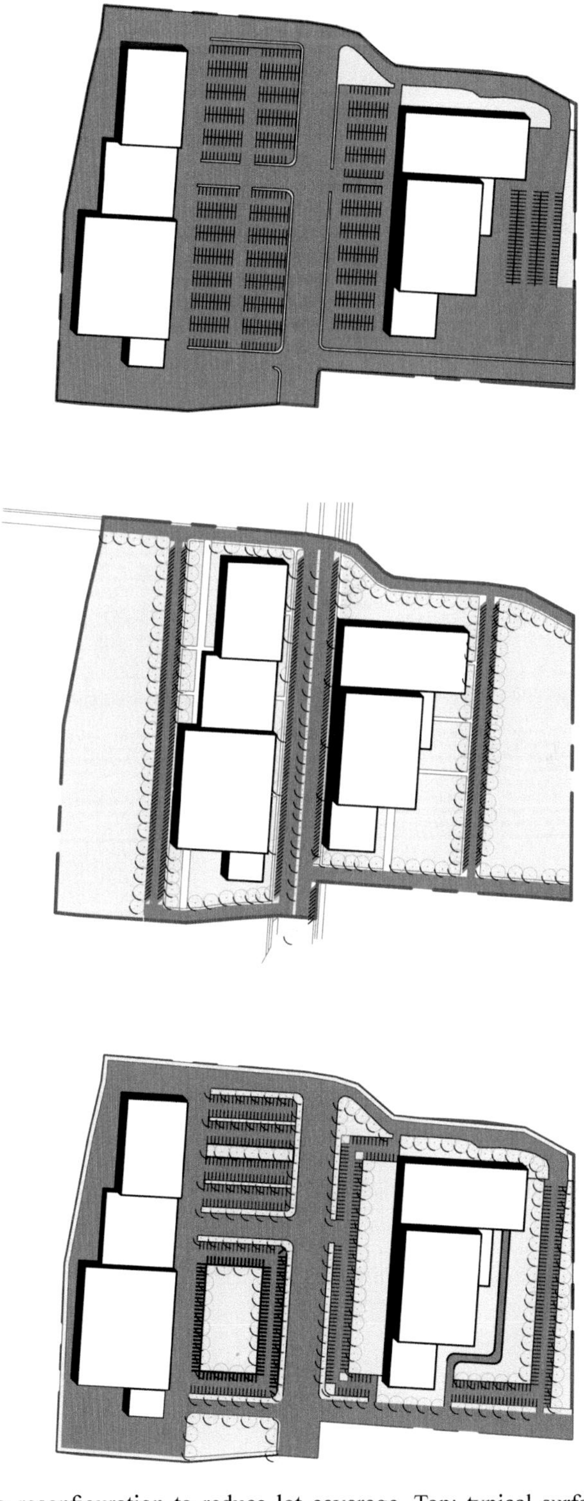

FIGURE 7.12 Parking reconfiguration to reduce lot coverage. Top: typical surface parking; middle: buildings moved to the street front with parking placed along the street and to the rear of the buildings; bottom: parking reconfigured with vegetated medians and pedestrian paths.

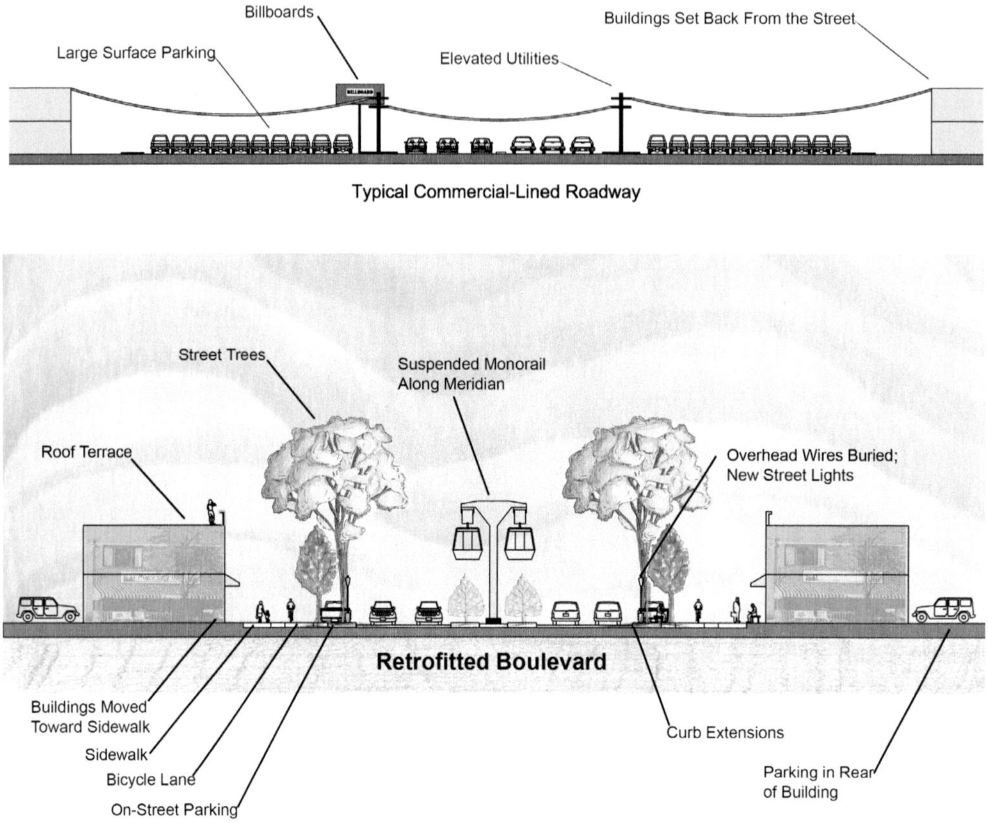

FIGURE 7.13 Top: typical commercial roadway; bottom: retrofitted boulevard.

and sidewalk. Buildings are required to have entrances onto the sidewalks, creating a mixed-use district with a downtown aesthetic.

Some commercial strip conversions are attempting to create village centers in towns that never had such in the first place because of being conceived by auto-centric design. The success of such "second cities" seems to be dependent on four factors: the proximity to other community services/attractions; providing a mix of different income range housing styles; the ability to incentivize businesses to relocate from their existing car-centric style plazas into the communities; and how well the community is integrated with the current street network and surrounding town. For the last of these, it seems some developers are not quite grasping the concept of infill development and leaving too much space between the existing street network and the new communities, resulting in insular parcels that are not connected to the community. Similar to mall designs (discussed in more detail in Chapter 10) that place the larger retailers as anchors at the ends of the malls, infill development also needs to anchor itself to the adjacent arterials to improve visibility (especially for the commercial spaces) and provide contiguity with the surrounding community.

FIGURE 7.14 Although the state of Vermont was one of the last states to be infused with corporate chains and box retail, there are some attempts that utilize Smart Growth and New Urbanism. In Williston, Vermont, some of the plazas are bringing the fronts of the buildings toward the sidewalks and streets, placing parking in the rear, and incorporating wide, tree-lined sidewalks with public sitting spaces. Some facades are being designed to mimic urban buildings to create the feel of a village center.

Route 20/Eastern Boulevard – Canandaigua, New York

Canandaigua, New York, is a small city along the northern shore of Canandaigua Lake. Given the proximity to the lake, the Bristol mountains, several wineries/vineyards, Sonnenburg Gardens and Mansion State Historic Park, two major hotels, and the Canandaigua Municipal Arts Center, the region has the potential to be a major destination. South of the city, the lake is lined with houses, cottages, condominiums, and townhouses and has few natural areas, riparian

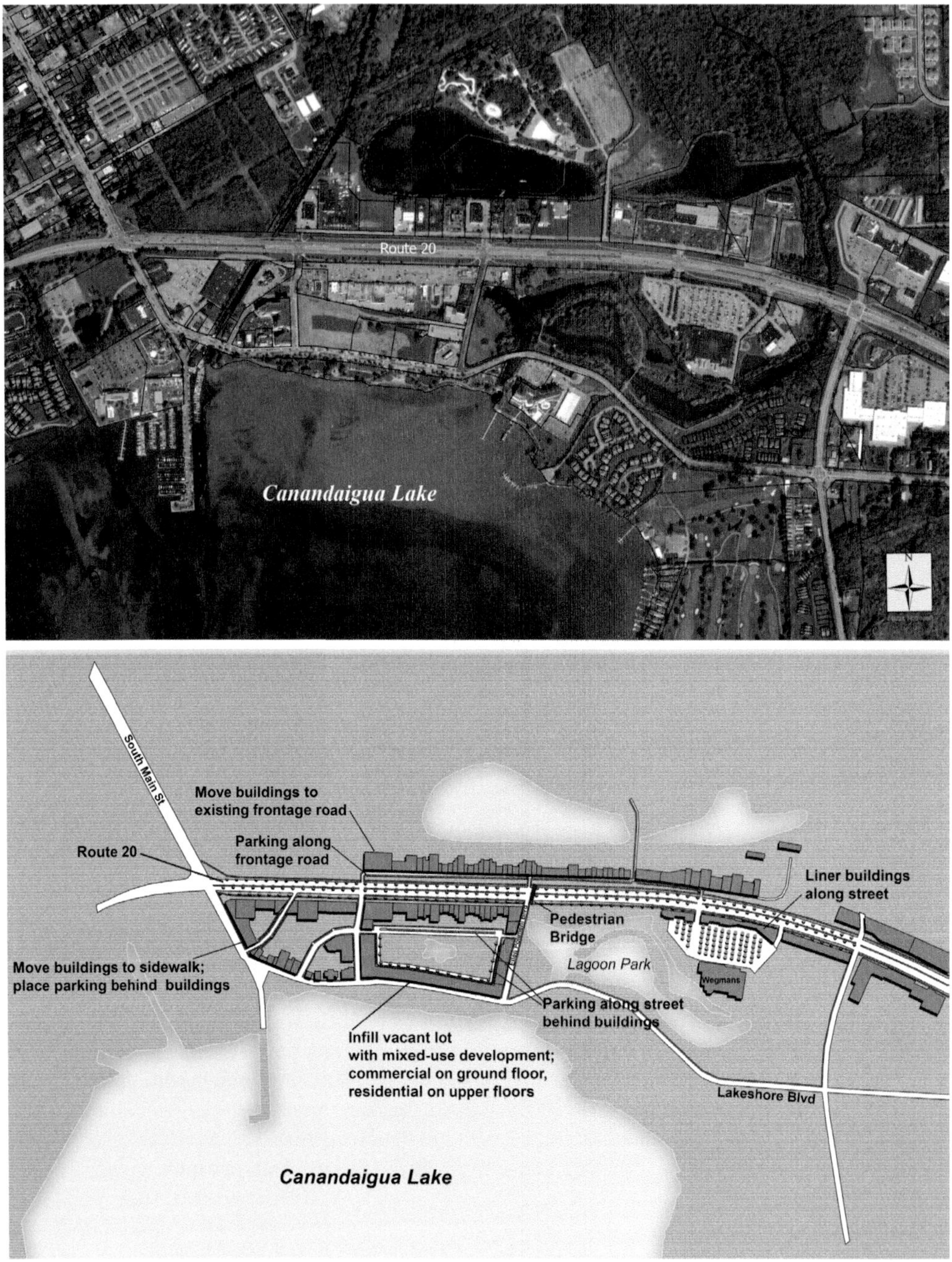

FIGURE 7.15 Top: existing Route 20/5 through Canandaigua, New York. Orthophotography courtesy of Esri, DigitalGlobe, GeoEye, i-cubed, USDA FSA, USGS, AEX, Getmapping, Aerogrid, IGN, IGP, swisstopo, and the GIS User Community; Bottom: Conceptual retrofit of Route 20/5.

buffers, parks, or public access points. Although Lakeshore Boulevard and Main Street have some degree of urban clustering, the city is bisected by Route 20 that contains the atypical, commercially lined, multi-lane road flanked by billboards, elevated utilities, large surface parking lots, and box retail. This stretch of road impairs the character, scale, and quaintness of the community. Using this stretch of road as a case study, Figure 7.15 demonstrates how a typical commercial arterial can be transformed into a more pedestrian-friendly, ecologically sensitive, multiway boulevard using the aforementioned roadway reduction strategies. Plazas and buildings offset from the street and surrounded by surface parking are brought to the sidewalk of the frontage street on the north side of Eastern Boulevard and the sidewalk on the south side of the boulevard. Liner buildings are placed between the boulevard and the parking lot for the Wegmans grocery market. The vacant lots along Lakeshore Boulevard are infilled with mixed-use development with commercial use on the ground floor and residential use on the upper floors. Surface parking is replaced by parking along the frontage street on the north side of the boulevard and parking along interior streets behind the buildings on the south side of the boulevard. Street trees are planted along the sidewalks and down the meridian and a pedestrian bridge crosses Eastern Boulevard near Ellen Polimeni Boulevard.

Reducing vehicles

The reduction of roadways and parking lots is obviously dependent on reducing the number of vehicles on the roads. To reduce vehicles, however, the auto-alternative transportation systems discussed in Chapter 6 and the pedestrian network of walking and bicycling routes discussed earlier need to be present. In addition to the trails themselves, infrastructure is needed for bicycle parking, storage, and renting along with park and ride locations, locker rooms, and showers for commuters. Other strategies to reduce vehicles include offering incentives not to drive, disincentives to drive or park, offering remote and/or hybrid work schedules, and staggering work schedules. While they do not necessarily reduce vehicle load or miles traveled, parking structures and smart parking apps also can distribute vehicles in a more efficient way, thereby indirectly reducing congestion by decreasing the amount of time a driver must scramble to find a space.

Incentives and disincentives

Employers can offer incentives to employees that encourage them not to drive, such as awards, gift certificates, or cash-outs of employee-paid parking. Some companies subsidize all or part of the costs of public transit, bike shares, or the purchase of a bike. There are also federal pre-tax incentives that allow employers and employees to deduct the cost of public transit, commuter-parking lot fees, or van pools. Some car insurance agencies are even starting to get on board to improve driver safety. Allstate Insurance currently offers the Drivewise program that monitors a user's driving behavior through a device that is plugged into the vehicle. The device tracks average speed, acceleration, deceleration (braking), total hours behind the wheel, and time of travel, to name a few. A user can then get cash rewards based on "safe" driving habits, such as slow, steady speeds with uniform acceleration/deceleration. The program is far from perfect, but it is a step in the right direction. The same principles can be applied to get people to drive less, but could be sponsored by employers, or by federal, state, or local government agencies. Such "the less you drive, the more you earn" programs could be cash rewards, gift cards, lower insurance/registration rates, or discounts for events or recreational activities.

Disincentives to drive, which can be somewhat contentious, to say the least, include limiting parking spaces, charging parking fees, raising federal and state gas taxes, or linking insurance rates/infrastructure tax to the amount people drive or to the type of vehicle they drive (e.g., a higher tax for less efficient vehicles). Such taxes can be used to fund more energy-efficient forms of transportation. These strategies are often easier to accomplish in high-density, urban areas where people can walk to destinations and/or have access to multiple modes of public transportation.

The single biggest driving disincentive may be something beyond our control: the cost of oil. Since oil is a finite resource and will run out sometime this century, it is inevitable that prices will start to escalate to a point where people will not drive. However, the data seems to be mixed as to where this upper limit will occur. According to Steiner (2009), Americans drove 100 billion fewer miles in 2008 than in 2007 when gas prices reached historical highs at $4 per gallon. However, with gas prices reaching the $5 per gallon threshold in 2022, data shows U.S. consumers only *slightly* reduced their fuel consumption as prices increased and either compensated or rationalized the additional cost (Barone, 2022, July 5). In many regions, there are no feasible alternatives to the car and people still must travel to school, work, or the grocery store. More so than rising gas prices or oil embargos, the COVID-19 pandemic seems to have had the most significant impact on driving. According to American Automobile Association (AAA), the average number of daily personal car trips in the United States plunged 45 percent in April 2020 and 40 percent for trips by all modes of transportation combined (Gross, 2021). Worldwide, lockdowns reduced the mobility of transportation by 74 percent, resulting in a 62 percent decrease in crashes and fatalities compared to pre-COVID-19 scenarios (Gupta et al., 2021).

Remote and hybrid work schedules

As a result of the COVID-19 pandemic, many workers have continued to work on a hybrid or fully remote work schedule. The Minnesota Department of Transportation conducted a study (Qian, 2022) to understand telecommuting impacts on traffic management and to no surprise, found drastic decreases in congestion because of remote work during the pandemic. However, with more people returning to work and/or working hybrid in-person/at-home schedules, it is unclear if there will be any significant lasting impact. Despite this uncertainty, the work-at-home model does give us a glimpse that it *can* be done. Flexible work hours and flexible workdays also may have the potential to stagger peak traffic flows both during the day and during the week, respectively. A demonstration project conducted in Honolulu, Hawaii, in 1988 (Giuliano & Golab, 1990) showed on average a three- to four-minute (79 percent) reduction in travel times, but questions remained as to whether this was meaningful or perceivable to commuters. The study also showed that given a choice, most workers would choose the earlier start time slot, showing that voluntary start times could simply shift the peak traffic periods and that mandating time slots might also result in decreased satisfaction rates among workers. The study also questioned if any time savings would be offset by latent-demand travel shifts. Simulations by the Vias Institute (Develtere & Leblud, 2018) showed staggering work schedules to have the greatest impact on journey times at peak rush hours with an average 25 percent reduction in journey times between 7:00 and 9:00 AM; unsurprisingly, the peak-time gains have an adverse effect on journeys made during all other time slots, particularly at 06:00, 10:00, 11:00 and 20:00 when journey times are almost doubled.

It seems more research is needed to determine how much distribution is needed to significantly reduce congestion and/or the impact of combining different start times with the increase of new hybrid/remote work schedules. In addition, there is little data in the United States that studies the impact of altering the Monday–Friday workweek system. As with staggering workday hours, there will certainly be specific five-day shifts that are more desired than others and further difficulties with life-family-work balance may arise. For example, one person may work Tuesday to Saturday, but his/her/their friend may work Wednesday through Sunday, leaving narrow or no amounts of time blocks to socialize. In addition, children's schedules – which seem to be more hectic in recent times – could add further complexity to time coordination. Regional coordination across state, province, or country lines would also be circuitous given the variance in laws, regulations, industry, and cultural holidays.

Parking structures

While parking structures do not necessarily reduce the total vehicle load, they do play a part in reducing the footprint of large, surface parking areas. Although underground parking and/or above-grade parking structures are ubiquitous in dense, urban areas where space is limited, they are not as common in suburban or rural areas because of the availability of land and the higher costs of excavation and/or for multi-story construction. However, in addition to the ecosystem benefits of minimizing large impervious surfaces, the higher initial cost for parking structures can yield economic benefits for property owners. Elevated structures can be built toward the streets and the ground level can contain space for office, retail, and services, thereby effectively increasing the square footage of rentable space. Parking structure rooftops can also contain gathering spaces, green roofs, recreation tracks, solar panels, and/or wind turbines and can connect to nearby businesses and residences with pedestrian bridges. Parking structures also do not have to be homogenous slabs of pre-cast concrete with dank interior spaces. The Greenway Self-Park in Chicago, Illinois, for example, is a ten-story, eco-friendly garage that contains a green roof, rainwater cisterns for irrigation, an array of vertical wind turbines that harness power back to the grid, high-efficiency glass, education kiosks with tips for sustainable living practices, electric car charging stations, a car share program, energy-efficient lighting, and 14,000 square feet of retail space on the ground level. The University of California Davis Medical Center garage uses a shade screen made up of over 3,000 aluminum fins that rotate and reflect natural light into the garage. The T3 parking structure in Austin, Texas contains both a green roof and a green wall that supports vines attached to a steel screen system.

Where there are height restrictions, underground parking can also be considered. Opponents of underground parking often cite concerns about safety, convenience, and energy requirements of keeping dark spaces adequately lit. Natural light can be incorporated by using skylights, light tubes, or, as in the case of the parking structure at the Cliniques Universitaires Saint-Luc hospital in Brussels, trenches that surround the structure. The garage contains three underground levels and two above-ground levels. The three underground levels are encircled by an open trench or "canyon" that allows natural light to enter, alleviating the electrical lighting demand and providing a more aesthetically pleasing environment. The canyon walls are planted with vegetation and the upper level contains a green roof that bridges different elevations at the street level. The facades are constructed from hardwood slats that are arranged at different angles that effectively stitch the structure into the site. Walls and ceilings are also an opportunity to provide space for local artists. This was employed at the 2KM3 parking garage at Saint Gervais in Mont Blanc, France. As far as safety is concerned, underground parking can be safer than large outdoor parking surfaces since access and egress are easier to monitor and/or control, not to mention it offers protection from harsh weather patterns.

Capping roadways

Existing roadways that are recessed below grade can be capped to reconnect communities that were fragmented by the infrastructure. Such caps can also be referred to as lids, decks, deck parks, or land bridges. The caps can range in size and shape from narrow strips to large, multi-acre spans that cover multiple blocks and/or interchanges. For example, Aubrey Davis Park over Interstate 90 in Seattle, Washington, Lake Place over Interstate-35 in Duluth, Minnesota (discussed in the following text), the Mt. Baker Lid over Interstate 5 in Seattle, Washington, Sam Smith Park over Interstate 90 in Seattle, Washington, and Penn's Landing Park over Interstate 95 in Philadelphia, Pennsylvania, are all over 10 acres. In Hamburg, Germany, a project is in progress to cap 2.1 miles (3.5 kilometers) of the A7 (Bundesautobahn 7) highway that will amount to roughly 30 acres (12 hectares) of green lids. The A7 caps – similar to the I-35 caps in Duluth – are being constructed simultaneously with the highway expansion. Another larger proposal is the Hollywood Central Park in Los Angeles, California, that is envisioned to be a 38-acre (15-hectare) park that stretches from Hollywood to Santa Monica Boulevard over the 101 Freeway. According to Friends of the Hollywood Central Park (n.d.), Hollywood has one of the lowest resident-to-park space ratios in California, and existing park space is disproportionately distributed toward the region's wealthier neighborhoods. The proposed park would reunite communities separated for over 60 years by the Hollywood Freeway, provide green space and recreation to more than 40,000 children, create more than 38,000 jobs, create transit-oriented development, and improve air quality. The proposed Rondo land bridge in Minnesota plans to cap Interstate 94 for five blocks between Chatsworth Street and Grotto Street and create a reconnected, vibrant, African American cultural enterprise district. Caps can also be networked with existing streets and parks to tie together disparate parts of a city. The proposed Stitch in Atlanta, Georgia, will create about 14 acres (5.7 hectares) of elevated parks, paths, development sites, green space, gathering spaces, and programmable spaces atop a three-quarter-mile platform between Ted Turner Drive and Piedmont Avenue. The connector cap runs at a roughly 45-degree angle relative to the orthogonal grid street system and extends its arms to network with the surface streets to unite all parts of Atlanta.

There are also numerous smaller caps less than 5 acres (2 hectares) or so, including five different lids in Seattle over several highways (Route 520, Interstate 5, and Interstate 590), the Downtown Loop proposal in Nashville, Tennessee, Kanawha Plaza in Richmond, Virginia, Klyde Warren Park in Dallas, Texas, Lytle Park in Cincinnati, Ohio, Memorial Park in La Canada Frintridge, California, the Sherman Heights-East Village lid and Teralta Neighborhood Park in San Diego, California, to name a few. One of the options for the Rethink 65/70 project in Indianapolis, Indiana proposes to remove the existing elevated inner loop that encircles the city and replace it below grade with "strategic" capping in specific areas that could contain structures, parks, and cap bridges to connect the streets. Some of the recessed sections would still be open air, which would give those driving natural light and glimpses of the city. Similar to the larger caps discussed earlier, the strategically placed lids can also provide space for infill development and community gathering that can provide opportunities for communities that were victims of social injustices when the highway was constructed.

Tunneling roadways

Underground engineering achievements can be dated back to the prehistoric caves of the Stone Age. Since then, various other sub-grade marvels have been achieved in projects such as the Christian Catacombs of Rome, the Roman Cryptoporticus, the 34-mile (54.7 km) Seikan Railroad Tunnel beneath the Tsugaru Strait, the New York and Harlem Railroad Tunnel, the Croton

FIGURE 7.16 Top two images: existing I65/70 through Indianapolis, Indiana; bottom two images: proposed caps with the bottom-most image showing a reconnection between the historic African American cultural center of the city to the new IU Health Campus and the rest of the city.

Source: Images courtesy of the Rethink Coalition, Inc./Rundell Ernstberger Associates, Inc.

FIGURE 7.16 (Continued)

Aqueduct, the various subways built in the nineteenth century, the Eisenhower-Johnson tunnels under the Continental Divide in Colorado, the Holland Tunnel under the Hudson River in New York, the Zhongnanshan Tunnel in China, and the Channel Tunnel under the English Channel, constructed in the 1980s. The tunneling of roads and highways through hillsides or mountains was traditionally viewed as a way of connecting two points via the shortest distance. The mountains and nature were thought of as a nuisance to the engineering process. In modern times, however, this paradigm has shifted. As greenspace and natural surroundings become increasingly endangered and as our planet continues to warm, engineers and designers are now looking at tunneling strategies from a different perspective. Existing and new roads can be tunneled underground and capped with deck parks, greenspace, surface boulevards, public squares, or even buildings. Although still complex, constructing a tunnel at the onset of a project where there is no existing infrastructure at the surface obviously presents fewer obstacles than having to tunnel an existing road while somehow keeping traffic at the surface still moving. But, as evident in the Big Dig case study discussed in the next section, it is not impossible.

Tunneling existing highways – the Rose Kennedy Greenway, Central Artery/Tunnel Project (the Big Dig)

When one thinks of burying highways, it is impossible to ignore the Central Artery/Tunnel Project (the Big Dig) in Boston, Massachusetts, in the 1990s. The project was not without its faults and drama, but much can be learned from the planning, design, and engineering of this undertaking.

Boston experienced the frenzy of post-World War II highway construction in 1951 with the construction of the John F. Fitzgerald Expressway – an elevated six-lane highway through the heart of the city. The 1.5-mile (2.4 km)-long Central Artery, as it came to be known, was originally intended to alleviate inner-city congestion. According to the Massachusetts Department of Transportation (n.d.), when the highway opened in 1959, it carried approximately 75,000 vehicles a day, but, by the 1990s, this had grown to 190,000 vehicles a day. Traffic jams were averaging 10 hours per day and were expected to rise to 16 hours per day by 2030.

Economists estimated that the congestion cost more than $500 million/year in fuel, accidents, late deliveries, and other traffic-related delays. In addition to congestion, the highway physically divided Boston's downtown from the waterfront, economically displaced businesses and the marketplace, and socially fragmented neighborhoods.

Construction of the Big Dig began in 1991. The project envisioned improving mobility, decreasing congestion, reconnecting neighborhoods to the harbor, and improving human livability by decreasing air and noise pollution and providing natural landscapes throughout the city and its surrounding communities. To date, the project is still the largest, most expensive, most complex, and most technologically challenging highway project ever attempted in American history. It spanned 7.8 miles (12.6 km) of highway and replaced the six-lane elevated I-93 artery with an eight- to ten-lane underground expressway known as the Thomas P. O'Neill, Junior Tunnel. To the north, the O'Neill Tunnel merges with the ten-lane, cable-stayed Leonard P. Zakim Bunker Hill Bridge and the Tobin Bridge via the City Square Tunnel across the Charles River. Toward the south of the city, the Massachusetts Turnpike (I-90) was extended underground to connect with the Ted Williams Tunnel under Boston Harbor. This section provides a direct, 3.5-mile (5.6 km) connection to Logan Airport east of Boston.

The Ted Williams Tunnel was completed in 1995, while the I-90 connector tunnel to it was opened in 2003; the I-93 tunnels were fully opened in 2005. The Rose Fitzgerald Kennedy Greenway now sits atop the former highway site and contains a series of gardens, promenades, and pedestrian pathways that have reconnected the communities of downtown Boston. The greenway occupies roughly 15 acres (6 hectares) of land, spans 1.5 linear miles (2.4 km), and includes Chinatown Park, Dewey Square Park, Fort Point Channel Parks, the Wharf District Parks, and the North End Parks.

While the new network of highways carries more traffic, congestion and traffic jams have decreased, thus decreasing idling and emissions. Gelinas (2007) summarized some of the positive results, including a decrease in carbon monoxide emissions of roughly 12 percent compared to prior to construction, a decrease in travel times between one-quarter to two-thirds, and an average increase of speeds in some sections between 10 and 43 mph. In addition, the

16 million cubic yards of dirt extracted during the project were used to create over 300 acres (121 hectares) of open land, including 27 acres (11 hectares) where the elevated highway used to span, 105 acres (42.5 hectares) at Spectacle Island where project dirt was used to cap brownfields, and an additional 40 acres (16.2 hectares) of parks throughout the city of Boston. Several other landfills were capped, including Quarry Hills, Gardner Street Landfill, and Granite Rail Quarry. The project also reclaimed the Rumney Marsh Wetland and created an artificial reef system in Boston Harbor with dredged soil from the tunnels. The shorter travel times and more efficient traffic flow have accounted for more than $200 million per year savings in time and fuel. Commercial properties along the old artery have increased in value by 79 percent in 15 years. New businesses have been established and existing businesses have reconfigured and renovated their spaces to adapt to the new greenway and connection to the harbor (Gelinas, 2007).

If you walk the streets of Boston and ask a hundred people how they feel about the Big Dig, you will get a hundred different answers – some positive, some negative, some combinations of the two. In 2005, Atlantic Research & Consulting conducted a survey about public perception of the Big Dig and asked the loaded question, "How has the Big Dig changed your life?" Forty-three percent claimed it had made the airport commute easier, 23 percent said it had made Boston prettier, 13 percent said it had made their commute shorter, 11 percent said it had made their commute longer, and 29 percent said it had not changed anything (Halpern, 2005). Much of the negative sentiment can be attributed to the plethora of problems that have plagued the project since the ground was broken in 1991. Roof leaks, construction errors, lawsuits, dubious accounting, and a fatal panel collapse due to inadequate epoxy-holding bolts have painted an ugly picture. Then, of course, many critics have also questioned the cost of the project, which by 2012 was around $24.3 billion, which was $21.5 billion over the original $2.8 billion estimate in 1982 (Goldman, 2012; Greiman & Warburton, 2009).

However, despite these negative aspects, much was and can be learned from the Big Dig in terms of engineering, project coordination, funding, and public policy. The technological advancements alone advanced the possibilities of architecture, design, and engineering and have inspired a legion of road tunneling projects. As mentioned earlier, the current highway carries more traffic than the old highway but is currently showing a reduction in total vehicle hours and fuel use. It will be interesting to see if this is sustained if auto use continues to rise due to induced traffic demand.

Tunneling roads to preserve smaller communities: U.S. Routes 2 and 7 in Vermont

Tunneling roads underground can also be done at smaller scales and used to preserve the historic character of small villages and urban centers plagued by traffic. The village of Shelburne, Vermont, for example, is currently bisected by U.S. Route 7 – a 308-mile (495.7 km) north–south highway that runs from Norwalk, Connecticut, to Highgate, Vermont. Due to the geography of Vermont – including Lake Champlain to the west and the Green Mountains to the east – Route 7 has become a major north–south artery for interstate trucking, tourists, and local commuter traffic. To accommodate the increase in traffic, the primarily two-lane road was expanded to four lanes in several sections, including the section north of Webster Road in Shelburne and South Burlington. Based on informal studies and observations from living on the road in the village for almost 20 years, these extra lanes have induced further traffic, which in turn has caused extreme bottlenecking in the two-lane sections that pass through the village. For residents and

FIGURE 7.17 Top: the elevated Central Artery through Boston, Massachusetts, in 2001; bottom: Rose Kennedy Greenway – textured water tables at North End Parks, 2016.

Source: Bottom image courtesy of Karen Breider

businesses along Shelburne Road, the noise is frequently at levels that make it difficult to have a conversation, listen to the radio/television, and/or sleep soundly through the night – not to mention air quality concerns, fuel consumption from idling, and safety issues/difficulties in navigating across the road on foot, bike, or wheelchair. Much of the village's small-town feel and historic character is essentially in jeopardy of becoming "extinct" or "endangered" due to the presence of this road. Although several traffic calming strategies have been proposed in the town's comprehensive plan, more ambitious strategies may need to be considered to mitigate the fragmentation caused by the road. Since there is no land suitable for a roundabout or frontage road adjacent to the village, the most effective solution (also most likely the most expensive solution) would be to tunnel a road for through traffic. Local traffic could still use the surface lane, where traffic calming strategies discussed earlier such as reduced speed limits and street parking, along with pedestrian-friendly bicycle and walking lanes and street landscaping, could be imposed.

Figure 7.18 demonstrates a conceptual proposal to bury 1.5 miles (2.4 km) of Route 7 from the jug-handle at Webster Road south through the village of Shelburne to the Bostwick/Marsett Road intersection. The tunnel would provide a fast, uninterrupted route through the village for through-traffic and interstate trucking. Local traffic could exit onto the existing surface road via feeders at the Webster Road jug-handle and at the Bostwick/Marsett Road intersection. A similar technique is also demonstrated for U.S. Route 2 in Burlington, Vermont, which is also plagued by traffic congestion and divides the University of Vermont's central campus from the residential/athletic campus. A land bridge in this area would unify the two sides of the university that are currently bisected by the road, decrease car and truck congestion by providing a more direct passage with fewer traffic lights, decrease noise, increase pedestrian safety by providing an uninterrupted connection, and create an aesthetically pleasing environment. Figure 9.15 in Chapter 9 also demonstrates the use of pedestrian bridges and bridging rooftops over the same section of the road.

Simultaneously tunneling and capping highways – I-35, Duluth, Minnesota

Interstate-35 project in Duluth, Minnesota, was unique in that it solved the quandary of expanding a highway and creating new greenspace. Unlike the Big Dig, the I-35 project had the benefit of simultaneously constructing the below-grade highways and the deck parks.

Interstate-35 was originally constructed through the city of Duluth in the late 1950s. Similar to Boston's I-93 situation mentioned earlier, I-35 quickly became overwhelmed with traffic that caused severe congestion at its terminus in the city center. In the early 1970s, a plan was drafted to extend the highway northward along Lake Superior with the intent of alleviating congestion in downtown Duluth by providing a bypass around the city. However, the extension was met with much opposition since its alignment created a barrier to the waterfront and would cause the fragmentation of Leif Erikson Park and the potential extirpation of several historical buildings in the vicinity. The Opposition groups Citizens for the Integration of Highway and Environment (CIHE) and the Stop the Freeway Action Group were formed, and, in 1975, Duluth Mayor Robert Beaudin created the I-35 Citizens' Advisory Committee that worked with the community, city and government officials, engineers, designers, and planners to devise a better solution. Kent Worley, a landscape architect and founder of CIHE, was one of the key contributors who envisioned sinking the road and enabling a pedestrian connection between the city and

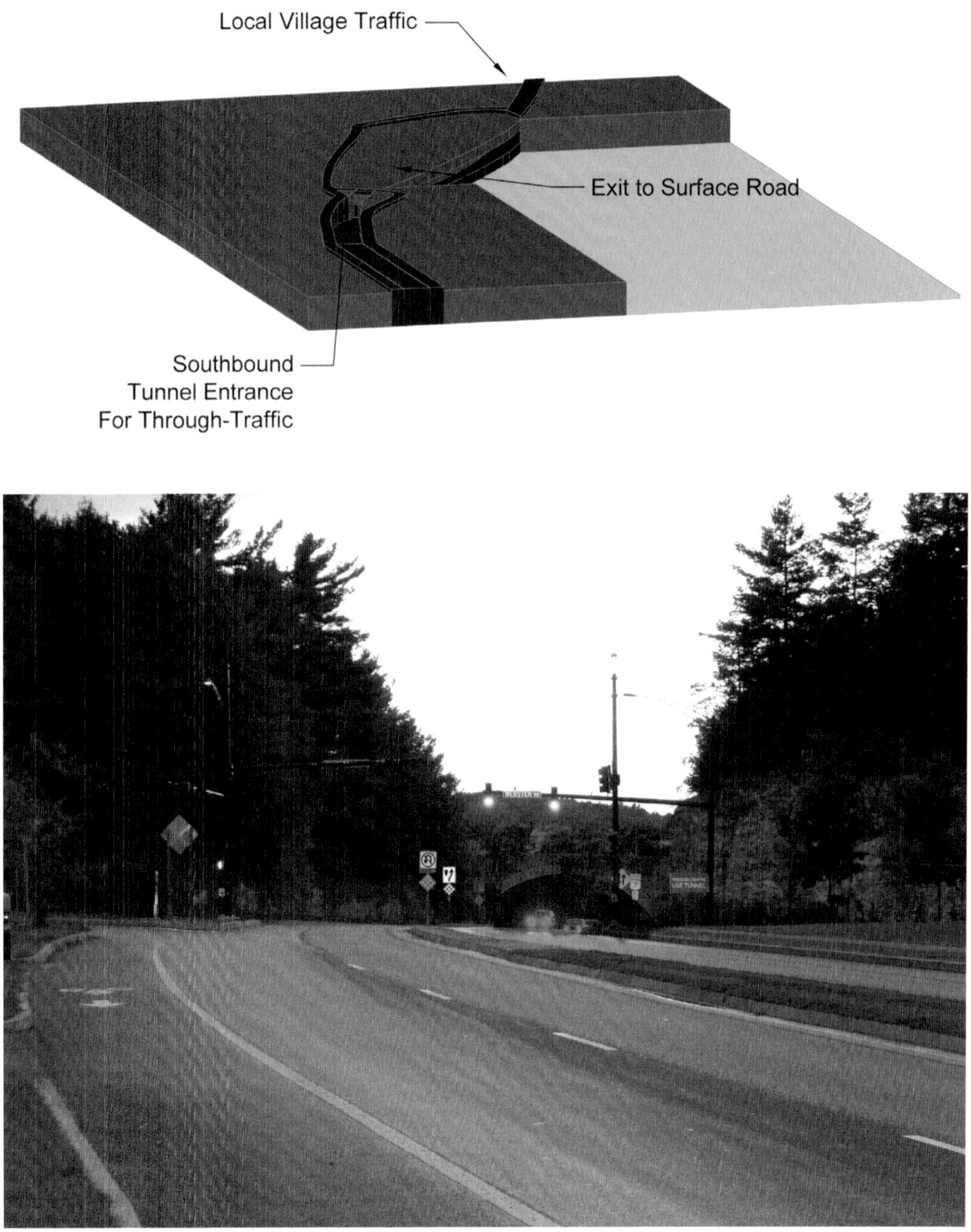

FIGURE 7.18 Top: conceptual massing model showing roadway and tunnel juxtaposition along U.S. Route 7 in the village of Shelburne, Vermont; bottom: conceptual rendering of the tunnel.

FIGURE 7.19 I-35 through Duluth, Minnesota.

Source: Image courtesy of the Minnesota Department of Transportation

the waterfront. The novel "cut and cover" plan enabled city and state officials to secure federal funding from the Federal Highway Administration.

Overall, roughly three miles (4.8 km) of highway were tunneled and a series of three parks was constructed, including the expansion of Leif Erikson Park and the creation of Lake Place that created new greenspace and connections from the city to the waterfront. Tunneling the roadways also protected the highways from icing from the lake and minimized air and noise pollution. In addition, excavated material from the tunneling was used to create new land along the shore of Lake Superior, the 2.6-mile (4.2 km) Lakewalk, and an offshore reef. The final phase of the project was completed in 1993 at a total cost of over $200 million (Gonzalez, 2006).

As of the writing of this book, the Duluth Waterfront Collective is advocating to replace a 1-mile stretch of Interstate-35 with a surface parkway that integrates the downtown and the waterfront. The proposal will free up 20 acres of land and incorporates many of the New Urbanist concepts mentioned earlier, including sidewalks and meridians lined by trees, a crosstown recreation trail, laned for bus rapid transit, and a rail corridor for tourist excursions.

Removing roadways

The complete removal of highways in favor of parks, greenspace, or at-grade boulevards is often met with much skepticism, as people fear other roads will become more congested and that commute and travel times will be longer. The truth of the matter is that much of the existing traffic would not even exist if the roads/highways were not constructed in the first place. As mentioned earlier, economists and transportation planners aptly refer to this as induced demand – the idea that, after supply increases, more of a good is consumed. New roads not only invite traffic that would otherwise seek different routes but also increase speed which initially decreases travel time but eventually increases the frequency of trips and the distance that people travel. Thus, in the long run, the amount of time people spend commuting remains relatively constant. This "fixed travel-time budget" was first pioneered by Yakov Zahavi when studying traffic patterns for the U.S. Department of Transportation in the 1970s and later advanced by physicist Cesare Marchetti in what was to become known as "Marchetti's Constant." Fixed travel-time budget centers on the notion "that regardless of culture, class, creed or access to advanced technology, the mean amount of time people all over the world spend in everyday transit is about an hour" (Turner, 2012). What this essentially means is, if you remove roads, people will eventually adapt their routes, routines, and modes of travel to stay within a "fixed" one-hour travel-time budget, thus ultimately reducing the demand on roads. Cairns et al. (2002) analyzed over 70 case studies of road space reallocation from 11 countries and found that significant reductions in traffic levels occurred after road removal, with an average reduction of around 11 percent.

The will of the community – Harbor Drive Freeway, Portland, Oregon

Harbor Drive was a four-lane highway that ran along the western edge of the Willamette River in Portland, Oregon. The freeway was completed in 1942 and the State Highway Department plans to widen it surfaced in 1968. However, the citizens' opposition to freeway expansion provoked Governor Tom McCall to appoint a task force to study other possibilities for Harbor Drive. In addition to widening proposals, the committee examined plans to relocate the highway or bury portions of it and cover it with a park. After much deliberation between citizens, private consultants, and traffic engineers, Governor McCall and Glenn Jackson of the Highway Department finally made the decision to remove the highway in 1974. Four years later, the freeway was replaced with Tom McCall Waterfront Park (renamed after the governor in 1984), which was expanded in 1999 through the addition of South Waterfront Park, ultimately creating roughly two miles (3.2 km) of the park along the waterfront in downtown Portland. The park revitalized downtown Portland, beautified the west bank of the Willamette River, and has been a catalyst for new development and more parks/greenways. For example, in December 2014, construction was completed of the 1.2-mile (1.9 km) South Waterfront Greenway in the Central District. As stated in Portland Parks and Recreation's South Waterfront Greenway Development Plan,

> the Greenway was designed to increase diversity of habitat types, create and enhance habitats that support native species while minimizing non-native species, provide, where feasible, off-channel habitat, such as coves and tributary mouths, protect and improve water quality in the Willamette River, provide stable riverbanks where needed to protect existing and future urban development and infrastructure, significant natural resources, and public safety, provide safe public access to the river where appropriate, and to enhance the aesthetic qualities of the river's edge.
>
> *(Portland Parks and Recreation, 2004)*

Portland was the first major city to remove an urban freeway. The project has served as a precedent for other cities planning highway removal and has spawned more highway

FIGURE 7.20 Top: Harbor Drive along the Willamette River in 1958; bottom: Harbor Drive removed and replaced with Governor Tom McCall Waterfront Park.

Source: Images courtesy of Portland Parks and Recreation, Portland, Oregon removal projects in the city itself. In the early 1990s, citizen advocate groups, particularly Sensible Transportation Options for People (STOP) and 1,000 Friends of Oregon, prevented the construction of the Western Bypass, which would have stretched from I-5 near I-205 south of Portland to I-5 north of Vancouver. The 1,000 Friends of Oregon proposed an alternative plan known as LUTRAQ (Land Use, Transportation, Air Quality) that included extension of the light rail and transit-oriented development rather than construction of the bypass. The new plan was shown to be superior to road expansion in terms of emissions/air quality and traffic congestion. It was implemented and became a major component of Portland's long-term land use master plan. As of the writing of this book, Riverfront for People is still advocating for the removal of I-5 along the east bank of the Willamette River and the No More Freeway Expansions Coalition is working to block the expansion of I-5 into Portland's Rose Quarter.

Mother Nature always wins – Embarcadero and Central freeways, San Francisco, California

San Francisco's Embarcadero and Central freeways were elevated, double-deck roads that were part of a larger concept for a series of interconnected freeways through the city known as the "Trafficways Plan." The Embarcadero extended 1.2 miles (1.9 km) along the bay, essentially dividing the downtown from the waterfront, while the Central bisected the Hayes Valley neighborhood. Opposition to the Trafficways Plan started to build before any freeway was even constructed and eventually mounted to a point that forced the cancellation of seven of the ten planned freeways, including the extension of the Embarcadero to the Golden Gate Bridge and the northern and western extensions of the Central.

Although there was ample support for the removal of the freeways in the 1980s (especially the Embarcadero, which was on prime real estate), it would take an act of nature – the 7.1-magnitude Loma Prieta earthquake of 1989 – to bring the Embarcadero and Central freeways to their knees. In addition to causing portions of the Bay Bridge and the Cypress Freeway in Oakland to collapse, the earthquake damaged both the Embarcadero and the Central to a point where they had to be closed. The damage, along with the fact that traffic seemed to be adapting without the freeways, spawned new support for demolishing the massive concrete freeways that were often compared to prison walls. In 1991, the Embarcadero Freeway was replaced with a surface boulevard – now known simply as "the Embarcadero" – that is flanked by a pedestrian promenade, parkland, open space, and recreational paths.

Replacing the double-decked freeway with a boulevard raised property values in the adjoining neighborhoods by 300 percent and stimulated development, including the restoration of the Ferry Building, the extension of the streetcar system to the waterfront, new retail and office space, and the restoration of park space (The Preservation Institute, n.d.). More importantly, San Francisco was reconnected with the waterfront, the very resource that influenced much of the city's early development. Portions of the Central Freeway were removed in 1992 and replaced with a surface boulevard (Octavia Boulevard); the final section was demolished in 2003. Octavia Boulevard contains four lanes for through-traffic and two service lanes for local traffic and pedestrians, all flanked by landscaping down the central meridian and along the edges, buffering the pedestrian paths and service lanes. Since the boulevard's completion, the Hayes Valley neighborhood has undergone a boom in new housing, business, retail, and park space.

Inner Loop, the "Big Fill," Rochester, New York

Rochester, New York, is a mid-sized city located 8 miles (12.9 km) south of Lake Ontario on the Genesee River. The first inhabitants in the area were the Seneca Indians, who were one of the Six Nations of the Iroquois Confederacy. The first permanent white settlement was in 1789 with the establishment of a grist mill on the upper falls of the Genesee River. Rochester was officially incorporated as a city in 1834 and prospered as a flour milling city in the mid-nineteenth century due to its proximity to rail lines, waterpower, and the Erie Canal (Britannica, 2023). When the flour milling industry moved west, Rochester became known in the nursery industry by producing and shipping seeds, hence changing its moniker from the "Flour City" to the "Flower City." In the late 1890s, George Eastman, John Jabob Bausch, and Henry Lomb developed photographic and optical technologies, subsequently forming Eastman Kodak and Bausch and Lomb, respectively. Between 1834 and 1900, Rochester grew from 12,000 to 163,000 (Grier, 2018), and after World War II, Rochester's population continued to grow to reach a peak in 1950 of 332,448 (BiggestUSCities.com, n.d.; U.S. Census Data, 2022).

As a result of the Federal Aid Highway Act of 1956, Rochester saw the construction of the Interstates 390, 490, 590, and the Inner Loop – a 2.68-mile (4.3 km) expressway encircling Rochester's central business district. To make room for the Inner Loop, officials removed 1,300 businesses and residences (FHWA, n.d.b). Although the highways were regarded as progressive at that time, history has demonstrated that the Inner Loop served as a major barrier between the city center and adjacent neighborhoods, including Upper Falls and Marketplace Heights to the north, Grove Place, East End, and Wadsworth Square to the east, South Wedge and Corn Hill to the south, and Mayor's Heights, Susan B. Anthony, Brown Square, and High Falls to the west. As an article from City Lab points out,

> Many places grappling with the damage from urban freeways remain stuck in the 1950s paradigm where planners and engineers misunderstood the relationship between the form of a city and its traffic. They thought they had to reconstruct the city to accommodate an anticipated flood of traffic in the future. What they didn't understand is that it was the very change that they advocated that would create the vehicular traffic their models predicted.
>
> *(Garrick, 2016)*

Many of the neighborhoods that were fragmented by the Inner Loop and the I-490 were already victims of the Federal Housing Administration (FHA) and the Home Owner Loan Corporation's (HOLC's) color-coding mapping practice that identified areas of "creditworthiness" for receiving mortgages. Red zones were deemed hazardous since they were infiltrated by the "undesirable population" and thus were precluded from the issuance of loans. Red zones also received significantly less funding for infrastructure, education, and public works projects. As evident in many cities, Rochester's red-lined neighborhoods – which included historic Ward 3 (west and southwest of the Inner Loop) and Ward 7 (north of the Inner Loop) – had the highest percentages of African American and foreign national citizens, respectively (Thompson, 2017). Over time, these neighborhoods fulfilled the prophecies that they were cast into by the discriminatory zoning practices, resulting in the decline of real estate values, decayed infrastructure, high rates of unemployment, poverty, crime, and a limited means of ever being able to relocate out of the neighborhoods (Thompson, 2017). Indeed, in addition to being "zoned out" of society, the construction of the Inner Loop and I-490 further exacerbated the isolation by creating a physical barrier between the adjacent neighborhoods and the city. The Inner Loop has often been described as a "noose around the neck" of the city, and this seems to be true in more ways than one. From its peak population of over 300,000 in 1950, Rochester's population decreased to just over 210,000 in 2020 (U.S. Census Data, 2022). Traffic studies showed the western portions of the Inner Loop had about 46,500 trips a day while the eastern portion only showed 10,500 trips per day (CNU, n.d.). Due to these low numbers, the eastern portion was the first to be planned for removal. Although planning to remove the eastern section began back as far as 1990, the city was awarded a $17.7 million Transportation Investment Generating Economic Recover (TIGER) grant in 2012 from the U.S. Department of Transportation (USDOT). Construction began in 2014 and was completed in 2017.

The project, which has become known as "the Big Fill," freed up 5.7 acres (2.3 hectares) of land and has been infilled with a mix of commercial and residential space (FHWA, n.d.b). South Union Street – once a one-way street that bordered the Inner Loop – was transformed into a walkable, two-way, tree-lined boulevard with designated bicycle lanes, fostering pedestrian connections between the eastern neighborhoods and the central business area. As of 2023, there have been more than 500 housing units constructed, 60 percent of which are occupied by people earning below the median income level of the area (City of Rochester, n.d.). Economically, the project created 170 permanent jobs, over 2,000 construction jobs, and over $229 million in economic development from new tax revenues and private investment (FHWA, n.d.b). As of 2023, plans for removing and redeveloping the northern portion of the Inner Loop are in design.

Niagara Scenic Parkway, Niagara Falls, New York

The Niagara River flows north for 37 miles (59.5 km) from Lake Erie to Lake Ontario, plunging about 165 feet (50 meters) at Niagara Falls. Although the falls, the river, and the waterfront have served as the nexus of the Buffalo – Niagara region and are the very foundation on which the region was built, commercialization to exploit the flood of tourists, the construction of hydroelectric power facilities, and the subsequent spread of industrial manufacturing have, to some extent, overshadowed the natural beauty and mysticism of the water.

The 2,400-megawatt Robert Moses Niagara Power Plant and Lewiston Pump Generating Station were constructed between 1958 and 1961. Robert Moses, then chairman of the New York Power Authority, was responsible for various highway and bridge projects from the Great Depression to the 1960s. In addition to several projects in and around the New York City/Long Island region, he constructed a four-lane parkway along the Niagara River at the same time as the power plant was being built. When it was finished, the parkway ran for almost 20 miles (32 km), beginning at the terminus of the La Salle Expressway, extending through the city of Niagara Falls, and terminating at Lake Ontario. In the late 1980s, a portion of the parkway was removed in the city of Niagara Falls to restore Frederick Law Olmsted's vision for Niagara Falls State Park; the remaining parkway was then split into two segments, with traffic rerouted through the city.

Although Moses envisioned a "drive-by" experience of Niagara Falls, the reality is that the parkway serves as a barrier to the river and the gorge. Throughout the 1980s and 1990s, various proposals and studies were conducted to restore the Niagara Gorge waterfront. The Niagara Heritage Partnership was formed with a mission of "preserving and restoring the Niagara

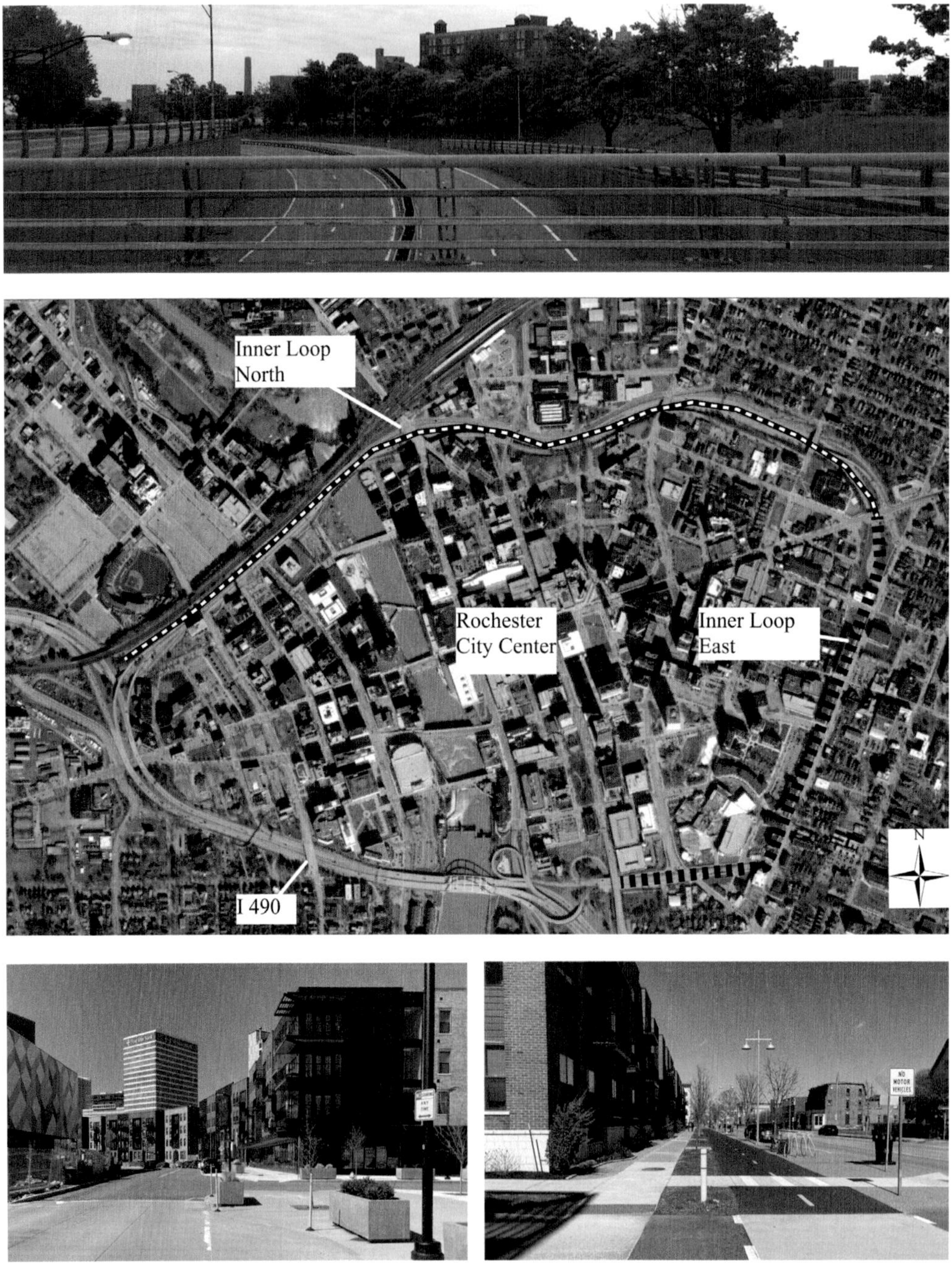

FIGURE 7.21 Top: existing image of the north section of the Inner Loop; middle: map on Inner Loop, imagery courtesy of Esri, DigitalGlobe, GeoEye, i-cubed, USDA FSA, USGS, AEX, Getmapping, Aerogrid, IGN, IGP, swisstopo, and the GIS User Community; Bottom: New infill mixed-use development/housing on removed east section of Inner Loop.

Region's natural environment and encouraging socially responsible development" (Niagara Heritage Partnership, n.d.). In 2000, the partnership advocated for the removal of 6.5 miles (10.5 km) of the parkway between the city of Niagara Falls and Lewiston, New York. In its place, it proposed a series of pedestrian/bicycling paths and revegetating the shore along the gorge. Also around this time, the city of Niagara Falls contracted the Urban Design Project at the University at Buffalo's School of Architecture and Planning, Foit-Albert Associates, and the Waterfront Regeneration Trust to assess the waterfront planning of previous years and implement a future plan. The Urban Design Project also had the opportunity to consider portions of the project as class design projects. In 2001, the state conducted a pilot program by closing the two lanes closest to the gorge to traffic but keeping them active as pedestrian paths. After this pilot program was shown to be a success, the state permanently closed two lanes from Cedar Avenue in Niagara Falls to I-190 in Lewiston.

In November 2014, ground was broken on an $18.8-million project to remove two miles of the southbound lanes along the Upper River Gorge from John B. Daly Boulevard to Old Falls Street. As of 2020, approximately 140 acres (56.7 hectares) of land was converted to park space, Whirlpool and Third Streets were reconfigured, the Niagara Gorge State Park was retrofitted,

FIGURE 7.22 Top left: southern portion of Robert Moses Parkway along Niagara River; top right: proposal to remove lanes and replace with a riparian buffer, pedestrian trails, and a monorail; bottom left: northern portion of Robert Moses Parkway lanes along Niagara Gorge; bottom left: completed removal of parkway along Niagara Gorge.

the Robert Moses Viaduct, abandoned railroad trestle, and pedestrian bridge were removed, and the area was renamed the Niagara Scenic Parkway. Neighborhoods on adjacent streets are now directly connected to the parkway and the gorge through a new trail network for hiking, biking, fishing, and cross-country skiing.

There's always potential – I-190, Buffalo, New York

The Niagara River is also separated from the community to the south of Niagara Falls, heading toward Buffalo, New York, where the I-190 occupies a 6-mile (9.7 km) stretch of land that was once the Erie Canal. I-190 was constructed in 1960 when shipping on the canal lost its competitive edge to interstate trucking and rail. Access to the river is blocked near Scajaquada Creek, where the I-190 and the I-198 intersect, forming an elevated spaghetti mess of feeder ramps and concrete. The I-198 (also known as the Scajaquada Parkway) occupies a portion of land that was once part of Frederick Law Olmsted's Emerald Necklace – a parkway in the shape of a wide arc across the northern portion of the city that linked three major parks: the Park (currently Delaware Park), the Parade (currently Martin Luther King Jr. Park), and the Front (currently Front Park). In addition to the I-190 and the Scajaquada Parkway, other portions of Olmsted's original conception have been interrupted by the Humboldt Parkway and portions of the Kensington Expressway (NYS Route 33). Riverside Park and Front Park are also disengaged from the river.

As of the writing of this book, Buffalo is the recipient of a $55,597,500 grant from the U.S. DOT's *Reconnecting Communities and Neighborhoods Grant Program* to cap portions of the Kensington Expressway (Route 33). In addition, the Scajaquada Corridor Coalition and the Buffalo Skyway Project have promoted and solicited designs for restoring the Scajaquada Parkway to realign with Olmsted's vision and repurposing the elevated portion of Route 5 known as the Skyway, respectively. In theory, all these projects could connect with the work done on the Robert Moses Parkway and could create an international, 37-mile greenway along both sides of the Niagara River, from Lake Erie to Lake Ontario (Richards, 2001).

For many years, Buffalo and the entire Niagara region have searched for a magic bullet that will rejuvenate their economy back to their heyday when shipping on the Erie Canal was king. Most of these proposed antidotes have failed due to a lack of capital, poor leadership, lackluster vision, and/or a lack of support. And shy of the Buffalo Bills winning the Super Bowl or the Buffalo Sabres winning the Stanley Cup, the region is going to have to look deeper for solutions. As discussed throughout this book, when in doubt, look to nature. The Niagara River has served as the region's jewel throughout history with respect to nature, development, culture, transport, and industry. Its role as an international nexus not only physically binds and defines the region but also provides a unique setting where different cultures interact, work, and celebrate. Regreening the Niagara River and the shores of Lakes Erie and Ontario would sustain the Buffalo–Niagara region in the twenty-first century and eliminate the poor, rust-belt perception that the region seems unable to shake off. Instead of being known for the aging industry, abandoned grain elevators, the Love Canal disaster, the twin-span failed proposal for the Peace Bridge, cold and snowy weather, gray skies, and Scott Norwood's infamous wide-right field goal attempt in Super Bowl XXV, the Niagara River Greenway has potential to rebrand the area as a region of parks. In addition to the environmental, social, and economic benefits of a north–south lineage connecting two of the Great Lakes, the greenway also has a

FIGURE 7.23 Top: I-190 along the Niagara River in Buffalo; bottom: conceptual rendering of the proposal to remove the I-190.

FIGURE 7.24 Top: existing intersection of I-190 and the Scajaquada Parkway (198) in Buffalo, New York. Bottom: proposed restored Scajaquada Creek Corridor. Orthophotography courtesy of Esri, DigitalGlobe, GeoEye, i-cubed, USDA FSA, USGS, AEX, Getmapping, Aerogrid, IGN, IGP, swisstopo, and the GIS User Community; bottom: proposal to remove I-190, courtesy of Buffalo Olmsted Parks Conservancy/Mark Nowaczyk, Cannon Design.

FIGURE 7.25 Proposal for a "Niagara River Greenway." Light Gray Canvas Map courtesy of Esri, DeLorme, HERE, MapmyIndia.

unique opportunity to forge an international partnership with Canada by networking greenspace across the border.

Invisible roads and flying cars

At the end of the movie, *Back to the Future*, the Doc proclaims to Marty: "Roads? Where we're going, we don't need roads." As with many of his movies, it seems Steven Spielberg foresaw the future. The Korea Advanced Institute of Science and Technology (KAIST) has developed a road-powered electric vehicle network called OLEV (Online Electric Vehicles). The road transfers energy to the vehicles via magnetic resonance from a series of cables buried under the surface. This technology could, in theory, power hover vehicles that could levitate over any surface, eliminating the need for any impermeable road surfaces at all – that is, "the invisible road."

In the second movie of *Back to the Future*, Marty McFly outruns Biff and his gang of thugs using a hover board that floats over the ground surface. I honestly never thought I would be alive to see this technology come to fruition, but, in 2014, ArxPax co-founders Greg and Jill Henderson demonstrated the Hendo Hover – the world's first hoverboard. The technology uses their patented Magnetic Field Architecture (MFA) that uses electromagnetic energy with two downward-firing engines that create an opposing magnetic field to allow the board to levitate off the ground. The science is based on Lenz's Law, and as described by Henderson:

> Lenz's Law states that if you move a magnetic field relative to a conductor, you generate an eddy current in that conductor. If you're moving it relative to a passive surface, you don't need magnetic floors. The primary magnetic field in our hover engines – and that's where the secret sauce is – focuses electromagnetic energy downward. . . . Our hover engine is over a conductive surface, which can be many different substances, and it generates a primary magnetic field. That field creates the eddy currents in the surface, and then the eddy current, following Lenz's Law again, generates a secondary magnetic field. That mirrors the primary magnetic field. And just like two north poles of two different magnets, they repel one another and generates lift.
>
> *(Micaleff, 2015)*

Interestingly, Henderson – who is an architect – originally researched the technology to hover buildings to make them more resilient to earthquakes. Lexus has also created a hover board called the Slide that uses two reservoirs of liquid nitrogen that keep the superconducting material at −197 degrees Fahrenheit. The board then floats over a track that contains magnets.

And then, of course, there is the prospect of flying vehicles. Before the Doc and Marty flew away at the end of *Back to the Future* in the retrofitted Delorean, images of flying cars/vehicles can be seen in *The Man with the Golden Gun*, *Blade Runner*, the *Fifth Element*, and of course, *Star Wars*, although these are technically referred to as "air speeders." Although there have been several iterations and designs dating back to the early twentieth century, there have been recent advances in both flying cars and electrically powered vertical takeoff and landing aircraft, also referred to as "eVTOL." Flying cars are a combination of a car and plane while eVTOLs generally do not have wheels and take off more like a drone or helicopter. The Jeston ONE eVTOL developed by the Swedish startup Jetson AB became the first of its kind to be certified for personal flight in uncontrolled air space in Italy. The PAL-V Liberty can be driven as a car or flown as a gyroplane simply by unfolding the rotors that collapse and are housed above the vehicle. When in drive mode, the vehicle can travel up to 100 mph (161 km/h) and has a range of 817 miles (1,315 km); in fly mode, the craft can travel 250–310 miles (402–499 km) at a maximum altitude of 11,000 feet (3,353 meters) (PAL-V, n.d.). Whether society is ready for another dimension of travel is still unclear. If flying vehicles become ubiquitous, are we just transferring congestion from one plane to another? Given that 1.35 million people are killed in vehicle-related accidents per year (CDC, n.d.), will flying vehicles add to this total, or will they provide congestion relief that makes travel safer? Or will it just be available to a select few who can afford it, creating more disparity and inequity that further fragments society?

References

Barone, E. (2022, July 5). $5 gas isn't changing how often Americans drive. *Time*. https://time.com/6193830/high-gas-prices-summer-travel/

Benedict, M. A., & McMahon, E. T. (2006). *Green infrastructure: Linking landscapes and community* (p. 258). Island Press.

BiggestUScities.com. (n.d.). *Rochester, New York population history 1840–2021*. www.biggestuscities.com/city/rochester-new-york

Blumgart, J. (2022, February 28). *Why the concept of induced demand is a hard sell*. Governing. The Future of States and Localities. https://www.governing.com/now/why-the-concept-of-induced-demand-is-a-hard-sell

Britannica. (2023). *Geography and travel: Rochester, New York*. www.britannica.com/place/Rochester-New-York

Cairns, S., Atkins, S., & Goodwin, P. (2002, March). Disappearing traffic? The story so far. *Municipal Engineer, 1*, 13–22.

Calthorpe, P. (1993). *The next American metropolis: Ecology, community, and the American Dream* (p. 110). Princeton Architectural Press.

Centers for Disease Control and Prevention (CDC). (n.d.). *Road traffic injuries and deaths – A global problem*. www.cdc.gov/injury/features/global-road-safety/index.html

City of Geneva Local Planning Committee. (2017). DRI Geneva.

City of Rochester. (n.d.). *Inner Loop East: From barrier to beautiful*. www.cityofrochester.gov/InnerLoopEast/

City Parks Alliance. (n.d.). *Portland's Green Streets program*. https://cityparksalliance.org/resource/portland-green-street-program/

The Confluence Project. (n.d.). *The Vancouver land bridge*. www.confluenceproject.org/river-site/vancouver-land-bridge/

Congress for the New Urbanism (CNU). (n.d.). *I-490 Inner Loop*. www.cnu.org/what-we-do/build-great-places/i-490-inner-loop

Curtis, P. D., & Hedlund, J. M. (2005). *Reducing deer – vehicle crashes*. Cornell University Cooperative Extension. http://wildlifecontrol.info/pubs/Documents/Deer/Deer-Vehicle_factsheet1.pdf

Develtere, A., & Leblud, J. (2018, September 26). *What effects would staggering work hours have congestion and road safety at peak times*. Vias Institute. https://www.vias.be/publications/De%20impact%20van%20flexibele%20werktijden%20op%20de%20congestie%20en%20de%20verkeersveiligheid%20tijdens%20de%20spitsuren/What%20effects%20would%20...%20(Summary).pdf

Dunham-Jones, E., & Williamson, J. (2011). *Retrofitting suburbia: Urban design solutions for redesigning suburbs*. John Wiley and Sons.

Federal Highway Administration (FHWA). (n.d.a). *Road function classifications*. https://safety.fhwa.dot.gov/speedmgt/data_facts/docs/rd_func_class_1_42.pdf

Federal Highway Administration (FHWA). (n.d.b). *Rochester Inner Loop East, New York, A freeway to boulevard*. Center for Innovative Finance Support. www.fhwa.dot.gov/ipd/project_profiles/ny_freeway_to_boulvard_rochester.aspx

Friends of the Hollywood Central Park. (n.d.). *Facts and benefits*. https://hollywoodcentralpark.org/home

Garrick, N. (2016, September 1). *Burying a 1950s planning disaster*. Bloomberg: City Lab. www.bloomberg.com/news/articles/2016-09-01/burying-rochester-s-inner-loop-a-1950s-era-planning-disaster

Gelinas, N. (2007). Lessons of Boston's Big Dig. *City Journal*. www.city-journal.org/article/lessons-of-bostons-big-dig
Giuliano, G., & Golab, T. F. (1990). Staggered work hours for traffic management: A case study. *Transportation Research Record, 1280*(46–58). https://onlinepubs.trb.org/Onlinepubs/trr/1990/1280/1280-006.pdf
Goldman, N. (2012, July 12). 7 Things that cost less than the Big Dig. *WBUR News*. www.wbur.org/2012/07/12/7-things-that-cost-less-than-the-big-dig
Gonzalez, P. (2006). *MNDOT celebrates interstate highway system's 50th anniversary*. Minnesota Department of Transportation. www.lrl.mn.gov/docs/2006/Other/060481.pdf
Greiman, V., & Warburton, R. D. H. (2009). *Deconstructing the big dig: Best practices for mega-project cost estimating*. Paper presented at PMI® Global Congress 2009 – North America, Orlando, FL. Newtown Square, PA: Project Management Institute.
Grier, E. (2018). *Negroes in five New York cities: A study of problems, achievements, and trends*. Forgotten Books.
Gross, A. (2021, July 15). *Pandemic pause: Americans cut driving nearly in half in early stage of COVID crisis*. American Automobile Association (AAA). https://newsroom.aaa.com/2021/07/travel-before-and-during-covid-july-2021/
Gupta, M., Pawar, N. M., & Velaga, N. R. (2021). Impact of lockdown and change in mobility patterns on road fatalities during COVID-19 pandemic. *International Journal of Transportation Research, 13*(5–6), 447–460.
Halpern, J. (2005, September). The next big thing. *Boston Magazine*. www.bostonmagazine.com/2006/05/the-next-big-thing/
Library of Congress (LOC). (n.d.). *Who invented the automobile*? www.loc.gov/everyday-mysteries/motor-vehicles-aeronautics-astronautics/item/who-invented-the-automobile/
Massachusetts Department of Transportation (MA DOT). (n.d.). *The big dig – project background*. www.mass.gov/info-details/the-big-dig-project-background
Micaleff, K. (2015). *How Arx Pax's magnetic field architecture can levitate hoverboards – and buildings*. Autodesk. www.autodesk.com/design-make/articles/magnetic-field-architecture-levitates-hoverboards-and-buildings
National Association of City Transportation Officials (NACTO). (n.d.). *Urban street design guide*. https://nacto.org/publication/urban-street-design-guide/street-design-elements/sidewalks/
The National Geographic Society (NGS). (1975). *We Americans* (p. 314). National Geographic Society.
National Wildlife Federation (NWF). (n.d.). *Critical paths*. www.nwf.org/RegionalCenters/~/media/PDFs/Regional/Northeast/NRC_CriticalPaths_Brochure.ashx#:~:text=of%20native%20species.-,Safe%20Crossings%20for%20Wildlife%20and%20People,and%20to%20minimize%20habitat%20fragmentation
Niagara Heritage Partnership. (n.d.). *Home page*. www.niagaraheritage.org/
Nogueira, X. R., & Mennis, J. (2019, October). The effect of brick and granite block paving materials on traffic speed. *International Journal of Environmental Research and Public Health, 16*(19), 3704. http://doi.org/10.3390/ijerph16193704
PAL-V. (n.d.). *Liberty*. www.pal-v.com/en/liberty
Portland Parks and Recreation. (2004, December 8). *South waterfront greenway development plan: Phase 1*. www.portlandoregon.gov/parks/article/175890
The Preservation Institute. (n.d.). *Removing freeways, restoring cities*. San Francisco, CA: The Embarcadero Freeway. www.preservenet.com/freeways/FreewaysEmbarcadero.html
Project for Public Spaces (PPS). (2008). *Traffic calming 101*. www.pps.org/article/livememtraffic
Qian, X. (2022). *Telecommuting during COVID-19: How does it shape the future workplace and workforce*. Minnesota Department of Transportation (MNDOT). https://www.dot.state.mn.us/research/reports/2022/202205.pdf
Richards, M. A. (2001). *Sustainable networks: Regreening the Buffalo-Niagara corridor* [Masters thesis, University at Buffalo]. University at Buffalo Abbott Library Architecture and Planning Thesis Collection, Thesis A7 2001 R52.
Seattle Public Utilities. (n.d.). *Street edge alternative project*. www.seattle.gov/utilities/neighborhood-projects/street-edge-alternatives
Smart Growth America. (n.d.). *Complete streets coalition*. https://smartgrowthamerica.org/program/national-complete-streets-coalition/
Steiner, C. (2009). *$20 per gallon: How the inevitable rise in the price of gasoline will change our lives for the better*. Grand Central Publishing.

Thompson, B. (2017, May 3). *Race and place in the flower city: A case study of perpetual marginalization through urban planning*. Medium. https://medium.com/@bt8002a/race-and-place-in-the-flower-city-300e8aac4d07
Turner, C. (2012, November 14). *For pedestrians, cities have become the wilderness*. Bloomberg City Lab. www.citylab.com/commute/2012/11/pedestrians-cities-have-become-wilderness/3878/
United States Department of Transportation (U.S. DOT). (2023). *Biden-Harris administration announces first-ever awards from program to reconnect communities*. https://www.transportation.gov/briefing-room/biden-harris-administration-announces-first-ever-awards-program-reconnect-communities
United Staes Department of Transportation (U.S. DOT). (2024). *Reconnecting communities and neighborhoods (RCN) announcement*. https://www.transportation.gov/grants/rcnprogram/rcn-announcement
United States Department of Transportation Bureau of Transportation Statistics (U.S. DOT). (2020). *Public road and street mileage in the United States by type of surface*. www.bts.gov/content/public-road-and-street-mileage-united-states-type-surfacea
U.S. Census Data. (2022). *Rochester, New York*. www.census.gov/quickfacts/fact/table/rochestercitynewyork/PST045222
Warren, R. (1998). *The urban oasis* (p. 18). McGraw Hill.
World Health Organization. (2020). *Number of registered vehicles*. www.who.int/data/gho/indicator-metadata-registry/imr-details/194

8

CORRIDORS AND GREENWAYS

National Geographic (n.d.) defines a landscape as part of the earth's surface that can be viewed at one time from one place that consists of geographical features that are characteristic of a particular area. Forman and Wilson (1995) define a landscape "as a mosaic of repeating patterns of patches and corridors and the matrix in which these elements sit." In landscape ecology, a patch is defined as a homogenous area of habitat differing from its surroundings; a corridor is a more linear strip of land that differs from the adjacent land surrounding it; and a matrix is the land cover that is dominant over the majority of the land surface that includes the collection of patches and corridors interconnected throughout the landscape. Corridors can be narrow lineages with no interiors, can be wider lineages with interior and edge conditions, or may consist of smaller, scattered stepping-stone patches of habitat (Hellmund & Smith, 2006, p. 46).

For the purposes of this book, greenways are corridors of land or water designed for multiple functions, including wildlife conservation, stormwater management, recreation, and social/cultural preservation. Greenways can vary in nomenclature and are often referred to as ecological/wildlife corridors, greenbelts, conservation corridors, vegetative/riparian buffers, parkways, and linear parks, to name a few. A greenway corridor can vary in scale from spanning entire regions to comprising no more than an alley between two buildings. Like roadways, parking surfaces, and rooftops, greenways can transform dead areas into multifunctional spaces that can unite fragmented habitats and/or provide uninterrupted conduits of land that can be connected to each other and/or connect with other parks and greenspace to form a contiguous network of green infrastructure on community, regional, and even continental scales.

Indeed, greenways can provide ecological, social, and economic benefits. Ecologically, greenways provide vegetation that can filter stormwater runoff, filter airborne particles and gases, and provide riparian buffers for wildlife. Socially, greenways provide community connections; space and pathways for physical activity, active transportation, and micromobility; and green space in dense urban areas that lack parks and open space. Economically, greenways can increase adjacent land value, attract tourists, create jobs, and save money by providing a cost-effective, natural way to clean the environment. However, it should be noted that increasing land values can be a double-edged sword. A property that is reassessed at a higher tax

DOI: 10.4324/9781032668666-8

rate could potentially price out and displace businesses and residents in some communities, particularly if people from higher-income brackets move into the neighborhoods and renovate or completely rebuild the property, thus further driving up values and fostering patterns of spatial-social restructuring and gentrification. Transparency, community involvement in the planning and decision-making process, and anti-displacement policies that ensure tenant protections and the availability of affordable housing should be implemented with the planning of any greenway.

Greenways can be designed as linear parks/parkways or beltways that encircle a city, town, or region and/or can be implemented in or along rail corridors, utility corridors, alley corridors, or waterway corridors (streams, lakes, rivers, canals, lagoons, oceans). Greenways can connect patches of habitat, parks, and natural areas to form a cohesive network of green space on community, regional, or even continental scales.

Parkways

An early form of the greenway was the parkway. In the 1860s, landscape architect Frederick Law Olmsted advocated for linear greenspace as a conduit to link parks and neighborhoods in urban areas. Between 1878 and 1890, he designed Boston's Emerald Necklace, which is an 1,100-acre (445-hectare) chain of parkways that link seven parks in Boston and Brookline, including Franklin Park, Arnold Arboretum, Jamaica Pond, Olmsted Park, The Riverway, Back Bay Fens, and Commonwealth Avenue Mall/Public Garden/Boston Common. In 1868, the U.S. District Attorney of Northern New York invited Olmsted to Buffalo to discuss the setting aside of land for public recreation. Olmsted, along with his partner Calvert Vaux, proposed three parks in the northern part of the city, including The Park (currently Delaware Park), The Front (currently Front Park), and The Parade (currently Martin Luther King, Jr. Park). The parks were connected through a series of parkways and avenues. The parkways were described as "sylvan tributaries" that enabled people to travel in a complete circle under a green canopy. Olmsted defined the parkways as "broad thoroughfares planted with trees and designed with special reference to recreation as well as for common traffic" (Kowsky, 1987). By 1870, there were close to seven park approaches, including Humboldt Parkway, which joined the Park to the Parade, Lincoln Parkway, Chapin Parkway, Bidwell Parkway, The Avenue (currently Richmond Avenue), Porter Avenue, and Fillmore Avenue. In addition to these more monumental approaches to the park, several existing city streets assumed new importance as residential avenues leading to the parkways. Joseph Ellicott's design for Buffalo's radial street grid combined with Olmsted's parkway system formed a symbiotic, spoke-and-wheel arrangement that demonstrated how an industrial city could be made more livable by establishing accessible thoroughfares and parks that permeated the existing residential and industrial grid, ultimately bringing spaciousness and the natural elements to the people throughout the city.

Later additions to the park system included Day's Park, The Terrace, Masten Place, Bennett Place, South Park, Cazenovia Park, and Riverside Park. In the 1960s, construction of Interstate 190 along the Niagara River and New York State 33 and 198 through the Scajaquada Creek Corridor and Humboldt Parkway interrupted the original Olmsted conception. Riverside and Front Parks now are currently disengaged from the river, and expressways for motorized vehicles currently dominate the once pedestrian-oriented parkways. However, as discussed in Chapter 7, there are currently plans underway to cap and/or remove portions of these highways.

FIGURE 8.1 Olmsted parks and parkways in Buffalo, New York, 1914.

Source: Map courtesy of University at Buffalo Libraries

Green belts

Another form of the greenways is green belts, which are bands of natural and agricultural land surrounding a metropolitan area that are conserved and delineate a growth boundary. As of 2022, England estimates there are 4,047,957 acres (1,638,150 hectares) of land designated as green belts, which accounts for approximately 12.6 percent of the land (GOV.UK, 2022). In addition to the London Metropolitan Green Belt discussed in the next section, other notable green belts in the United Kingdom include the North West Green Belt, the South and West Yorkshire Greenbelt, and the West Midlands Green Belt. Other countries also have notable green belts, including the Ottawa and Golden Horseshoe green belts in Ontario, Canada, the Royal National City Park in Stockholm, Sweden, the European Green Belt along the route of the former Iron Curtain, and the German Green Belt along the former border of East and West Germany, to name a few. Although parkways and urban growth boundaries are more common in the United States, there are still several corridors that use the "green belt" nomenclature, including the three green belt cities created from Franklin D. Roosevelt's New Deal (Greenbelt, Maryland, Greenhills, Ohio, and Greendale, Wisconsin), along with Boulder, Colorado, and 70,000 acres of land conserved by the Greenbelt Alliance in the San Francisco Bay area in California.

London Metropolitan Green Belt

Although the notion of preserving land around towns in the United Kingdom can be traced back to the Old Testament and to Elizabeth I in 1580, Ebenezer Howard's Garden City movement in the late nineteenth century catalyzed a series of proposals and the birth of regional planning, including the Town and Country Planning Association in 1919 and the Greater London Regional Planning Committee in 1927. After reviewing the merits of the proposals, concerns of open space being lost, and political change, the London County Council green belt scheme was launched in 1935 with a vision "to provide a reserve supply of open spaces and recreational areas and to establish a green belt or girdle of open lands, not necessarily continuous, but as readily accessible from the urbanized area of London as practical" (Thomas, 1963, pp. 17–18). The London County Council made loans available to county authorities that enabled them to purchase land. The Green Belt Act of 1938 allowed permanent protection of the purchased land, and the Town and Country Planning Act of 1947 permitted local authorities to designate land as part of the Green Belt in their town plans (Bridgers, 2017). Today, the London Metropolitan Green Belt (Green Belt) is 1,200,932 acres (486,000 hectares) and contains portions of the outer London Boroughs, Hertfordshire, Essex, Kent, Surrey, Berkshire, Buckinghamshire, and Bedfordshire (LGBC, n.d.). Twenty-six percent of the land is environmentally protected, including recreational land, parks, and woodlands, and roughly two-thirds is agricultural (Bridgers, 2017).

As of 2019, London's population reached 8.982 million, and it is expected to reach 11 million by 2050. Although many people in London agree that more housing is needed to meet the population increase, opinions are divided as to where the new housing should be built. London's Mayor Sadiq Khan and the London Green Belt Council are committed to preserving and enhancing the Green Belt due to its critical role in preventing sprawl, mitigating climate change, controlling stormwater, and providing habitat, agriculture, and recreation. Although the boroughs can de-designate portions of the Green Belt in their local plans, both the London Plan and the National Planning Policy Framework emphasize that this should only be done in exceptional circumstances. The London Plan also states that roughly 300,000 new homes could be built by utilizing brownfield sites in and around London without having to encroach on the Green Belt (London Assembly, 2021). The Ministry of Housing, Communities, and Local Government, on the other hand, is pressuring local councils to release Green Belt land for development. This could be compounded if housing targets are increased by Parliament. The business campaigning group BusinessLDN (formerly London First) has advocated for a more nuanced approach that preserves the Green Belt but re-evaluates portions of the land that are not as environmentally sensitive and/or sites near transit stations as potential sites for new development. BusinessLDN claims that 49,000 to 62,000 new homes per year are needed to meet the growing demand for housing between 2015 and 2036. However, since housing targets were introduced in 1984, London has been hovering around the 14,000–15,000 range per year. With demand exceeding supply, the cost of housing is increasing faster than wages which could make it difficult for people to afford to work and live in London (London First, n.d.).

In addition to the polarization between environmentalists and capitalists, many sustainable strategies create chasms within the environmentalist community. This is often seen with renewable energy proposals and historic preservation projects. For example, wind turbines and solar panels – while ultimately good for reducing carbon emissions – also impact aesthetics and wildlife; likewise, while restoring existing buildings can preserve culture and conserve land by preventing new buildings on undeveloped sites, some historic preservation boards may not allow retrofits to the buildings that make them more energy efficient. London's Metropolitan Green Belt is currently dealing with this paradox – more affordable housing contributes to social sustainability, but preserving the Green Belt contributes more to ecological sustainability. As discussed in Chapter 2, the three Es of sustainability can overlap or can be nested, depending on which "E" is deemed most important. Using the Green Belt as an example, an eco-centric nested model might favor land preservation on the Green Belt at all costs, while a more equitable or economic model might favor establishing more housing on the Green Belt to prevent displacement and/or migration of people/businesses out of the region. Indeed, this case study beckons one of the fundamental questions this book is trying to answer: how can we meet society's physical spatial needs while still conserving open land and natural resources?

Rail corridors

In the late nineteenth and early twentieth centuries, the United States witnessed a surge in railroad construction that produced a network of tracks that enabled people and freight to be transported between communities, cities, and states. In 1860, the United States was crisscrossed with 30,626 miles (49,288 km) of railroads; by 1890, there were 200,000 miles (321,869 km). The railroad was a dramatic, visible symbol of national union. After 1869, a person could travel anywhere through all the states and territories, coast to coast, north to south (NGS, 1975, p. 178). In the mid-twentieth century, railroads began declining because of passenger vehicles, interstate trucking, air travel, and buses. The rail industry in the United States peaked in 1916 when there were roughly 275,000 miles (442,570 km) of freight and passenger railroads (AAR, 2024). Today, there is less than half of this original network remaining, as many lines have been abandoned, fragmented, or sold to developers (Ryan, 1996).

Such rapid contraction of rail lines forced Congress to amend the National Trails System Act and create the Railbanking Program. Railbanking is a method by which rail lines proposed for abandonment can be preserved for future rail use through interim conversion to a trail. If a line is railbanked, the corridor is treated as if it had not been abandoned, preventing any reversions that could break it up into small pieces. As of 2006, there were 256 railbanked corridors, accounting for about 4,628 miles (7,448 km). Of these, 102 corridors (2,451 miles/3,945 km) were open as trails and 112 corridors (1,683 miles/2,709 km) were under development (RTC, 2006; RTC, n.d.b). Often, portions of lines – railbanked or not – are sold, leased, and/or donated. In some states, if the corridor is not railbanked or transferred to a different owner, the land often reverts to adjacent landowners. The railroad company may apply to the Surface Transportation Board to resume rail service, at which time the trail use would be vacated. Often, however, the right of way sits in stalemate and the corridor becomes a site of urban blight.

Fortunately, the Rails to Trails Conservancy (RTC) was founded in 1986 with the goal of promoting and stimulating trail creation and connectivity. In 1999, the RTC worked with the U.S. Department of Transportation to establish the Millennium Trails Program. This designates trails into three categories – National Millennium Trails, Millennium Trails, and community trails – and aims to advance trail-based greenways by increasing public awareness and inspiring future greenway development. As of the writing of this book, the RTC has helped to create 2,403 rail trails amounting to close to 26,000 miles (41,843 km) that serve over 100 million users per year. RTC is currently working on the Great American Rail Trail – a 3,700-mile (5,955 km) network of rail trails connecting Washington, D.C., to the state of Washington. In addition, RTC has created the TrailNation collaborative that brings together projects, leaders, and resources to network and connect rails and has created the TrailLink website and mobile app which helps users find and map over 40,000 multi-use trails across the country more efficiently (RTC, n.d.a).

In addition to converting abandoned rail corridors to greenways, design parameters can be initiated that allow active rails and trails to coexist. The Rails to Trails Conservancy refers to such design as "rails with trails," as opposed to the more familiar "rails to trails," discussed earlier. The trails can be located within or directly adjacent to active freight or passenger rail corridors. In the United States, there are currently 442 active rail/pedestrian trails with a total trail length close to 6,000 miles (9,656 km), with the rail with trail shared portions amounting to over 1,000 miles (1,609 km) (RTC, 2023). The exclusive right of way offers the same benefits for pedestrians as it does for the transit system – an uninterrupted line of travel separated from roadways and auto traffic. The shared use also promotes the idea of multi-modal forms of transportation, which can ultimately increase train ridership, walking, bicycling, and micromobility, thus decreasing personal automobile use. The RTC conducted a survey of 88 rails with trails in 2013 and found that every trail was open to walking, while approximately 95 percent also allowed bicycling, skating, or skiing; 13 percent allowed equestrian use; and 3 trails allowed limited motorized use, such as ATV or snowmobile (RTC, 2013).

The two biggest concerns of rails with trails are safety and liability. The design must provide an adequate buffer between the pedestrians and the trains. This can be achieved by fencing, vegetation, berms, ditches, grade separation, and spacing. The parameters used are often dependent on the frequency and speed of the trains. In the 2013 RTC survey, 60 percent of trails were 30 feet (9 meters) or less from the railroad tracks, and more than a quarter reported a minimum distance of 11–20 feet (3.4–6 meters); some trails were as close as 2 feet (0.6 meters) to the tracks (RTC, 2013). Nevertheless, data shows rails with trails to have an exemplary safety record, with only one fatality recorded since 1992 (RTC, 2013). In comparison, data from the World Health Organization (WHO, 2018) reported 1.35 million road traffic deaths in 2016, of which more than half were among pedestrians, cyclists, or motorcyclists.

Although many of the current rails with trails have lease agreements between the trail manager and the railroads, rail companies are often apprehensive to enter into such agreements because of liability. Indeed, they often require trail managers to indemnify them against any liability from accidents. State recreational use statutes (RUSs) can also protect landowners from liability. The state of Maine, for example, amended its recreational use statutes to protect railroad and utility owners in the same way as it protects private landowners.

The Island Line Trail – a rail to trail and a rail with trail

The 12-mile (19.3 km) Island Line Trail along the shoreline of Lake Champlain in Burlington, Vermont is a good example of a rail to trail *and* a rail with trail. The entire Island Line includes both the Burlington Bikeway and the Colchester Causeway and sits atop the former site of a rail line and rail yard constructed by the Central Vermont Railroad in the late nineteenth century. The rail line ran 30 miles (48.2 km) north to Canada, connecting the islands with a series of marble-banked causeways. It was abandoned in 1964 and the state of Vermont obtained the right of way in the early 1970s.

Planning for the rail-to-trail conversion began in 1973, and the Burlington Waterfront Park and Bikeway – which are situated on the site of a reclaimed brownfield – were opened in 1986. The latter runs 7.6 miles (12.2 km) from the southern end of Oakledge Park, through the city of Burlington, and north to the Winooski River. In 2004, the 7.6-mile (12.2 km) stretch was linked

FIGURE 8.2 Top left: Island Line Trail near Colchester Causeway; top right: Burlington Bikeway through Burlington along Lake Champlain; bottom left: Burlington Bikeway coexisting with active rail line (rail with trail); bottom right: Auburn Trail, a rail to trail in Victor, New York; note: the former train station and caboose have been retrofitted and currently house the Finger Lakes Cannabis Company.

to the 4.4-mile (7 km) Colchester Causeway via a truss bridge over the Winooski River/Delta Park Flood Plain. The causeways were originally built on the ridge of a sandbar, and today this section of the path offers a dramatic and scenic voyage into the middle of Lake Champlain with panoramic views of the Adirondack Mountains to the west and the Green Mountains to the east. In addition to Oakledge Park and Burlington Waterfront Park, the trail links Battery Park, North Beach Park, Leddy Park, Starr Farm Beach/Park, Delta Park, and Colchester Bog Natural Area along with several cultural attractions, businesses, recreational facilities, and restaurants. The Island Line Trail and the waterfront redevelopment rejuvenated Burlington's environmentally, socially, and economically and networked greenspace along the shoreline of the lake. The path draws roughly 150,000 people a year and has served as a catalyst for additional recreation, businesses, festivals, concerts, entertainment, and tourism (Ryan, 1996, pp. 191–195).

Elevated rail corridors – the High Line, New York

Elevated rail corridors in urban areas can also be reallocated for pedestrian and ecological use. The High Line in New York City is currently taking advantage of an abandoned elevated railway that weaves through the alleys of the meat-packing district in the Lower West Side. The High Line was constructed in the 1930s as part of the West Side Improvement Project. The aim of the project was to reduce fatalities and congestion on the streets by elevating the rail track 30 feet (9.1 meters). The 1.45-mile-long (2.3 km) elevated line ran from the meat-packing district through the West Chelsea neighborhood and allowed freight rails a direct passage into buildings. It was used heavily for about 20 years but suffered a large loss in usage because of the increase in interstate trucking in the 1950s.

The line was last used in 1980 and was subsequently abandoned and slated for demolition. However, this was halted because of resistance from Chelsea residents. In 1999, the Friends of the High Line community-based organization was formed with the mission to preserve and reuse the High Line. By 2002, Friends of the High Line had gained support from the city council and the Federal Surface Transportation Board. An open design competition was held in 2003 that attracted 720 teams, and in 2004, the landscape architecture firm James Corner Field Operations and the architecture firm Diller, Scofidio, and Renfro were selected to design the project. A pedestrian path lined with vegetation, fixed and movable seating, and LED lighting occupies the elevated line, offering unique views of the Hudson River and the city skyline while providing pedestrians with separation from the congestion on the streets below. CSX Transportation donated the High Line to the city in 2005. Section 1 from Gansevoort Street to 20th Street opened to the public in 2009, and Section 2 (20th Street to 30th Street) opened in 2011.

Utility corridors

Utility corridors represent another circumscribed lineage of space that can double as greenspace for wildlife or pedestrians but are often unusable, hazardous, and aesthetically unappealing. It is estimated that there are more than 200,000 miles (321,869 km) of high-voltage transmission lines and 5.5 million miles (8,851,392 km) of local distribution lines across the United States (Weeks & The Daily Climate, 2010). Elevated lines are frequently plagued with malfunction and maintenance issues due to damage from wind, ice, and snow. Trees around overhead wires and utility poles are often destroyed to keep the lines clear of debris. Ideally, more buildings and homes would be self-sustainable and generate their own power through renewable sources

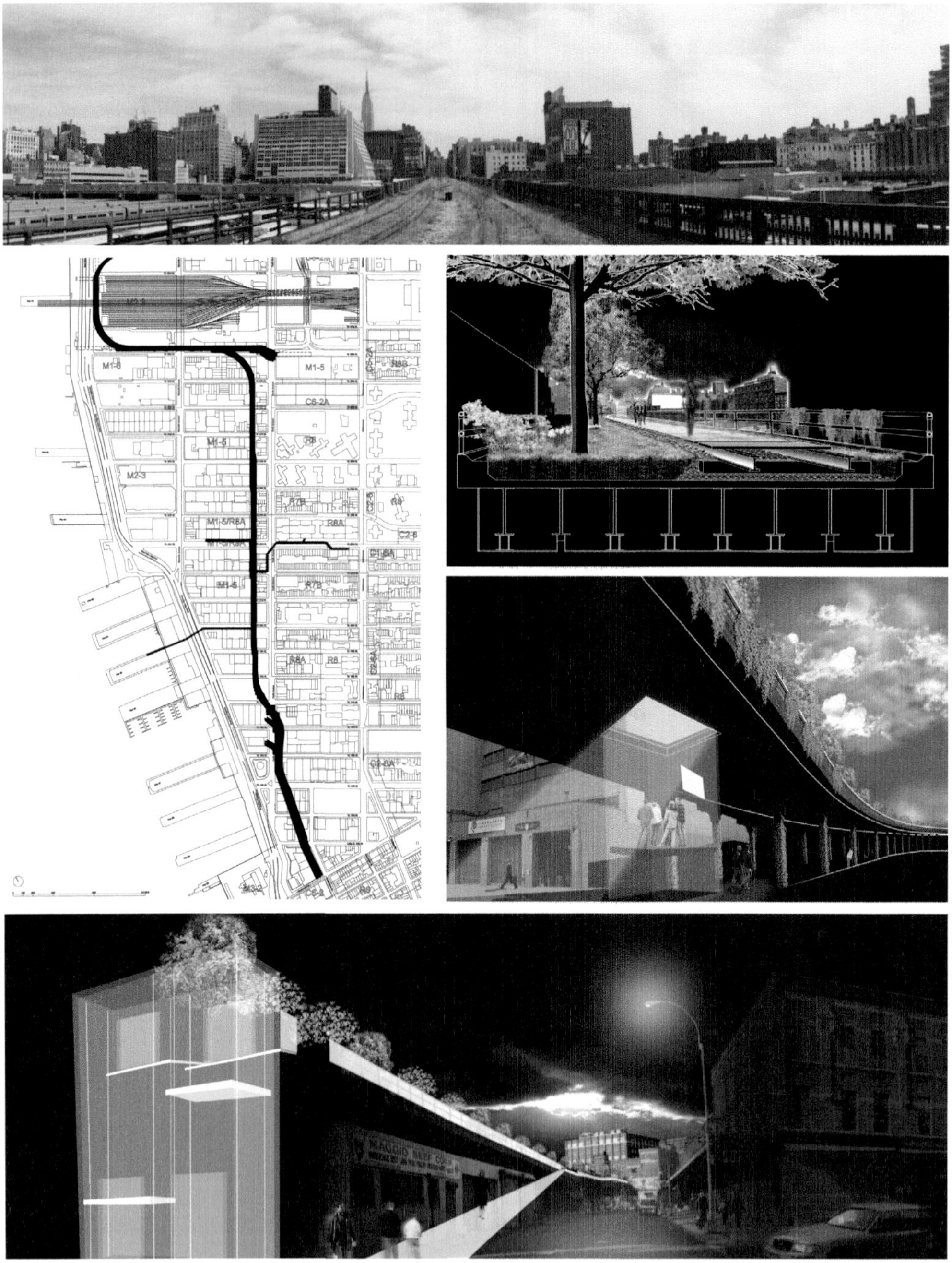

FIGURE 8.3 Top: the High Line in disrepair in 2003 looking north near 30th Street; left: map of the High Line; middle and bottom: design proposal from the 2003 competition by Vladimir Levin and Alex Bitterman showing pedestrian walkway, light cascading to the ground level, and ramp access near Gansevoort Street.

Source: All images and designs courtesy of Vladimir Levin and Alex Bitterman of energy, reducing the need for grid expansion. However, until self-sustainability becomes a reality, the existing grid of elevated poles and cables – many of which need replacing – should be placed underground, camouflaged, or relocated to minimize the impact on the environment.

Burying lines not only improves the aesthetics of a community but also provides more reliable service due to fewer malfunctions compared to overhead lines, improves safety by reducing collisions with poles and accidents from downed wires, reduces potential health risks associated with electromagnetic radiation, and increases land values. Rights of way can be used as wildlife corridors and as pedestrian spaces where voltage levels are safe.

The burial of elevated lines has sparked much debate across the country between utility companies, municipalities, zoning boards, and taxpayers. The lines can be buried at the same cost as constructing elevated lines when excavation is already under way or planned for new construction. However, the cost of excavation and burial of *existing* lines – and who should bear that cost – often waylays the sub-grade utility option. Nationally, the cost of burial is about ten times higher than for elevated lines, equating to $1 million or more per mile (Roe, 2007). However, the cost of burial should also be weighed against payback and maintenance and outage costs. A 2013 report by the U.S. Department of Energy and the President's Council of Economic Advisors estimated the total annual cost of weather-related power outages to be between $18 billion and $33 billion. Scenic America estimates undergrounding could save $17.1 billion per year in lost economic activity due to power outages (Scenic America, n.d.).

Decisions are also affected by who owns the right of way and/or the conditions of any easements given to the utility company. If the municipality foots the bill, it invariably finances the project by increasing taxes. If the utility pays for the burial, it customarily increases its rates, once again deferring the cost to the consumer. Land developers will often bail out of a bid if they are forced to finance the burial. Other opponents of utility burial claim underground lines are more difficult to reach for maintenance and repair, and that flooding, corrosion, collision with hidden lines during excavation, and uprooting trees can cause just as much damage as the construction of elevated lines. However, burying lines in concrete vaults with access panels can protect them from natural elements, provide easy access for maintenance, and prevent accidental collisions when the ground is excavated.

Often, state and/or federal agencies will partner with the utilities and/or the developers to reduce the financial burden on the taxpayer. For example, Sound Transit and the city of Seattle partnered to finance the burial of lines in the Ranier Valley as part of a proposed light rail project in 2002. Sound Transit financed portions along residential neighborhoods while the city financed the commercial portions. Another financing option is to spread the cost over a period of time akin to a tax increment financing strategy to keep the financial burden on individual households as low as possible. In San Diego, California, the Utilities Undergrounding Program has been funded by the state of California, and by 3.5 percent surcharge increase customers have been paying since 2003 to AT&T and San Diego Gas & Electric. The goal of the project is to bury 15 miles (24 km) per year, and with roughly 1,000 miles (1,609 km) of lines left to bury, the city estimates all residential areas will be buried within the next 54 years (City of San Diego, n.d.).

Federal funding and grants can also catalyze burial projects. The Infrastructure Investment and Jobs Act signed into law in 2021 includes three provisions for communities interested in burying their utility lines, including a $5 billion program to enhance the resiliency of the electric grid, amendments to the Robert T. Stafford Disaster Relief and Emergency Assistance Act to utilize disaster mitigation programs to underground wires, and the National Highways

Performance Program that makes the undergrounding of utility wires an eligible expense (Scenic America, 2022).

If it is not feasible to bury lines due to financial constraints or rugged terrain, co-locating utilities with other utilities or rail lines in locations that are offset from residential neighborhoods and vistas can be considered. Landscaping can also improve the aesthetic of lines that cannot be moved. Utility poles can be treated with an exo-skeleton frame and cloaked with plants (similar to the concept employed in green walls). In addition, vegetation and tree species can be strategically planted so they camouflage the wires without interfering with the mechanics of the utility. The mature height, shape, and spread of a tree or shrub species determine the spacing and juxtaposition relative to the wire. Shrubs and trees that grow to a maximum height of less than 15 feet (4.6 meters) and require minimal crown reduction can be planted under most lines. Trees above 15 feet with narrow crowns can be planted in rows along the sides of lines. Trees that grow to significant heights with wide crowns should be planted at an adequate distance to minimize potential damage to the utility lines and the need for frequent pruning.

With the endless debate over where to construct renewable energy projects, utility rights of way provide ideal sites for wind turbines, micro-hydroelectric penstocks, and/or solar panels. This averts the need to develop additional land and displace vegetation and wildlife. Overhead utilities might be better accepted if they were coupled with green energy projects. Merely seeing the green energy sources in the vicinity of the wires may change the perception of utilities from mere nuisance to facilitator of clean power. Given adequate dampening from vibration, wind turbines could potentially double as utility poles and support wires, and/or photovoltaic panels could be mounted on the tops of the poles. Overhead utilities can be co-located with wind turbines, railways, and pedestrian paths to foster multi-use corridors.

Alley corridors

In dense, urban areas, alleyways between buildings can be retrofitted from dark, dead spaces into functional greenspace. As introduced in Chapter 2 and discussed in more detail in Chapter 9, green roof technology can be applied vertically to walls, creating a contiguous blanket of vegetation extending from the roof, down exterior walls, and connecting to green space in alleys. The ground between alleyways can be designed as pedestrian space or made into a working space, such as an agricultural or stormwater collection area. The alleyway garden and/or raised garden beds are common throughout much of Europe in dense urban areas where open space is scarce and where people want to grow their own produce. In some communities, alleyways are utilized for parking and/or as driveways. Instead of paving edge-to-edge, driveways can be constructed with permeable pavers, porous asphalt, pervious pavers, turf-block systems, or can be constructed as ribbon driveways (also known as Hollywood driveways) where two parallel hard-scaped strips are separated by a strip of grass/vegetation (see Figure 2.5).

Spaces between and/or adjacent to buildings are also ideal for vegetated swales, infiltration trenches, and/or rain gardens. Runoff from rooftops that would otherwise hit the street surface and carry pollutants and sediment into waterways can be absorbed and filtered in an alleyway. Channeling rainwater can also become a determinant of form in landscape design and an effective teaching tool that demonstrates real-time runoff patterns. Observers can visually follow the path of the rainfall along the watercourse to the terminus at the collection pond. Seating areas, interpretation stations, and walking paths can be juxtaposed along the watercourse, ultimately forging a veritable coalescence of ecological function and landscape design.

FIGURE 8.4 Conceptual renderings of multi-use corridors. Top: utilities, wind turbines, and grade difference to separate pedestrian path from commuter rail; bottom: vertical-axis wind turbines, solar panels, utility poles with green wall system, and fencing between commuter rail and pedestrian path.

FIGURE 8.5 Water course between two buildings of the University Heights Residential Complex at the University of Vermont.

Riparian corridors

Riparian corridors (also known as riparian buffers) are natural or engineered lineages of land and vegetation that border the banks of both standing and flowing bodies of water. The buffers preserve the natural characteristics of the water body they serve by filtering sediment and pollutants from runoff before it is channeled into streams, protecting water banks from erosion, providing a sink on the flood plain and thereby regulating the hydrologic cycle/flooding, providing food and habitat for wildlife, and controlling water temperature. The passage of the National Wild and Scenic Rivers Act in 1968 and the Clean Water Act in 1972 catalyzed the conservation of river corridors, wetlands, and riparian zones along waterways.

Although zoning regulations often dictate minimum setback dimensions – which are often around 100 feet/30.5 meters from any watercourse or wetland edge – the size of riparian buffers should be determined by the characteristics of both the water body they border and the larger watershed they serve (Watson & Adams, 2011). For example, watersheds with steeper slopes will produce more sediment and water runoff than flat slopes and therefore require wider riparian zones. An examination of the flood plain can reveal patterns of flooding that are likely to recur over specific time intervals (i.e., the 100-year flood plain). Given the current and projected increase in extreme weather events, melting polar ice, and subsequent sea-level rise, it may be prudent to expand the setback of all riparian buffers to cover a 100-year flood, given the increased likelihood of such an event.

Tom Schueler at the Center for Watershed Protection offers eight criteria to determine the size, management, and crossings of stream buffers, which include using a three-zone buffer system; restoring buffers to predevelopment vegetation; allowing for buffer expansion and contraction; clear delineation of buffer dimensions; identifying provisions for infrastructure that crosses the buffer; utilizing low-impact development/best management practices/green stormwater infrastructure outside of the buffer to maximize stormwater capture; educating the public and ensuring there is no encroachment on the buffer; and making buffer ordinance flexible through conservation easements, density compensation, and variances to protect the rights of the property owners (Otto et al., 2004).

The three-zone system has been adopted by several agencies, and although there will be some variation based on geography and climate, the system can provide a template that can be adapted to a specific region. The New York State Department of Environmental Conservation (NYSDEC, n.d.), for example, describes the three zones as the following: Zone 1, also referred to as the streamside/riverside/lakeside zone, is the zone closest to the water body. It is ideally occupied by mature, water-tolerant trees and shrubs that require little or no harvesting. The vegetation provides a bank stabilization and cools the water, providing better conditions for cold water fish species. The minimum recommended width is 15 feet (4.6 meters). Zone 2, also referred to as the middle zone, should be planted with native, shade-tolerant trees and shrubs that are smaller and faster growing than those in Zone 1. The plantings allow the infiltration and filtration of water runoff and the uptake of nutrients. The recommended width is 20–60 feet (6–18 meters). Zone 3, which is the furthest from the stream and the closest to development, should be 15–60 feet (4.6–18 meters) and be planted with native grasses, wildflowers, or other herbaceous plants that can slow fast-moving water and filter nutrients.

Although some studies have examined optimal corridor width based on specific species, variables such as the length of the corridor and the quality and continuity of habitat can cause variance between sites. The U.S. Department of Agriculture (Bentrup, 2008), however, has developed some general guidelines based on the research that is available, including the larger the species, the wider the corridor will need to be to facilitate movement and habitat; the width of the corridor needs to increase incrementally with increases in length; wider corridors are needed in landscapes dominated by human use and development; and corridors that need to function for long periods (decades or centuries) need to be wider to accommodate slow-moving organisms, gene flow, and range distribution due to climate change.

Although the current riparian zone surface area in the United States is between 37 million and 123.6 million acres (15–50 million hectares), it is estimated that 70–90 percent of the natural riparian vegetation in the United States has been lost or degraded due to human activities (NRC, 2002; Doppelt et al., 1993). In addition to irresponsible land development and urbanization,

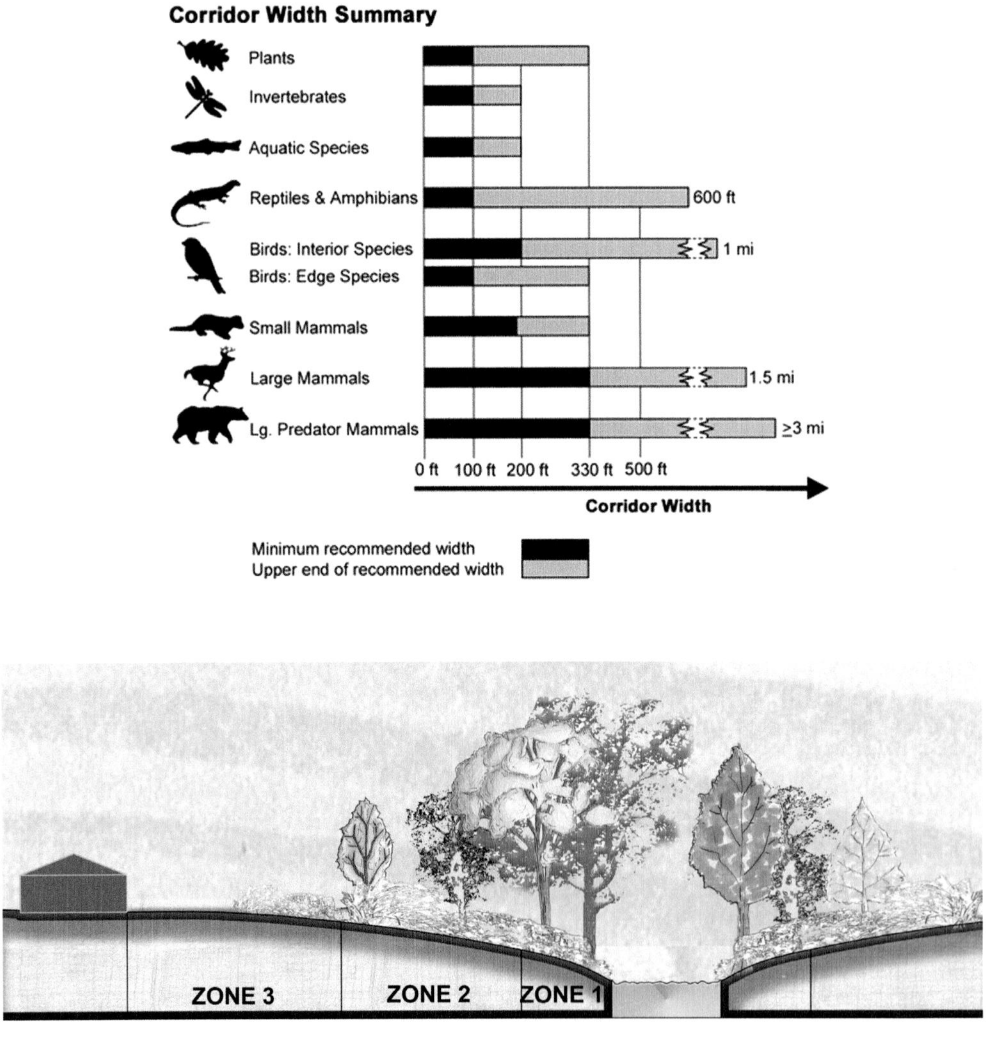

FIGURE 8.6 Top: recommended corridor widths by species, courtesy of USDA/Gary Bentrup; bottom: typical three-zone riparian buffer cross section.

other human impacts from agriculture, livestock grazing, forestry, and flood control have severely degraded riparian corridors.

In recent times, many regions have recognized the severity of this degradation and implemented large-scale restoration projects. For example, the Great Rivers Greenway District is an organization in Missouri that was created in 2000 as a result of the passage of the Clean Waters, Safe Parks, and Community Trails initiative. The district has developed a master plan known as the River Ring that envisions an interconnected web of 45-plus greenways that will cover over 600 miles (965.6 km) along the Mississippi and Missouri rivers, ultimately connecting St. Louis City, St. Louis County, and St. Charles County. The greenway network includes natural, undisturbed, and restored wildlife and conservation areas, parks for active and passive recreation, residential neighborhoods, river towns and cities, agricultural and natural open space, and commercial and industrial operations. The guidelines in the River Ring plan recognize that ecological and economic goals are mutually beneficial. In addition to protecting and restoring the natural features and functions of the rivers, flood plains, and wetlands, the plan emphasizes relationships between the natural features and the human realm by highlighting distinguishable characteristics of the Mississippi region with respect to history, industry, ecology, and culture. The park and trail system continues to expand and provides direct access to the waterfront and a vast array of recreational activities that stimulate regional economic growth, preserve natural resources, and generate a new sense of cultural heritage.

Since its inception in 1960, the Southeast Wisconsin Regional Planning Commission (SEWRPC) has identified and protected a regional system of environmental corridors across the seven counties of Kenosha, Milwaukee, Ozaukee, Racine, Walworth, Washington, and Waukesha. SEWRPC has identified primary environmental corridors that are at least 400 acres (162 hectares), 2 miles long (3.2 km), and 200 feet (60.9 meters) wide; secondary environmental corridors that are at least 100 acres (40.5 hectares) in size and one mile in length; and isolated natural resource areas that are at least 5 acres (2 hectares) and 200 feet (60.9 meters) wide. In 2000, the corridors totaled roughly 600 square miles (1,554 square km), which is 22 percent of the region, and roughly 87 percent of primary corridors had received some level of protection. The SEWRPC recommends that all lowland corridors and isolated natural areas that are floodplains and wetlands should be kept free of development and should not be filled or drained; upland portions of all primary corridors should not be developed except for residential use at one household per five acres; and local governments should consider measures to protect upland portions of secondary environmental corridors and isolated natural resource areas (SEWRPC, n.d.).

In New York State, several land acquisitions facilitated the formation of Genesee Valley Greenway State Park (GVGSP), which can be included in various sections of this chapter seeing it is a canal to rail, a rail to trail, and a riverside greenway. Indeed, GVGSP runs along the former sites of the Genesee Valley Canal (1840–1878) and the Pennsylvania Railroad, Rochester Branch (1882–1963), and portions of the Genessee River. After rail operations ceased in 1963, Rochester Gas and Electric (RGE) purchased 80 miles (128.7 km) of the corridor in Livingston, Wyoming, and Alleghany Counties. Between 1967 and 1978, the New York Office of Parks, Recreation, and Historic Preservation (OPRHP) purchased roughly 260 acres (105 hectares) that were added to Letchworth State Park and 118 acres/10 miles (48 hectares/16 km) from Penn Central Corporation, the latter on a quit claim deed. In 1990, the New York Parks and Conservation Association and RGE initiated efforts to open 50 miles (80.5 km) of the trail between Genessee Valley Park and Letchworth State Park, and in 1992, the first two miles of the corridor opened in Livingston County. In 1994, a public–private partnership was formed between

OPRPH, the New York State Department of Environmental Conservation (NYDEC), and the Friends of the Genessee Valley Greenway (FOGVG) – a not-for-profit established in 1993. In 2000, New York State purchased the segments still owned by RGE. Federal grants, including those from the Federal Transportation Enhancements Program and the Recreation Transportation Program, along with donations and fundraising through the Adopt a Trail Program, were used for the land acquisitions and maintenance of the corridor (FOGVG, n.d.).

As of the writing of this book, GVGSP extends 90 miles (144.8 km) through 16 towns and six counties from Genessee Valley Park in Rochester, New York, south to the village of Cuba in Allegany County. The GVGSP connects to the Finger Lakes Trail, the Erie Canal Recreation Way (discussed later), the Erie-Attica Trail, the Lehigh Valley Trail, and the Genessee Riverway Trail, which connects to several trails directly or indirectly linking Ontario Beach Park, Durand Eastman park, Turning Point Park, Seneca Park, Maplewood Park, Lower Falls Park, High Falls Terrace Park, Genessee Gateway Park, Bauch and Lomb Riverside Park, and Genessee Valley Park. There are plans to extend the corridor to Hinsdale in Cattaraugus County (OPRHP, n.d.).

Canals as greenways – the New York State Canal System

Many historic towpath canals – once the heart of shipping and transport – have been transformed into multifunctional greenways, including the Royal Canal Greenway between Dublin and Galway in Ireland, the High Line Canal in Denver, Colorado, the Pittsburgh–Harrisburg Main Line Canal in Pennsylvania, the Cross Florida Greenway, the Ohio Canal, and, as discussed in more detail next, the Erie Canal in New York State (the New York State Canal System). The dedicated rights of way constructed for the canals can provide the same community connections they did in their heyday. Repurposing canals as multifunctional greenways can preserve history, provide recreational opportunities, provide ecological benefits, and revive old rust-belt port cities.

After the War of 1812, development in New York State was catalyzed by the construction of the Erie Canal, which was also known as "Clinton's Ditch" after Governor Dewitt Clinton who was an early proponent of the canal. Construction began in 1817 and was completed in 1825, forming a 363-mile (584.2 km), east-west canal connecting the Niagara River in Buffalo, New York, to the Hudson River in Albany, New York, which ultimately linked the Atlantic Ocean to the Great Lakes. Shortly after the completion of the Erie Canal, Canada initiated construction of the Welland Canal that eventually connected Lakes Erie and Ontario west of the Niagara River, and the 66-mile (106.2 km) Champlain Canal was also constructed connecting the Hudson River to Lake Champlain. According to the Erie Canalway National Heritage Corridor (n.d.), 83 locks were originally constructed to navigate the 571-feet (174-meter) elevation change between Buffalo and Albany. Between 1826 and 1836, New York State added to the Erie Canal and constructed the Oswego, Cayuga-Seneca, Chenango, Crooked Lake, and Chenango canals. The original canal was 4 feet (1.2 meters) deep and 40 feet (12.2 meters) wide but was expanded between 1836 and 1862 and again between 1903 and 1918, the latter being dubbed as "Barge Canal," which expanded the Erie Canal, the Champlain Canal, the Oswego Canal, and the Cayuga-Seneca Canal. The Barge Canal project increased the depth to 12–23 feet (3.7–7 meters) and the width to 120–200 feet (36.6–60.9 meters).

The canal system established New York as a leader in population and industry and established several thriving port cities – including Buffalo, Rochester, Syracuse, Utica, and Albany – that became the center of trade between the Atlantic Ocean and New York City to the east and the

FIGURE 8.7 Left: map of Genessee Valley Greenway, courtesy of Friends of the Genessee Valley Greenway; top right: Genessee River Trail through boardwalk at Turning Point Park; Middle: Genessee Valley Greenway through Letchworth State Park; bottom: trail along Erie Canal through Pittsford, New York.

Great Lakes and the agricultural heartlands to the west. Before the construction of rail and highways, the canal was the fastest and cheapest way to transport freight and people. In addition to enhancing the entire region, the canal made New York City the commercial capital that it is today and established the city of Buffalo as one of the busiest ports in the world. New York City's population quadrupled between 1820 and 1850 and Buffalo's population increased from 2,500 to 74,000 between 1825 and 1855 and to 352,387 in 1900. By 1853, the canal carried 62 percent of the nation's trade (Herzberg, 2001; History Channel, 2018).

Although the construction of railways initially complemented the canal, by the mid-twentieth century, the expansion of rail combined with highway construction and the opening of the St. Lawrence Seaway caused a sharp decline in the canal for both shipping and transport. Since the 1990s, the canal has been mostly used for recreation, and in 2000, Congress designated it a National Heritage Corridor. It has since been rebranded as the New York State Canal System.

In 2019, the Reimagining the Canals Initiative was launched to examine how the canal can be reimagined in the twenty-first century. The initiative had five objectives, including identify potential new uses for the Erie Canal aimed at improving the quality of life for New Yorkers; evaluate how the Erie Canal can support and enhance economic development along the canal corridor; find new opportunities to enhance recreation and tourism along the Erie Canal; assess how the Erie Canal can help mitigate impacts from flooding and ice jams to improve resiliency and restore ecosystems in canal communities; and identify opportunities for using Erie Canal infrastructure to expand irrigation for Western New York farms (Reimagine the Canals, 2020). Some of the themes that came out of the sessions and the design competition included creating walkable canal-side neighborhoods and business districts; making paths more accessible; the need for strategies addressing water quality, flooding, stormwater management, invasive species, and ecosystem protection; improving recreational infrastructure; improving wayfinding; and a need for a better understanding, inventory, and preservation of historically and culturally significant sites, many of which belonged to natives of the Haudenosaunee (Iroquois) Confederacy – including the Mohawk, Oneida, Onondaga, Cayuga, and Seneca nations – that were displaced as a result of illegal treaties, the invasion of settlers on native land, and the disruption of natural water flows caused by the dredging and damming of rivers and watersheds.

Another ensuing issue with the canal has been over tree clear-cutting practices along the embankment. In 2017, the New York Power Authority (NYPA) – which owns the New York State Canal Corporation (NYCC) – presented plans to clear cut trees and vegetation along the inside slope and crest of the embankments, claiming such removal was necessary to repair leaks, prevent embankment collapse, and to ensure safety. However, some of their vegetation removal practice conflicted with the Canal Recreation Way Commission's 1992 vision of the canal as a "greenway." This prompted the formation of the citizen advocacy group called Stop the Canal Clear-cut, and in 2018, the towns of Pittsford, Perinton, and Brighton took the case to court and obtained an injunction for the Canal Corporation to stop clear cutting and requiring the NYPA to complete an environmental review under state law and a proposed guide book that is subject to approval by the town of Pittsford's Planning Board and pursuant to Pittsford's Local Waterfront Redevelopment Plan. Similar to some of the adaptive reuse projects discussed in Chapter 10, historic preservation can often conflict with goals of sustainability and ecological protection where a benefit to one is detrimental to the other. Often, compromise is the only solution. As of the writing of this book, NYCC has released a plan that has more selective cutting and proposals to plant trees with shallower root systems, plants, pollinator areas, and grasses on the outer slope.

Continental-scaled trails and greenways

Greenways can be planned between states, regions, and countries to produce a contiguous network at continental and even global scales. To date, some of the grandest-scaled lineages have come in the form of trails. The National Trails System Act of 1968 called for the "establishment of trails in both urban and rural settings for people of all ages, interests, skills, and abilities" (PCTA, n.d.). Since its inception, roughly 88,600 miles (142,588 km) of National Trails have been created, 11 of which are National Scenic Trails, 19 National Historic Trails, 1,300 National Recreation Trails, and seven side and connecting trails. The Appalachian Trail (the AT) on the East Coast and the Pacific Crest Trail (PCT) on the West Coast were the nation's first National Scenic Trails. American forester, planner, and conservationist Benton MacKaye published his idea for the Appalachian Trail in 1921 in the Journal of the American Institute of Architects after being inspired by a hike up Stratton Mountain in Vermont. His original vision proposed a trail from Mt. Washington in New Hampshire to Mt. Mitchell in North Carolina with shelter communities for hikers, small clusters of homes/camps for recuperation and study, and agricultural settlements near the trail (Hanson, 2014). The trail was completed in 1937, and it currently spans 2,198.4 miles (3,538 km) from Springer Mountain in Georgia to Mt. Katahdin at Baxter State Park in Maine. The AT passes through 14 states, six national parks, eight national forests, and two wildlife refuges (ATC, n.d.). The AT also connects with the Long Trail in Vermont that spans 273 miles (439.4 km) along the spine of the Green Mountains. Vermont is also home to the 300-mile (482.8 km) Catamount Trail, which is a backcountry skiing trail that also runs the length of the state.

On the west coast of the United States, the Pacific Crest Trail spans 2,650 miles (4,265 km) from Mexico to Canada. Clinton Clarke promoted the idea for the PCT in the early 1930s,

TABLE 8.1 Prominent continental-scaled trails

Trail	*Region*	*Length (miles/km)*
Appalachian Trail	Georgia–Maine	2,198/3,537
Pacific Crest Trail	Mexico–Canada	2,650/4,265
East Coast Greenway	Florida–Maine	3,000/4,828
The Great Trail	Newfoundland–Yukon	16,777/27,000
The Great Western Trail	Western United States	6,875/11,064
American Discovery Trail	California–Delaware	6,800/10,944
Eastern Continental Trail	Key West, FL–Belle Isle, Canada	4,400/7,081
The North Country Trail	Vermont–North Dakota	4,600/7,403
Grand Italian Trail	Italy	3,831/6,165
The Continental Divide	Mexico–Canada	3,100/4,989
Hokkaido Nature Trail	Hokkaido, Japan	2,848/4,583
The English Coastal Path	United Kingdom	2,795/4,498
The Great Himalayan Trail	Western Nepal	2,800/4,506
European Long-Distance Path	Europe	2,900/4,667
Te Araroa Trail	New Zealand	1,900/3,058
Eurovelo	Europe	55,923/89,999

Sources: Poirot, L. (2024). *30 longest hiking trails in the world, ranked.* Far & Wide. https://www.farandwide.com/s/longest-hiking-trails-world-008dd8ef3c094041; World Atlas. (n.d.). *Longest hiking trails in the Unired States.* https://www.worldatlas.com/articles/longest-hiking-trails-in-the-united-states.html; East Coast Greenway. (n.d.). *About the East Coast Greenway.* https://greenway.org/about/the-east-coast-greenway

and shortly after, Warren Rogers and the YMCA scouted the route. The PCT was designated a National Scenic Trail in 1968 and it was substantially completed in 1993 (PCTA, n.d.). The PCT connects to the Great Western Loop – a 6,875-mile (11,064 km) trail that, in addition to the PCT, connects the Pacific Northwest Trail, the Continental Divide Trail, the Grand Enchantment Trail, and the Arizona Trail.

East Coast Greenway

Construction of a 3,000-mile (4,828 km) greenway/shared-use trail system from Calais, Maine, to Key West, Florida, began in 1991. The project was conceived by a group of enthusiastic cyclists who eventually formed the East Coast Greenway Alliance (ECGA). The north–south spine will be networked with 1,000 miles (1,609 km) of complementary routes that provide connections to parks, urban centers, greenways, and bicycle routes. The principal goal of the project is to separate the greenway from motorized traffic wherever possible. This includes utilizing both active and abandoned rail lines, grade-separation techniques, and overpasses/tunnels, as discussed in previous chapters. As of the writing of this book, 35 percent of the total route is now on firm surface trails protected from motorized traffic (ECGA, n.d.).

The design criteria for the permanent route include connecting the route to towns, cities, population centers, housing, jobs, and essential services; providing trails that are open, free of cost, and contain amenities and services users will need, such as restaurants, lodging, restrooms, picnic areas, health services, and bike repair shops; using grades and trail widths that conform to the American Association of State Highway and Transportation Officials (AASHTO) guidelines; providing firm, navigable surfaces; ensuring safety from traffic, crime, and discrimination; and utilizing traffic-separating barriers through horizontal spacing and vertical elements (ECGA, 2023).

When complete, the East Coast Greenway will link 15 states, 25 major cities, and 450 communities along the Eastern Seaboard to provide a truly "alternative" way of traveling down the coast. Indeed, it might be ideal for those who are afraid to fly, dislike driving, and do not have the patience for the train.

References

Appalachian Trail Conservancy (ATC). (n.d.). *Explore the A.T.* https://appalachiantrail.org/explore/

Association of American Railroads (AAR). (2024). *Chronology of America's freight railroads*. www.aar.org/wp-content/uploads/2020/07/AAR-Chronology-Americas-Freight-Railroads-Fact-Sheet.pdf

Bentrup, G. (2008). *Conservation buffers: Design guidelines for buffers, corridors, and greenways*. Gen. Tech. Rep. SRS-109. USDA, Forest Service, Southern Research Station. www.fs.usda.gov/nac/buffers/index.html

Bridgers, L. (2017). *The Green Belt: What is it and why does it matter*? High Living. www.highlivingbarnet.com/green-belt-matter/

City of San Diego. (n.d.). *Utilities underground program: Overview/History*. www.sandiego.gov/undergrounding/overview

Doppelt, B., Scurlock, M., Frissell, C., & Karr, J. (1993). *Entering the watershed: A new approach to save American river ecosystems*. Island Press.

East Coast Greenway Alliance (ECGA). (n.d.). *Frequently asked questions*. www.greenway.org/faqs

East Coast Greenway Alliance (ECGA). (2023). *Design guide*. https://greenway.org/design-guide

Erie Canalway National Heritage Corridor. (n.d.). *Historical timeline*. https://eriecanalway.org/learn/history-culture/timeline

Forman, R. T., & Wilson, E. O. (1995). *Land mosaics: The ecology of landscapes and regions*. Cambridge University Press.
Friends of the Genessee Valley Greenway (FOGVG). (n.d.). *GVG history*. https://fogvg.org/gvg-history-2/
GOV.UK. (2022). *Local authority green belt – England 2021–2022 statistical release*. www.gov.uk/government/statistics/local-authority-green-belt-statistics-for-england-2021-to-2022/local-authority-green-belt-england-2021-22-statistical-release#:~:text=1.-,Main%20points,2021%20and%2031%20March%202022
Hanson, A. (2014). The founder of the Appalachian Trail imagined something even grander: Utopian vision of a Harvard forestry grad. *Humanities*, *35*(6). www.neh.gov/humanities/2014/novemberdecember/statement/the-founder-the-appalachian-trail-imagined-something-even
Hellmund, P. C., & Smith, D. S. (2006). *Designing greenways: Sustainable landscapes for nature and people* (p. 46). Island Press.
Herzberg, D. (2001). The Niagara frontier: A border zone or middle ground? In *Rethinking the Niagara Frontier: A report on the continuing bi-national forum* (pp. 36–53). The Urban Design Project. University at Buffalo School of Architecture and Planning.
History Channel. (2018). *The Erie Canal*. www.history.com/topics/landmarks/erie-canal
Kowsky, F. R. (1987). *Municipal parks and city planning: Frederick Law Olmsted's Buffalo park and parkway system*. Buffalo Architecture and History. https://buffaloah.com/h/kowsky/kowold/
London Assembly. (2021, June 24). *London's Green Belt: QA session with Mayor Sadiq Khan*. www.london.gov.uk/who-we-are/what-london-assembly-does/questions-mayor/find-an-answer/londons-green-belt
London First. (n.d.). *The Green Belt: A place for Londoners*? BusinessLDN. www.businessldn.co.uk/sites/default/files/documents/2018-05/Green-Belt.pdf
London Green Belt Council. (n.d.). *Information*. https://londongreenbeltcouncil.org.uk/information/
National Geographic. (n.d.). *Landscape*. https://education.nationalgeographic.org/resource/landscape/
National Geographic Society (NGS). (1975). *We Americans* (p. 178). National Geographic Society.
National Research Council (NRC). (2002). *Riparian areas: Functions and strategies for management*. The National Academies Press. https://doi.org/10.17226/10327
New York Office of Parks, Recreation, and Historic Preservation (OPRHP). (n.d.). *Genessee Valley Greenway Park*. https://parks.ny.gov/parks/geneseevalleygreenway/
New York State Department of Environmental Conservation (NYSDEC). (n.d.). *Riparian buffers*. www.dec.ny.gov/chemical/106345.html
Otto, B., McCormick, K., & Leccese, M. (2004). *Ecological riverfront design: Restoring rivers, connecting communities*. American Planning Association. PAS Report Number 518–519.
Pacific Crest Trail Association (PCTA). (n.d.). *America's national trails system*. www.pcta.org/our-work/national-trails-system/
Rails to Trails Conservancy (RTC). (2006). *Railbanking and rail-trails: A legacy for the future*. www.railstotrails.org/resourcehandler.ashx?name=railbanking-and-rail-trails-a-legacy-for-the-future&id=3489&fileName=Railbankingreport_7-06_lr.pdf
Rails to Trails Conservancy (RTC). (2013). *America's rails with trails: A resource for planners, agencies, and advocates on trails along active railroad corridors*. www.railstotrails.org/resourcehandler.ashx?id=2982
Rails to Trails Conservancy (RTC). (2023). *Rails with trails list*. www.railstotrails.org/resourcehandler.ashx?name=americas-rails-with-trails-rail-with-trail-list&id=32593&fileName=Rail-with-TrailList%20spreadsheet_Sept2023.pdf%20(1)
Rails to Trails Conservancy (RTC). (n.d.a). *About rails to trails conservancy*. www.railstotrails.org/about/
Rails to Trails Conservancy (RTC). (n.d.b). *Railbanking*. www.railstotrails.org/build-trails/trail-building-toolbox/acquisition/railbanking/
Reimagine the Canals. (2020). *Task force report*. www.canals.ny.gov/reimagine/TaskForceReport.pdf
Roe, A. (2007, January 9). Burying power lines: Is it worth the cost? *Seattle Times*. www.seattletimes.com/seattle-news/burying-power-lines-is-it-worth-the-cost/
Ryan, K. L. (1996). *Great rail trails in New York and New England* (pp. 2, 191–195). Rails to Trails Conservancy.
Scenic America. (2022). *Underground resources in the infrastructure investment and jobs act*. www.scenic.org/2022/05/20/undergrounding-resources-in-the-infrastructure-investment-and-jobs-act/
Scenic America. (n.d.). *Burying utilities for safety, resiliency, and scenic beauty*. www.scenic.org/why-scenic-conservation/energy-infrastructure-and-equity/undergrounding-utility-infrastructure/

Southeast Wisconsin Regional Planning Commission (SEWRPC). (n.d.). *Land use: Environmental corridors*. www.sewrpc.org/SEWRPC/LandUse/EnvironmentalCorridors.htm
Thomas, D. (1963). London's Green Belt: The evolution of an idea. *The Geographical Journal*, *129*(1), 14–24, 17–18. https://doi.org/10.2307/1794894
Watson, D., & Adams, M. (2011). *Design for flooding: Architecture, landscape, and urban design for resilience to climate change* (pp. 119–121). John Wiley & Sons.
Weeks, J., & The Daily Climate. (2010). U.S. electrical grid undergoes massive transition to connect renewables. *Scientific American*. www.scientificamerican.com/article/what-is-the-smart-grid/
World Health Organization (WHO). (2018). *Global status report on road safety 2018*. www.who.int/publications/i/item/9789241565684

9

BUILDINGS AND ROOFTOPS

One of the major culprits impacting global climate change and environmental health is buildings. According to the U.S. Energy Information Agency, nearly 29 percent of total U.S. end-use energy consumption in 2022 was consumed in residential and commercial buildings or about 22 quadrillion British thermal units (EIA, 2022). The World Green Building Council estimates that buildings are currently responsible for 39 percent of global energy-related carbon emissions, including 28 percent from energy for heating, cooling, and electricity and 11 percent from materials and construction (WGBC, 2019). In urban areas, buildings – along with the other vast amounts of concrete and asphalt for roads, parking, and paths – can create urban heat islands that put further strain on energy consumption and degrade air quality and human health.

Although concepts of utilizing the sun and the earth as building elements date back millennia, the modern notion of green buildings or "bioclimatic" architecture gained momentum with the environmental movement of the 1960s and 1970s and the OPEC oil embargo of 1973. Architects, engineers, environmentalists, ecologists, and landscape architects began to coalesce around notions of conserving energy and reducing the footprint of buildings. The U.S. Environmental Protection Agency (EPA) defines green building as

> the practice of creating structures and using processes that are environmentally responsible and resource-efficient throughout a building's life-cycle from siting to design, construction, operation, maintenance, renovation, and deconstruction; green buildings are designed to reduce the overall impact of the built environment on human health and the natural environment by efficiently using energy, water, and other resources; protecting occupant health and improving employee productivity; and reducing waste, pollution and environmental degradation.
>
> *(EPA, n.d.a)*

Green building rating systems

In the late 1980s and early 1990s, green building policies, oversight agencies, and standards started to formalize. The American Institute of Architects (AIA) established the Committee

DOI: 10.4324/9781032668666-9

on Environment in 1989 and went on to publish the *Environmental Resource Guide* in 1992. That same year, the U.S. Environmental Protection Agency (EPA) and the U.S. Department of Energy established Energy Star – a voluntary program that helps businesses and individuals save money and protect our climate through superior energy efficiency. In 1990, the Building Research Establishment Environmental Assessment Method (BREEAM) was created in the United Kingdom. This was followed by the establishment of the U.S. Green Building Council (USGBC) in 1993, and its first iteration of the building certification program known as Leadership in Energy and Environmental Design (LEED). Since that time, the World Green Building Council (WGBC) has close to 80 established and prospective members globally and there have been hundreds of green building rating systems (GBRSs) developed across the world, some of which are listed in Table 9.1. As defined by Marchi et al. (2021),

> Green Building Rating Systems (GBRSs) are typically third-party, voluntary, and market driven standards that measure buildings' sustainability level by multi-criteria assessment, and encourage the adoption of environmentally, socially, and economically sustainable practices in design, construction, and operation of buildings (or neighborhoods). GBRSs aim at guiding and assessing the project throughout all its life cycle, thus limiting the negative impact on the environment, as well as on the building occupants' health and well-being, and even reducing operational costs.

Some GBRSs are more design-based, some are more performance based, and others do some degree of both. Several studies have attempted to compare the GBRSs quantitatively and qualitatively. A study by He et al. (2018) comparing LEED, ASGB, and GS found LEED and ASGB to be more design focused and Green Star to be more performance based. A study by Song et al. (2023) comparing LEED, Green Mark (GM), ASGB, WELL, LBC, and the ASHB showed that LBC gave the highest weight to architectural design, followed by LEED, ASGB, ASHB, GM, and WELL. Another study by Assefa et al. (2022) compared ITACA, Green Star, DGNB, CASBEE, HQE, BREEAM, and LEED to the International Standardization of Organization (ISO) sustainable building criteria and found DGNB to be the strongest in covering most of the ISO indicators, BREEAM to be the only rating system that covers all economic aspect indicators of the ISO, LEED and ITACA to most emphasize environmental aspects, CASBEE to emphasize the local scenario, and HQE to be the only rating system that evaluates all outdoor environment-related ISO indicators. A study by Mattoni et al. (2018) compared CASBEE, Green Star, BREEAM, LEED, and ITACA based on seven macro-areas, including site, water, energy, comfort and safety, materials, and outdoor quality. Quantitatively, energy scored the highest importance for all except CASBEE, which gave the highest importance to comfort and safety; water had the lowest impact on scores followed by materials and outdoor quality. In addition, CASBEE was found to consider the highest number of sustainability issues within each of the seven macro-areas while ITACA was found to consider the least amount of sustainability issues.

BREEAM and LEED – the two longest-running systems – are still the most widely used, with BREEAM having 535,000 certified buildings and over two million registered for certification in the United Kingdom and 74 other countries and LEED having about 100,000 certified projects in more than 160 countries and territories (BRE Group, n.d.b; Verdinez, 2023; Stanley, 2018). However, some newer systems are starting to gain popularity. In 2013, the International Well Building Institute (IWBI) created the WELL Standard that focuses more on buildings' impact on human health and well-being. Well version 2 centers around six overarching principles: equity

TABLE 9.1 A sample of green building rating systems

System	*Name*	*Initiator*	*Region*	*Created*	*Certified projects (approx)*
ASHB	Assessment Standard for Healthy Buildings	Architectural Society of China	China	2017	N/A
BEAM/BEAM Plus	Building Environmental Assessment Method	Hong Kong Green Building Council (HKGBC)	Hong Kong	1996	1,500
BREEAM	Building Research Establishment Environmental Assessment Method	Building Research Establishment (BRE)	UK	1990	535,000
CASBEE	Comprehensive Assessment System for Built Environment Efficiency	Japan Sustainable Building Consortium (JSBC)	Japan	2004	541
DBJ	Development Bank of Japan	Development Bank of Japan	Japan	2011	624
DGNB	German Sustainable Building Council	German Sustainable Building Council (GSBC)	Germany	2007	4,800
FITWEL	Fitwel	U.S. CDC and U.S. General Services Administration	United States	2016	950
GB/T	Evaluation Standards for Green Buildings	Ministry of Housing and Urban-Rural Development of People's Republic of China (MOHURD)	China	2006	10,636
GBI	Green Building Index	Greenbuildingindex Sdn Bhd (GBI)	Malaysia	2009	500
GG	Green Globes	Green Building Initiative (GBI)	Canada	2000	900
GM	BCA Green Mark	Building and Construction Authority (BCA)	Singapore	2005	3,408
GS	Green Star	Green Building Council Australia (GBCA)	Australia	2003	4,466
GSAS	Global Sustainability Assessment System	Gulf Organization for Research and Development (GORD)	MENA Region	2007	950

(*Continued*)

TABLE 9.1 (Continued)

System	*Name*	*Initiator*	*Region*	*Created*	*Certified projects (approx)*
HQE	Haute Qualité Environnementale	Association pour la Haute Qualité Environnementale (ASSOHQE)	France	2004	N/A
IGBC	Indian Green Building Council	Confederation of Indian Industry (CII)	India	2001	3,000
IgCC	International Green Construction Code	International Code Council	United States	2009	Adopted by 13 states and Washington, D.C.
ITACA	Innovation and Transparency of Procurement and Environmental Compatibility	Sustainable Building Council (SBC)	Italy	2003	N/A
LBC	Living Building Challenge	International Future Living Institute (IFLI)	United States, Australia, Europe	2006	250
LEED	Leadership in Energy and Environmental Design	U.S. Green Building Council (USGBC)	United States	1998	100,000
Miljobyggnad	Swedish Environmental Certification for Swedish Conditions	Sweden Green Building Council (SGBC)	Sweden	2009	1,000
Minergie	Minergie	Minergie Association	Switzerland	1998	55,000
NGBS	National Green Building Standard	National Association of Home Builders	United States	2012	315,000
WELL	WELL	International Well Building Institute (IWBI)	United States	2013	25,550

Note: Not all systems are listed.

Sources: HKGBC, n.d.; BRE Group, n.d.a; IBEC, n.d.; DBJ, n.d.; IEA, n.d.; GSBC, n.d.; Fitwel, n.d.a; MOHURD, n.d.; GBI, n.d.; GG, n.d.; BCA, n.d.; GBCA, n.d.; GORD, n.d.; ASSOHQE, n.d.; IGBC, n.d.; CII, n.d.; ICC, n.d.; SBS, n.d.; ILFI, n.d.; USGBC, n.d.; SGBC, n.d.; Minergie, n.d.; NAHB, n.d.; IWBI, n.d.; Song et al., 2023; Tang et al., 2020.

to disadvantaged and vulnerable groups; interventions with global relevance; evidence-based research across multiple disciplines; technically robust best practices and rigorous third-party verification; a customer-centric, intuitive platform; and resiliency to keep pace with technological advances and changes in society. Building evaluation is based on ten concepts: air, water, nourishment, light, movement, thermal comfort, sound, materials, mind, and community (IWBI, n.d.).

The U.S. Centers for Disease Control and Prevention and the U.S. General Services Administration also recognized the importance of health and well-being at both the building and community levels and developed Fitwel in 2016. Fitwel addresses health as an interconnected system and focuses on seven health impacts, including strategies that impact surrounding community health, strategies that reduce morbidity and absenteeism, strategies that support social equity for vulnerable populations, strategies that instill feelings of well-being, strategies that enhance access to healthy foods, strategies that promote occupant safety, and strategies that increase physical activity (Fitwel, n.d.b).

Although some programs such as Japan's CASBEE are mandated, most GBRSs are still voluntary, which begs the question if the standards would have more impact if they were integrated with building codes or zoning regulations. The International Code Council's International Green Construction Code (IgCC), for example, is designed to be incorporated into a jurisdiction's codes and ordinances and function as an overlay to other International Code Council codes adopted by the jurisdiction. Likewise, many jurisdictions have also adopted the American Society of Heating, Refrigeration, and Air-Conditioning Engineers' ANSI/ASHRAE/USGBC/IES Standard 189.1 standard and the Americans with Disabilities Act (ADA) standards into their building codes. Jurisdictions that do adopt standards also have the option of making the code more stringent. Adopting green building standards into the building codes and/or adopting policies to make all buildings carbon neutral or net-zero energy buildings (NZEB) may be the most effective way to meet the Net Zero Emissions by 2050 Scenario (NZE) by 2050 Scenario developed by the International Energy Agency (IEA), the Sustainable Development Goals (SDGs) established by the United Nations, the World Green Building Council's Advancing Net-Zero program, and the International Panel on Climate Change's emissions reductions recommendations to limit global temperature rise to 1.5 degrees Celsius. As stated by Maclay (2014),

> We can make significant headway by ensuring that every new building lives within the planet's flow of energy. We can decide to build and rebuild only buildings that are net producers of energy, either with renewables on-site or powered with nearby renewable community systems. From there, we can make our communities net producers.

Since their inception, many professionals have been critical of GBRSs, citing concerns about buildings that achieved high ratings but overlooked embodied carbon and energy efficiency.

For example, Seven Bryant Park in New York was certified LEED Gold in 2016, but the city's Building Energy Efficiency and Rating program (BEER) gave it a grade of "C," which was 60 out of 100 points (Millard, 2021). Much of this was due to older LEED requirements that examined forecasted energy models rather than ongoing operational energy. The USGBC has since rectified this by raising the threshold for energy performance in LEED v4, requiring building owners to report their annual energy efficiency, and adding a mechanism for specifically evaluating and incentivizing greenhouse gas emissions reductions (USGBC, 2023). In addition, many GBRSs have evolved to better align with the United Nations' Sustainable Development Goals and have subsequently expanded the range of project types and assessment criteria. Buildings, neighborhoods, townships, communities, and infrastructure have certifications for the planning, design, construction, operation, and performance phases of a project and are being assessed with criteria such as energy efficiency, carbon footprint, emissions, site sustainability, land use, construction management, water efficiency, material efficiency, indoor environmental quality, transportation, waste, innovation, resilience, innovation, equity, and health and well-being, to mention a few.

Another criticism of the GBRSs is the required training and associated course and exam fees that are required to become certified to be eligible to submit a project. Although professional certifications are becoming more ubiquitous, it seems more professionals in architecture, landscape architecture, planning, engineering, or construction management would participate in a GBRS submittal if the process was intuitive and could be completed with their previous academic training and without additional training, which, hopefully at this point in history, would include a comprehensive understanding of the principles of sustainability. An alternative would be for schools to incorporate GBRSs into their curriculum so students have the credential when they graduate.

Love them or hate them, GBRSs are here to stay. Informally, many professionals may agree that no GBRS system is perfect, but any system is better than no system. Indeed, although certification does not always directly sync with what some define as "sustainable" or "resilient," GBRSs created a standard where there was none before. Whether they are used to greenwash or not, GBRSs have brought many notions of the green movement mainstream. Green roofs, living walls, active and passive solar design, geothermal heat pumps, on-site water treatment, the use of local materials, climate control, health, comfort, well-being, and equity are just a few strategies that have become more ubiquitous since GBRSs were launched in the 1990s. As to what is truly "green" or "sustainable," Mother Nature will ultimately have the last word, but until then, professionals will have to continue to morph the assessment standards to adapt to new challenges our planet faces.

Smart buildings

Since it opened in 1975, the Carousel of Progress attraction at Walt Disney World's Tomorrowland showcased the evolution of technology in a typical household. The last rotation features a glimpse into the future where the family is seen using voice-automated appliances, virtual reality, and computerized controls of the appliances and equipment in the house. In the twenty-first century, it seems technology has finally caught up to Walt's original vision. We are now able to communicate, access any information desired, and control things remotely with a device slightly larger than a deck of cards. The days of worrying about whether you turned off an appliance, fed your cat, locked the door, turned down the heat, watered your plants, or if there is an intruder in your house when you are away are quickly starting to fade.

Although there is currently no internationally agreed definition of what makes a building "smart" or "intelligent," at its root, a smart building is a building that leverages best practices, technology, and automated processes to control, monitor, and improve building systems, operations, and occupant well-being. A review of nomenclature by Dakheel et al. (2020) found four major themes common to all definitions of smart or intelligent buildings: achieving the net-zero energy (NZE) standard, the building's response to external conditions, the building's response to users' needs, and the utilization of building energy management systems – also known as building automation systems (BASs) or building management systems (BMSs) – to provide monitoring, control, and supervision. BMSs connect to and collect information from physical devices and equipment fitted with sensors, software, and network connectivity, which are collectively referred to as the internet of things (IoT). IoT devices include sensors that monitor temperature, humidity, air quality, water usage, motion, proximity, lighting, fire, and carbon monoxide; digital/LED screens to display signage, mapping, and floor plans; indoor positioning systems to track user circulation, occupancy, and usage patterns; smart meters that allow bidirectional communication with the energy grid; cameras for security; smart door-access control systems; smart tinting on windows; and energy storage systems.

BMS also contains some type of software, app, or user interface, or, in the case of larger facilities/complexes, an Integrated Work Management System (IWMS) software that collects the data, generates reports, and provides a central repository for information. The IWMS can have a mobile interface for technicians and occupants that they can access on their smart devices. Building information modeling, computer-aided design, and geographical information systems can be integrated with the IWMS to give users real-time information on assets inside and outside of buildings.

Sustainability Institute Hall – which houses the Golisano Institute for Sustainability at the Rochester Institute of Technology (RIT) – is used as a testbed for several smart building and sustainability technologies, including 1,200 sensors embedded throughout the building that monitor and can automatically adjust temperature, light, occupancy, and energy generation. The building generates more power than it consumes by using a 400-kilowatt fuel cell as its primary energy source, three vertical-axis wind turbines, 144 roof-mounted solar panels, and an eight-well geothermal heating/cooling system, all of which are monitored by an intelligent microgrid tracking system. The building also utilizes a high-performance building envelop with transparent, heat-mirror glass that contains coatings that conduct electricity and can be heated to room temperature; a stationary, louvered sun shade system on the south facade; an interior green wall that contains 1,776 plants; an extensive and intensive green roof; a rainwater collection and recycling system; low-flow fixtures; electric vehicle charging stations; regional and recycled materials; bike commuter infrastructure; an energy-regenerating elevator; and water efficient landscaping that contains native, drought-tolerant plants. The building is LEED Platinum certified and serves as a real-time teaching and research tool (Kiley, 2014).

The same intelligent monitoring systems can also be applied at larger scales in the form of "smart cities" where infrastructure, transportation, energy, emissions, air and water quality, waste management, and pedestrian circulation can be tracked, monitored, and controlled. In Zurich, Switzerland, smart streetlights (also known as Smart City Towers) can control the brightness of streetlights in response to traffic flow, supply power to electric cars, identify free parking spaces, provide WiFi, and determine the fullness of garbage containers. In Seoul, Korea, a BMS was implemented that monitors vehicle positioning and communicates with information panes at bus stops as well as to applications on the internet or mobile devices. Seoul is also implementing AI surveillance cameras that can detect specific movements such as falling, physical assault, or disasters and transmit the images to police, fire, or other emergency management agencies. Given these technologies, it may not be long before AI or an algorithm will be able to predict crime before it happens – that is, Steven Spielberg's movie *Minority Report* or Harold Finch's "machine" in the show, *Person of Interest*.

Green roofs

Green roofs – also referred to as eco-roofs or living roofs – generally contain a waterproof membrane, a drainage layer, a filter, and a substrate of growing medium/soil and vegetation that are applied to the structural base layer of rooftops. Green roofs can reduce the urban heat island effect, provide energy efficiency, manage stormwater, improve air quality, reduce noise, improve aesthetics, improve biodiversity, and create new amenities for gathering, agriculture, and education. Green roofs are classified by their soil depth. Extensive systems have 3–5 inches (7.6–12.7 cm) of soil depth, are relatively low-maintenance, contain hardy vegetation such as mosses, sedums, succulents, and some grasses and can support around 15–25 pounds/square foot (73–98 kg per square meter); semi-intensive systems are generally in the 5–7-inch (12.7–17.8 cm) depth range, contain perennials, sedums, grasses, and small shrubs, and can support between 25 and 40 pounds per square foot (122–195 kg per square meter); intensive systems have soil depths in the 7–24-inch (17.8–61 cm) range, can support 80-plus pounds per square foot (390 kg per square meter), and can be planted with perennials, shrubs, trees, and vegetables; earth-sheltered structures have 3 feet (0.9 meters) or more soil depth and contain three or more surfaces that are buried by the earth (floor, wall, roof); and subterranean structures are completely submerged underground.

Rooftops are a major culprit of the urban heat island effect. A typical asphalt, tar, or gravel rooftop absorbs around 85–90 percent and reflects 10–15 percent of solar radiation, subsequently making urban areas an average of 2–8 degrees Fahrenheit warmer than surrounding areas (Earth Pledge, 2005). The higher temperatures also create formidable conditions for the formation and trapping of smog and air pollutants. Data from several test sites shows that greening rooftops can effectively combat the perils of urban heat islands. On a summer day, a typical asphalt roof can reach up to 160 degrees Fahrenheit (71 degrees Celsius), while a green roof rarely exceeds 80 degrees Fahrenheit (26.6 degrees Celsius) (Earth Pledge, 2005). Environment Canada estimates that greening just six percent of Toronto's rooftops (70 million square feet) would decrease summer air temperatures in the city by between 1.8 and 3.6 degrees Fahrenheit (equivalent to a 5–10 percent reduction) (Liu & Baskaran, 2003). Air quality is also improved by the vegetation's ability to remove particulate matter through deposition and absorbing greenhouse gases and carbon dioxide. In 2018, the U.S. Environmental Protection Agency (EPA, 2018) conducted a study in Kansas City, Missouri, and found that the over 700,000 square feet of green roofs installed between 1999 and 2020 would result in a decrease of 384 pounds of nitrogen oxide, 734 pounds of sulfur dioxide, and 269 tons of carbon dioxide in 2020 which equate to monetized health benefits estimated between $35,500 and $80,500.

In addition to outside air, green roofs can insulate a building to moderate the temperature of the building and decrease heating and cooling losses. The National Research Council of Canada has shown green roofs to decrease the average daily energy in a 400-square-foot test building by over 75 percent (Liu & Baskaran, 2003). The decrease in energy use ultimately leads to reduced greenhouse gas emissions. Green roofs also have the potential to balance the hydrologic cycle by absorbing rainwater, filtering pollutants, and mitigating the spillage issues associated with combined sewer overflows (CSOs), where sewage, industrial wastewater, and stormwater are channeled into a single wastewater treatment plant. Extensive green roof systems can retain 50 percent of a one-inch rainfall and 50–75 percent of all rainfall annually, and they can delay runoff by about 4 hours and slow it by 42–96 percent (Earth Pledge, 2005).

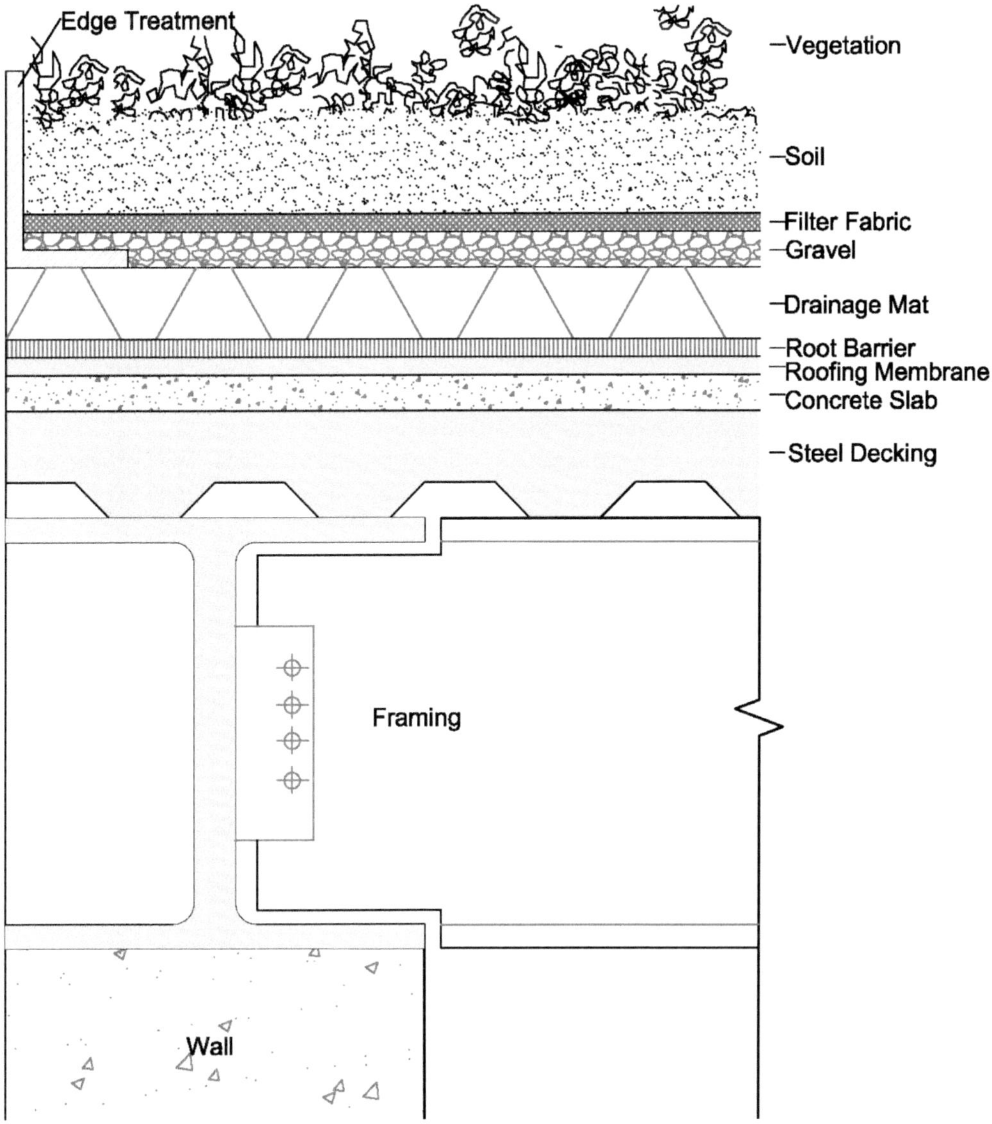

FIGURE 9.1 Typical section of a green roof.

Note: Modular systems are available that combine the root barrier, drainage mat, gravel, filter fabric, soil, and vegetation.

Green roofs can also be integrated into a larger ecological framework. Green rooftops on various buildings can serve as stepping-stone green spaces that connect to green walls, alley gardens, parks, natural areas, or riparian corridors to create a cohesive network of greenspace in urban community, or even regional scales. This can improve biodiversity, aesthetics, and human health by cleansing the lungs of urban areas. In addition, green roofs can provide spaces for agriculture that provide citizens in urban areas access to local, healthy foods and can serve as gathering areas for socialization, education, events, relaxation, or recreation.

Extensive green roof systems

As compared to intensive green roofs, extensive green roofs are lighter, less expensive, and require less maintenance. They are also easier to implement on existing buildings because of the lower impact on the roof's structural loading capacity. An extensive green roof was installed on the George D. Aiken Center at the University of Vermont as part of a project to green the entire building. The Aiken Center was built in 1981 and has since housed the Rubenstein School of Environment and Natural Resources. Ironically, the interior space did not share the same "green" themes that were taught in many of the environmental courses offered in the building. Students complained of dark, dingy, and windowless spaces with poor air quality and circulation. In 2010, the university began a major gutting of the building that became known as "the Greening of Aiken." William Maclay Architects was chosen to lead the design. Construction was completed in 2012 and the renovated building also includes a high-performance building envelope and windows that are 62 percent more energy efficient than the original; waterless/low-flow fixtures that use 50 percent less water than those in a similar-sized, conventional building; natural ventilation and lighting; environmental/energy-monitoring systems; a multi-purpose solarium on the south facade; and local, and renewable/recycled building materials (UVM, n.d.a). The extensive green roof contains eight different micro-watersheds – six with vegetated trays and two control sites without vegetation. In addition to species that are commonly used in green roofs, such as sedum and chives, researchers decided to test native Vermont plants and the use of biochar – a carbon-rich soil additive produced without the creation of carbon dioxide emissions. The green roof is monitored by tipping buckets that are visible on the third floor to monitor water absorption of the different mediums and the impact of each watershed on stormwater runoff. The building was awarded LEED Platinum certification in 2013 and officially became an NZEB in 2023 (White, 2023).

In Vermont, pre-vegetated sedum trays used for extensive green roofs were also successfully installed on the roof of the Heritage Aviation Hangar, atop the parking garage at the Burlington International Airport, and atop the Squash Center at Middlebury College. Heritage Aviation has an upper 30,000-square-foot green roof that retains the first inch of water in a rainfall event. After this, excess water is drained to a lower green roof, and, if needed, to a 34,000-gallon underground storage tank. If the tank fills, the excess water is drained to an on-site bioretention basin. The parking lot is also constructed of pervious concrete. The green roof, collection tank, and pervious pavement combine to decrease stormwater runoff for roughly three acres of impervious surface area. A 29,400-kilowatt-hour (kWh) photovoltaic array is also mounted on the roof's southern edge. Twelve solar hot-water panels heat food-grade glycol fluid that is circulated into a holding tank and used to heat the building's domestic water supply. In addition, the site contains a 100,000-kWh wind turbine (enough to supply energy to 16 average houses) (Heritage Aviation, n.d.). This is an example of how multiple green technologies can be combined at a site to increase the overall impact on the environment. The rooftop of the parking garage at the Burlington International Airport also combines an extensive green roof with a large solar array.

The Verdant Laboratory at the California Academy of Sciences also utilizes an extensive green roof that is a demonstration of the academy's mission to protect the natural environment. The laboratory was designed by the Renzo Piano Building Workshop and contains a museum, a planetarium, an aquarium, and several laboratories, collections, and offices. The green or "living" roof contains a central glassed piazza that is flanked by seven vegetated domes, two of which are 90 feet (27.4 meters) in diameter and house a planetarium and a rainforest. The domes were strategically placed to funnel cool air into the central piazza. The rainforest dome is dotted with several skylights that

FIGURE 9.2 Extensive green roof systems. Top left: vegetated trays at the Aiken Center, University of Vermont; top right: Heritage Aviation in South Burlington, Vermont, with wind turbine and roof-mounted solar panels in the background; bottom left: Squash Center, Middlebury College, Vermont; bottom right: roof of the parking garage at Burlington International Airport.

allow ample light for the plants and can be opened or closed by the BAS to facilitate further natural ventilation. The BAS contains sensors on the interior and exterior of the building that monitor air quality and temperature. In addition to windows and vents, the canopy is made up of 60,000 photovoltaic cells that are estimated to produce 213,000 kilowatt hours of clean energy a year that prevents the release of 405,000 pounds of greenhouse gas emissions annually (CAS, n.d.).

The 2.5-acre (1-hectare) green roof is composed of seven layers. From lowest to highest, these layers consist of a concrete slab, a waterproofing layer, an insulation layer, a drainage layer, an erosion control blanket, 3 inches (7.6 cm) of soil, and the coconut-husk trays that contain the plants and another 3 inches of soil. There is also a grid of rock gabions that keeps the soil in place and provides walking access. The massive roof contains enough vegetative space to absorb 100 percent of excess storm water, which is about 3.6 million gallons of water per year (CAS, n.d.). When saturated, the runoff is channeled into an underground chamber where it disperses into the surrounding landscape. Vegetation was selected to attract bird and insect species, but it also had to be able to survive the roof's minimal maintenance/irrigation schedule, shallow soil depth, and microclimate. Nine native California species were chosen, including beach strawberries, selfheal, sea pink, stonecrop, tidy tips, miniature lupine, California plantain, and California goldfield plants. Overall, 1.7 million plants were planted in the 50,000 biodegradable coconut husk trays (CAS, n.d.). The green roof keeps the surrounding air 40 degrees Fahrenheit cooler and interior temperatures 10 degrees Fahrenheit cooler than a standard roof. In addition to providing insulation for heating and cooling, the green roof reduces low-frequency noise in the interior by about 40 decibels (Greenroofs.com., n.d.).

Extensive green roofs can also be used on water reservoir covers. To protect the public water supply from disease-causing microorganisms and contaminants (especially *Cryptosporidium*) and acts of terrorism, the U.S. Environmental Protection Agency developed the Long Term 2 Enhanced Surface Water Treatment Rule (LT2 rule) in 2005 to improve the quality of drinking water. The rule requires that uncovered finished water reservoir systems be either covered or retreated to inactivate 4-log virus, 3-log *Giardia lamblia*, and 2-log *Cryptosporidium* (EPA, n.d.b). While covering previously uncovered reservoirs may detract from the aesthetics of the landscape surrounding the water source, the covers can employ the same extensive green roof technology that is used for building rooftops, depending on the structure of the lid. The tops of reservoirs can often provide new open space in dense, urban areas that are land-locked and pressed for greenspace. The city of Seattle, for example, is in the process of adding 76 acres (30.8 hectares) of new park land atop reservoir decks; the deck parks will be "intensive" and include running tracks, sports fields, picnic grounds, and unprogrammed lawns (Harnik, 2010, pp. 116–117). Similar to green roofs on standard buildings, vegetation has to be selected that will not penetrate the structure and jeopardize the water supply. Extensive tray systems can also be employed on reservoir covers that are constructed with lighter materials.

Semi-intensive green roofs

As stated earlier, semi-intensive green roofs are in the 5–7-inch (12.7–17.8 cm) depth range. At Middlebury College in Vermont, a semi-intensive, sloping green roof was installed on the main portion of the oval-shaped dining hall roof of Atwater Hall. The poured-in-place concrete roof was covered by a loose-laid waterproofing membrane and 7 inches of engineered soil. As opposed to an extensive system of pre-vegetated trays, Atwater's roof was planted in place. Designed by Kieran Timberlake Associates, the concept of the green roof was to utilize native vegetation to tie in with the surrounding landscape that includes the Green Mountains in the background and upland meadows and rock outcrop plant communities in the foreground. Species with bulbs, tubers, or shallow roots were selected because they are more adaptable to the relatively thin soil depth. To mimic vegetation in a flowering meadow, vegetation was planted in bands that vary in color in each season.

Intensive green roofs

As mentioned earlier, intensive green roof systems have deeper growing depths that facilitate the planting of larger trees and bushes and can also provide usable space for agriculture, recreation, events, or socialization. In addition to the extensive green roof on Aiken mentioned earlier, the University of Vermont has three other intensive green roofs: the 18,000-square-foot (1,672 square meters) roof over the loading dock on the west end of the Dudley H. Davis Center, and two smaller roofs on the terraces of the University Heights Complex. The Davis Center green roof is planted with a variety of drought-resistant grasses, with the taller ornamental grasses around the central lawn comprising a fescues/ryegrass blend that requires little maintenance or irrigation. The soil depth varies from 6 to 21 inches (15–53 cm) and can absorb between 12 and 42 pounds of water per cubic foot. Overall, the roof can hold up to 80 pounds per cubic foot (1,281 kg per cubic meter), including the soil itself, the grasses, and stormwater (UVM, n.d.b).

FIGURE 9.3 Left: Verdant Laboratory; right: close-up of the vegetated dome.

Source: Mages courtesy of Renzo Piano Building Workshop

FIGURE 9.4 Top: smart building, Sustainability Institute Hall, RIT; bottom: semi-intensive green roof of Atwater Hall, Middlebury College.

The University of Vermont Medical Center also has an intensive green roof on the wing of the Radiation/Oncology Unit. The space was designed by Wagner–Hodgson Landscape Architects and consists of a recreational lawn, a healing garden planted with native shadblow serviceberry and ferns, a plaza, and vegetable gardens. Rainwater is harvested in a garden shed and used to irrigate the plants and vegetables. The vegetables are used to supplement the hospital's food service, in line with the ethics of "responsible site development." Keith Wagner, the principal of

FIGURE 9.5 Intensive green roofs. Top left: loading dock, Dudley H. Davis Center, University of Vermont; top right: seating area, Radiology/Oncology Unit, University of Vermont Medical Center; bottom left: meadow grasses, Radiology/Oncology Unit; bottom right: water-harvesting shed (right rear), raised vegetable gardens (right), and perennial/herb bed (left), Radiology/Oncology Unit.

the firm, describes the design as an "abstraction of the aerial patterns of Vermont fields, meadows and villages . . . the space is both therapeutic and an educational tool for the patients of the Center to learn about nutrition and healthy food production" (Monty, 2014).

Intensive green roofs can also support agricultural gardens that can provide a source of local food for people in urban areas with little or no access to farms and/or space to plant a garden. Top Leaf Farms in Oakland, California offers design, build, and consultation services for farming systems, components, and lifestyles with a vision of creating 1000 new urban rooftop farms, street farms, and community food hubs by 2030 (Top Leaf Farms, n.d.). The firm has several urban and rural projects, including the Garden Village student apartment complex that contains 18 modular buildings that contain roof farms that are linked by a network of exterior walkways. In addition to the 12,000 square feet (1,115 square meters) of roof farms, green space is weaved in common areas between the buildings at the ground level, a concept that breaks the mold of the monolithic form that is often associated with large apartment buildings. The Dutch firm MVRDV has also broken ground on the Tainan Xinhua Fruit and Vegetable Market in Taiwan, which will contain an undulating roof garden used for public gathering and growing crops.

Blue, brown, white, solar, and wind roofs

Blue roofs, also referred to as blue-green roofs, are designed to increase the water detention of a standard green roof through the use of detention ponds or tanks that can be open or placed under vegetation (hence the "blue-green" nomenclature). The tanks can temporarily store the water and slowly release it to control flooding and stormwater runoff. Water can be used for irrigation or used as design elements to enhance the roofscape on rooftops that allow public access. Blue roofs can also cool and decrease energy consumption for a building. Water from the detention/ponds/tanks can also be sprayed onto other surfaces to enhance cooling.

Brown roofs – also known as biodiverse roofs – are similar to semi-intensive and intensive green roofs but are focused more on replicating the pre-construction ecology of the site. Soil excavated from the site is used as the substrate and native vegetation is planted to attract and encourage specific habitats. Over time, the roof is left to self-colonize by wind, birds, and insects, thus producing a more complex and diverse range of habitats than green roofs that can include wetland areas, landforms, logs, boulders, and stones.

Where there is insufficient budget for a green, brown, or blue roof, painting the roof with a reflective color (often white) is a less expensive alternative to foster cooling. A study by Sharma et al. (2016) conducted in Chicago, Illinois, found white or "cool" roofs to decrease the daytime surface temperature for core high-intensity urban areas by 7–8 degrees Celsius, which was approximately 1 degree Celsius better than a green roof applied to 100 percent of the roof surface. A 2014 study conducted by the Lawrence Berkeley National Laboratory found white roofs to be the most cost effective compared to a green or black roof over a 50-year life cycle cost and to reflect roughly three times more sunlight back to the atmosphere than green roofs (Sproul et al., 2014). White roofs can also be combined with green, brown, and blue roofs where stormwater management and/or biodiversity are needed for the site.

It is also possible to combine solar and wind installations with green, brown, blue, and white roofs. Such systems are often referred to as solar garden roofs, solar green roofs, biosolar roofs, wind roofs, or "yellow" roofs. Design considerations should ensure vegetation does not grow over the solar panels and that the panels do not interfere with the infiltration of the growing medium. Small-scale rotary and vertical-axis wind turbines can also be installed on rooftops. Although adjacent buildings can block a consistent stream of wind on rooftops in dense, urban areas, new technologies are emerging that may be able to harness wind on rooftops more efficiently. PowerNEST/IBIS Power in the Netherlands is developing vertical turbines that sit below a roof of solar panels, Aeromine Technologies in Houston Texas is working on a bladeless system that funnels wind into hollow airfoils, and RidgeBlade in Canada is developing a horizontal system that is mounted along the ridge line of pitched roofs.

Green walls

Green walls are essentially a vertical manifestation of green roofs. There are four major categories of green walls: naturally occurring green walls where vegetation grows directly on the wall or climbs up from the ground without the use of a support structure; facade systems that contain trellises and/or planter boxes that allow plants to attach and climb up and/or cascade down the support structures; modular or "living wall" systems where pre-vegetated panels, bags, mats, or blankets with or without irrigation systems are affixed to the wall or a frame; and retaining walls that are engineered structures to stabilize slopes and prevent erosion. Similar to green roofs, the deeper mediums support a larger variety of plant species.

In addition to aesthetical improvements, green walls can decrease the urban heat island effect in urban areas, improve exterior air quality by absorbing particulates and gases, improve energy efficiency, decrease noise levels, increase biodiversity, and can be planted to support fruits and

FIGURE 9.6 Green wall systems. Top left: modular green wall system at Golisano Institute for Sustainability at RIT; top right: Trellis interior green wall system; bottom: exterior cable system on Jeffords Hall at UVM allows plants to grow vertically on walls without damaging the facade.

vegetables. Systems using support structures or modules can prevent structural damage to the wall and can provide an air space between the wall and vegetation that can increase the efficiency of the vegetation's insulating properties. A study by Minke and Witter (1982) found a 1.6-inch (4 cm) gap between the wall and a 6.3-inch (16 cm) green wall system could raise the thermal efficiency of the wall up to 30 percent. A study at the University of Plymouth compared a west-facing facade of a building with and without a living wall system and found the amount of heat loss to be 31.4 percent lower for the living wall compared to the original structure (Fox et al., 2022). An analysis by Hadba et al. (2017) showed living wall systems have the highest installation and maintenance costs but can provide more vegetation variety and better noise reduction and thermal comfort than facade or direct systems. Façade and direct systems had roughly the same energy efficiency and air quality impacts, with facade systems causing less damage to the wall than direct systems.

Earth-sheltered structures

Using the earth to shelter portions of a structure is essentially like a large-scale intensive green roof system. However, as mentioned earlier, if three or more surfaces of a structure (ground, roof, and wall) are buried in the earth, this book classifies it as an earth-sheltered structure rather than an intensive green rooftop. Structures can be bermed with earth, stepped into a hillside, or completely submerged underground. The earth acts not only as a means of support but also as a thermal mass that can be used to insulate the structure. Since the earth has a high density, it takes longer to heat in the summer and longer to cool in the winter. This "thermal lag" makes it more efficient to maintain constant internal temperatures. In addition to energy savings, earth-sheltered buildings provide privacy, soundproofing, and protection against high winds, tornadoes, and hurricanes. Concrete construction supplemented with waterproof membranes and insulation provides both adequate structure and protection against moisture and humidity. Earth tubes can also be used to provide fresh air and passively heat and cool the interior. Fresh air from the surface is filtered and enters underground tubes where heat is either captured from or dissipated to the ground, depending on the season. As discussed in Chapter 5, this geothermal heat exchange is a component of passive design and can either supplement or be used as an alternative to conventional heating and cooling systems.

In regions with cold winters, the exposed wall should ideally face south (in the northern hemisphere) and be angled to maximize winter sunlight and provide design parameters (such as overhangs and smart glass) to provide shading in the summer. Smart glass systems allow the glass to adapt to be either more translucent or more transparent, ultimately controlling how much light is absorbed or reflected. Some technologies automatically transform when exposed to heat or light, while others can be controlled either mechanically or through electric voltage. In regions with mild winters and hot summers, the exposed wall should ideally face north (in the northern hemisphere).

In general, the steeper the slope, the less excavation is required. Flat topography requires more excavation than sloped topography and can thus be more expensive and labor intensive. Sloped topography can also provide cascading terraces that can be programmable space. The terraces can be used for agriculture (roof gardens), gathering spaces, or even as seating for events taking place on the ground level. In the event of an emergency, the terraces can provide an alternative means of egress to the outside. Earth-sheltering on sloped topography can also minimize the visual impact of a structure. This is evident in the addition to Lycée Jean Moulin in Revin, France by Duncan Lewis Scape and OFF Architecture. From a distance, the building seems to cascade down the hillside in a geometrically organic fashion. A similar technique

FIGURE 9.7 Terracing roofs into sloped topography. Top: Lycée Jean Moulin, Revin, France; bottom: Janelia Research Campus, Howard Hughes Medical Institute, Ashburn, Virginia.

Top image courtesy of Duncan Lewis, Jean De Giacinto, OFF Architecture, Egis Grand Est–BET TCE, Egis Concept Elioth–BET HQE, Echologos–Acousticien Base–Paysagiste, and Region Champagne Ardenne; bottom image courtesy of Paul Fetters.

FIGURE 9.8 Earth-sheltered buildings. Top: Campus Center for EWHA Women's University, Seoul, South Korea; middle: Larner Learning Commons, University of Vermont; bottom: Peter Vetsch's Nine Houses development.

Source: Top image courtesy of Gaëlle Lauriot-Prévost, Dominique Perrault Architecture; Bottom image courtesy of Christine Thuring

FIGURE 9.8 (Continued)

was employed at the Howard Hughes Medical Institute's Janelia Research Campus in Ashburn, Virginia. The building is 900 feet long (274 meters) and 270 feet deep (82 meters), and roughly 180,000 square feet of roof space is bermed into the earth. Rafael Vinoly, the architect, states, "The landscape is the building – it creates a counterpoint between technology and nature that enriches the research performed there. It is a highly technical structure in terms of equipment and flexibility, yet the entire composition looks like a natural thing" (Vinoly, 2014).

The slope of the site does not even need to be that extreme to employ a subterranean or earth-sheltering technique. In Dominique Perrault's Campus Center for EWHA Women's University in Seoul, South Korea, the building is constructed underground with a large linear rift or "valley" down the center that provides passage through the campus. Overall, six levels are built underground, the bottom two of which are used for parking. The green rooftops are lined with pedestrian paths, vegetation, and seating areas and are gradually sloped along the rift, thereby providing a functional connection through the campus. This project erases the stigma of the building as an isolated object on the landscape that is often an obstacle to circulation patterns and/or to views of the surrounding landscape. The Larner Learning Commons at the University of Vermont also took advantage of a moderate change in topography. The roof consists of gardens, terraces, paths, and roadways and relates to the roof of the underground parking facility.

The building–land fusion need not be exclusive to large-scale commercial or public architecture, but can also be done at smaller, residential scales. Architect Peter Vetsch's Nine Houses development in Dietikon, Switzerland represents a radical approach to housing. The cave-like houses – which have been compared to a Hobbit community – are positioned around a small lake and covered with 3.3–6.5 feet (1–2 meters) of earth. The roof construction creates an insulation layer that eliminates the need for air conditioning or significant heating. When heat is required, it comes from a geothermal system that pumps heat up through 300-foot-deep (91-meter) underground ducts.

TABLE 9.2 Comparison of green roofs

	Extensive	*Semi-Intensive*	*Intensive*	*Earth Sheltered*	*Subterranean*
Soil depth	3–5 inches	5–7 inches	7–24+ inches	>36 inches, surrounded by ground on 3 or more sides	>36 inches; structure completed below grade
Vegetation	Mosses, sedums, herbs, succulents, grasses	Perennials, sedums, grasses, small shrubs	Perennials, shrubs, trees, vegetables, paths	Perennials, shrubs, trees, vegetables, grasses, stones, paths, driveways	Perennials, shrubs, trees, vegetables, grasses, stones, paths, driveways
Load capacity	15–25 lbs/sf	25–40 lbs/sf	80+ lbs/sf	80+ lbs/sf	80+ lbs/sf

Sources: Minnesota Stormwater Manual. (n.d.). *Types of green roofs*. https://stormwater.pca.state.mn.us/index.php/Types_of_green_roofs; Earth Pledge. (2005). *Green roofs: Ecological design and construction*. Schiffer Publishing; Green Roof Technology. (n.d.). *Green roof types all under one tent*. https://greenrooftechnology.com/green-roof-finder/green-roof-types/; Department of Energy and Environment. (n.d.). *Green roofs*. https://doee.dc.gov/sites/default/files/dc/sites/ddoe/publication/attachments/Section%203.2%20%20Green%20Roofs.pdf

Buildings that create green space

When it is not topographically possible to build into the earth, there is always the option to construct buildings to be their own, self-contained "mountains." This was the concept behind Vo Trong Nghia Architects' design for FPT University in Ho Chi Minh City, Vietnam. Given the pace of infrastructure expansion, cities in Vietnam are experiencing environmental stress, energy shortages, pollution, and reduced green space. The design of FPT University seeks to mitigate these impacts by creating an undulating building akin to a forested mountain that will disrupt the verticality that dominates the city's skyline. The central building's floors are staggered and integrated with pockets of trees and vegetation. Greenery from the levels is connected to a central courtyard and around the perimeter of the building. (VTN, 2016).

In the Oersted City neighborhood in Copenhagen, the architecture firm BIG (Bjarke Ingels Group) designed a stepping, landscaped, ten-story building for 80 apartments known as the Mountain. The parking forms the base of the terraced housing and is integrated into the design by utilizing ramped, cathedral-like spaces that are lined with art and coordinated by color. The north and west facades of the parking base are covered by perforated aluminum plates that filter air and light. The plates display a mural of Mount Everest in the day and are illuminated from the interior at night. Each apartment has direct access to a terrace that is surrounded by wooden, vegetative planters that double as parapets for the apartments on the level below. The wooden parapets and turf surrounding the deck are connected at the edges to form a contiguous array that makes the structure look like a mountain of gardens. Rainwater is collected in a central tank and distributed to the planters. More than 20 different plant species that blossom at different times of the year are planted in the planters, adding to the living hillside aesthetic.

BIG is also working on a project for the redevelopment of 120 Fleet Street in London. The proposal is for a roughly 20-story building that contains a sloped, terraced roof from the tip to

FIGURE 9.9 FPT University, Ho Chi Minh City, Vietnam.

Source: Image courtesy of Vo Trong Nghia Architects

FIGURE 9.10 The Mountain in Copenhagen with close-ups of terraces.

Source: Images courtesy of Bjarke Ingels Group (BIG)

the base. The vegetated terraces will create usable space and make the building look essentially like a "green mountain."

Rooftops for recreation and circulation

In addition to the earth-sheltered terraced roofs discussed earlier, freestanding buildings can utilize terracing and ramping of rooftops for amenities, recreation, navigation, and connection. In Copenhagen, a rooftop activity park was designed on the sloped roof of the Amager Bakke (also known as CopenHill) waste-to-energy plant. The activity park contains a 1,640-feet (500-meter) ski/snowboard trail, hiking/running trails, playgrounds, a toboggan area, and fitness structures. The ski/snowboard surface is made of green plastic (Neveplast) and can be used year-round.

FIGURE 9.11 CopenHill. Images courtesy of Bjarke Ingels Group (BIG).

Source: Photographer: Rasmus Hjortshoj

One of the vertical walls of the building also features a 279-feet (85-meter) climbing/rappelling wall. Beneath the activity park, turbines convert 440,000 tons of waste annually into enough clean energy to deliver electricity and district heating for 150,000 homes (Baldwin, 2019). As stated by Bjarke Ingels – the founder of the firm BIG (Bjarke Ingels Group) that designed the building:

> CopenHill is a blatant architectural expression of something that would otherwise have remained invisible: that it is the cleanest waste-to-energy power plant in the world. As a power plant, CopenHill is so clean, that we have been able to turn its building mass into the bedrock of the social life of the city – its façade is climbable, its roof is hike-able and its slopes are skiable. A crystal-clear example of hedonistic sustainability – that a sustainable city is not only better for the environment – it is also more enjoyable for the lives of its citizens.
>
> *(Ingels, 2019)*

BIG also utilizes rooftops in several other projects, including the Audemars Piguet Hotel in Le Brassus, Switzerland, the VLTAVA Philharmonic in Prague, and the Gastronomy Open Ecosystem building in San Sebastian, Spain, to name a few. At the Audemars Piguet Hotel, five zig-zagging slabs provide continuous interior and rooftop circulation that connects the village to the valley. The offsetting of the slabs forms accessible terraces that weave the building into the topography and frames unique views of the landscape. Interwoven wood louvers provide shading for the facade, cool the building, and are integrated with the slabs, further blending the building with the pastoral landscape. The energy performance is also assisted by 86 photovoltaic panels that provide a portion of the building's energy use. The ecological and socially sustainable design construction, design, and performance have earned the building Minergie-ECO certification. A similar zigzag array is also being utilized in their project for Google Caribbean in Sunnyvale, California. At the VLTAVA Philharmonic, the corners of the buildings were lowered and raised at strategic touch points, allowing a contiguous connection from the public plaza, the interior space, the rooftop, and the Vltava River. At the Gastronomy Open Ecosystem building, BIG solved the issue of a 33-feet (10-meter) height difference by creating terraces in the building that gently rise from the ground and serve as an extension of the landscape. The building creates a public plaza that connects the Camino de Santiago pilgrimage route and the Cantabrian Sea.

Infiltrating buildings with greenspace

Another building design strategy that can preserve the landscape is to allow vegetation and trees to grow through the space. Early renditions of this concept date back to ancient Greece and Rome, where open, peristyle courts were placed in the centers of the spaces to funnel natural light and the collection of rainwater. Courtyards and atria can break up a structure with a large footprint and can decrease the overall density of a project if strategically designed.

As a graduate student in architecture, one of our projects was to design an environmental education center on a forested coastline near Whidbey Island in Puget Sound. After analyzing the site, it was my belief that the best way to educate people about the environment was to not remove any trees or vegetation, build nothing, and let people explore nature as it was created.

FIGURE 9.12 Audemars Piguet Hotel. Images courtesy of Bjarke Ingels Group (BIG).

Source: Top and middle left images courtesy of Audemars Piguet Hotel

The professors didn't go for this at all. One of them asked me, "If you don't want to build anything, why are you even in architecture school?" Point taken. My next proposal was a series of pathways with interpretation kiosks, but once again, I was told this didn't meet the interior spatial program requirements. After several rudimentary sketches (I am not that great of a designer, hence why I write books), I came up with something that looked like the Ewok Village on Endor in *Star Wars: The Return of the Jedi*. However, for the sake of not stealing George Lucas's idea, I came up with an undulating design of post and beam, curvilinear sections that contained voids in the roofs and floors to allow the trees to grow through the structure. The roofs of each level contain usable space and are connected by a cascading ramp down the center of the structure that links the forest and the beach with an underwater section, giving the user a face-to-face learning experience with all the ecosystems on the site.

Similar void methods to preserve existing trees and to have the trees interact with the living space were used by Paz Arquitectura in the Corallo House, located in a forest near Guatemala City, and by Travis Price Architects in the Hayes Residence, Berkeley Springs, West Virginia.

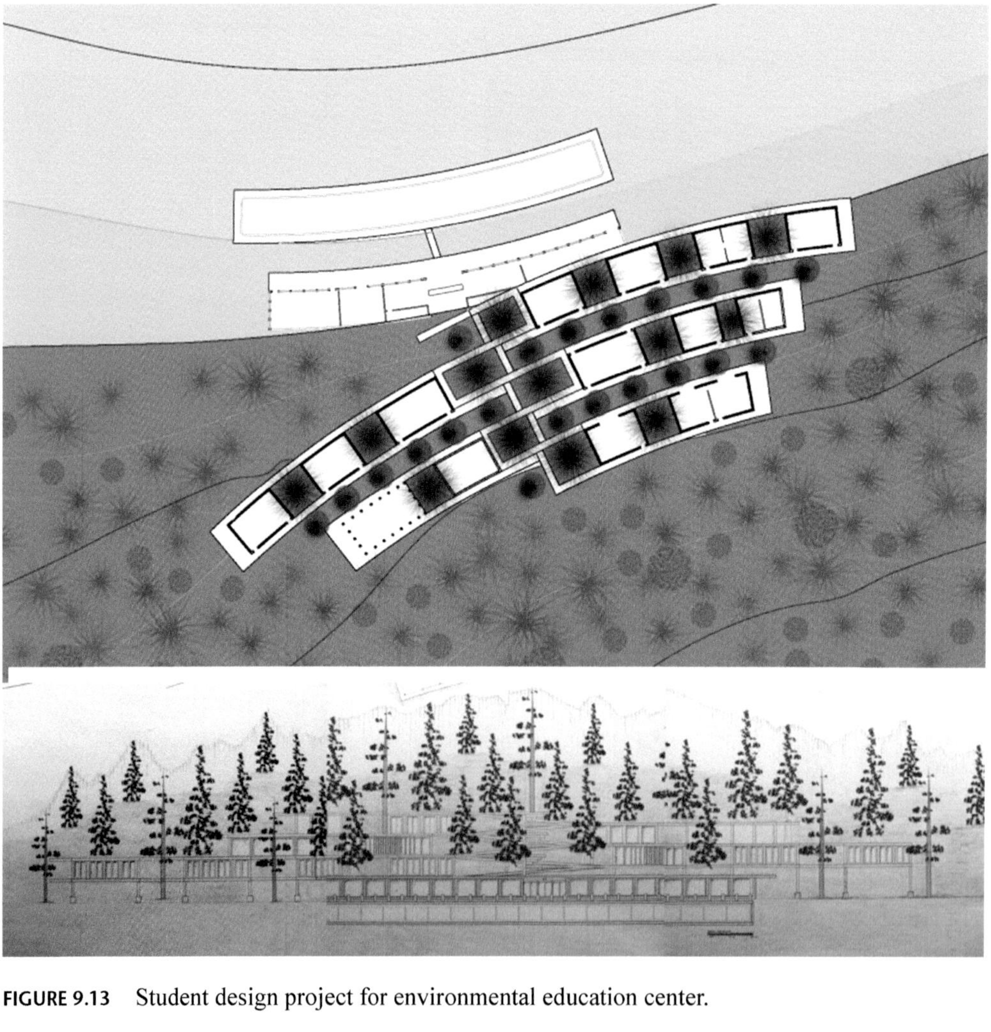

FIGURE 9.13 Student design project for environmental education center.

FIGURE 9.13 (Continued)

In the latter, voids are incorporated in both the exterior central courtyard and the interior living spaces. Trees puncturing the interior of the space are encased in glass boxes, allowing the trees to thrive while simultaneously shielding the interior from the outside elements. The exposed timber plank ceilings and parallel strand lumber beams further connect the structure to the wooded landscape.

Structures can also be built *within* the trees themselves. Although such designs are usually associated with small forts constructed for children, they can also be utilized for habitation on larger scales. Designs can be constructed around a single tree or may use multiple trees. The most efficient designs are flexible in that they allow the trees and beams to move independently to allow for tree growth. The goal is to provide the most structural support while minimizing damage to the tree. Designs range from small, intimate spaces, such as the Yellow Treehouse Café in Aukland, New Zealand or the cable-suspended, "levitating" 4Treehouse by Lucasz Kos in Ontario, Canada, to large, multi-towered resorts, such as Ariau Amazon Towers Hotel in Brazil or the Tree House Lodge in Limon, Costa Rica.

Bridging rooftops and buildings

As evident in the Garden Village catwalk-connected roof farms discussed earlier, buildings and rooftops can be bridged together to link attractions and provide an alternative site for programmable space that is separated from the ground plane. Skybridges have been implemented at the InTempo skyscraper in Benidorm, Spain, the Copper residential buildings in New York, the Sky

FIGURE 9.14 Top: trees growing through glass boxes at the Hayes Residence, Berkeley Springs, West Virginia; bottom: Community Treehouse at Oakledge Park, Burlington, VT.

Source: Top image courtesy of Travis Price Architects and Ken Wyner

Habitat residential development in Singapore, the Petronas Towers in Malaysia, and the Cross Towers in Seoul, Korea, and the Sky Park observation deck that bridges the roof tops of three buildings at the Marina Bay Sands resort in Singapore, to name a few. Dutch architects MVRDV also have a proposal for Seoul, Korea called the Cloud where two skyscrapers are joined by a pixelated, lego-like matrix at the 27th floors that will accommodate an atrium, restaurants, gym facilities, a conference center, and office-hotels.

Bridging buildings can also reroute pedestrians off crowded streets and provide a safer, weather-protected route for circulation. The Minneapolis Skyway System is a 10-mile (16 km) network of climate-enclosed skyways, corridors, and tunnels that connect 150 buildings across 80 city blocks. The first skyway was constructed in 1962 between the Northstar Center building and the Northwestern Bank Building. Although this section was demolished in the 1980s, it served as an impetus for the construction of more skyways and the official launch of the network in 1963. In 1974, the construction of the IDS building incorporated skyways on all four facades, effectively tying together several disparate systems. Skyways are now included in nearly all new buildings constructed in the central city. In addition to traffic flow, safety, and weather protection, the skyways have increased property and rental values due to the additional retail and service space.

Biomorphic and biomimetic design

As introduced in Chapter 2, biophilic design focuses on relationships between humans and nature, biomorphic design is design that is inspired by naturally occurring geometries and patterns found in nature and living organisms, and biomimicry or biomimetic design learns from or mimics the systems, patterns, and ecological processes of nature to solve human design challenges.

At the Mushroom House or "Pod House" in Pittsford, New York, architect James H. Johnson used the umbels of the wildflower Queen Anne's Lace as inspiration for the design for the attorney-artist couple Robert and Marguerite Antell. The residence consists of five circular pods that range between 30 and 34 feet (9–10 meters) in diameter that sit on reinforced concrete stems that range from 14 to 20 feet (4–6 meters) in height and taper from about 5 feet (1.5 meters) in diameter at the base to about 3 feet (~1 meter) in diameter where they attach to the pods. The pods consist of the foyer/family room/kitchen, the living/dining area, master bedroom, guest suite, and an outdoor deck. Molds of concrete and polyurethane were constructed on-site for both the bases and the roofs and were then reinforced with steel and raised into place. An earth-sheltered great room was added to the original design and consists of a den, a tree-form sculpture, fireplace area, and a pool table. The great room is connected to an outdoor terrace with a hot tub that conjures up the feeling of basking in a grotto. The interior truly feels as if you were living inside of the Queen Anne Lace, with the stems organically rising from the floor to the plant or fossil-like sculpted ceilings. Over 20 artists worked on the details, including sculptures, custom lighting, a water fountain, and painted tile floors by Marguerite Antell.

Although the Mushroom House effectively blends with its natural surroundings, other biomorphic designs may abstract the forms, patterns, and geometries inspired by nature to produce a design that complements the surroundings without fully committing to either contrast or blend with them. For example, although Chicago's Aqua Tower draws geometric inspiration from striated limestone outcroppings, the building is not meant to blend, but rather to create a sense of organic form in a dense, urban environment dominated by rectangular geometries. Likewise,

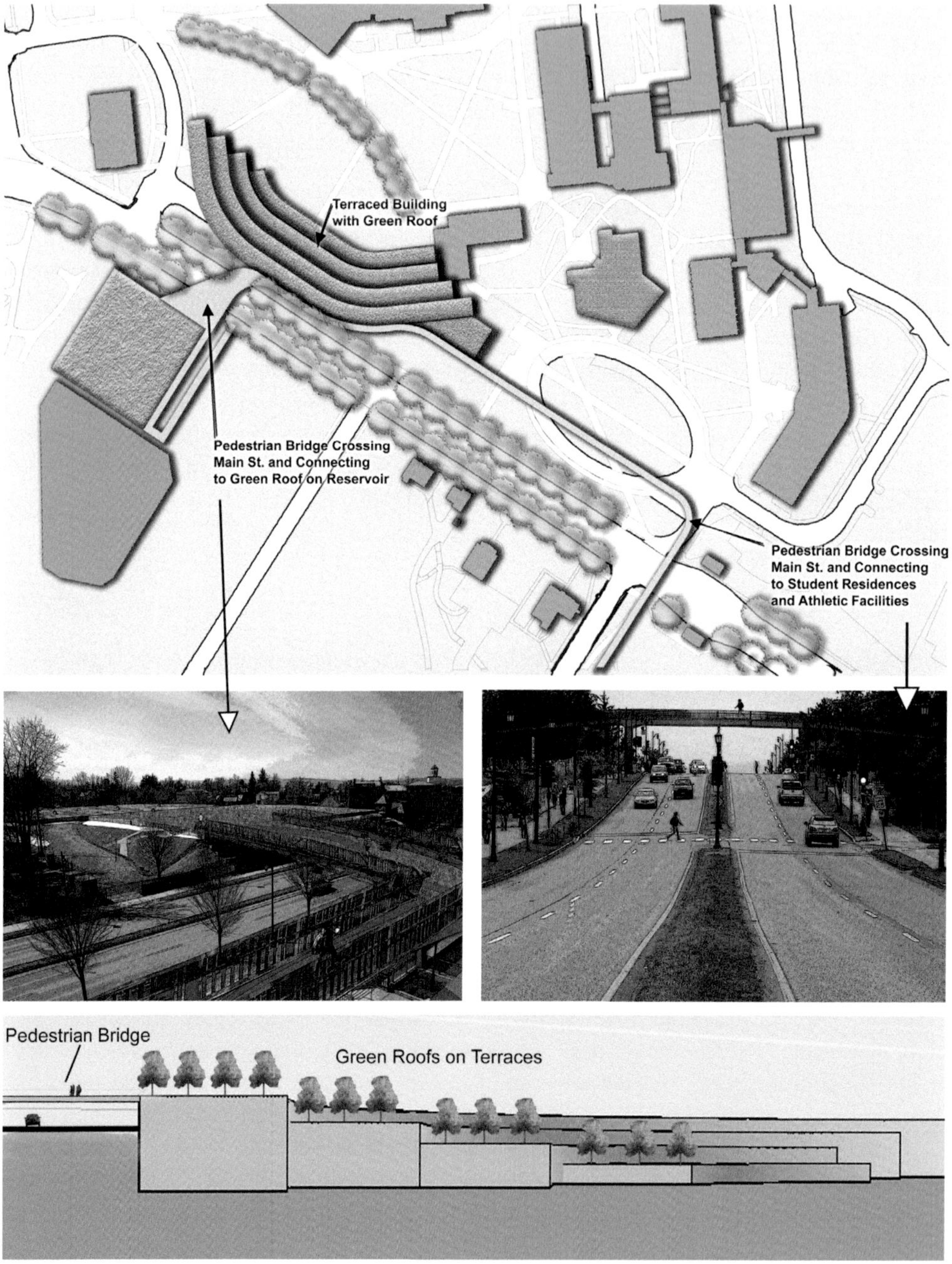

FIGURE 9.15 Conceptual proposal for a student center that combines several strategies: four curvilinear, sloped slabs with green roof terraces can be used for agriculture and gathering. The structure forms a natural amphitheater where people can gather on rooftops and watch concerts or ceremonies taking place in the central courtyard at the ground level. A pedestrian bridge connects the upper terrace to a proposed green roof on the city's water reservoir to the south. The pedestrian bridge ramps down to the ground heading east and ascends back up to cross over Main Street to connect the central campus to the residential and athletic campus.

FIGURE 9.16 Mushroom House in Pittsford, New York.

Toyo Ito's Grin building in Fukuoka, Japan also provides undulating, elliptical forms that break the grid of the island's rectilinear infrastructure.

Some critics argue that biomorphic design is just another form of modernism where designers are merely engaged in a form-making process with no regard for harmonizing with the surroundings or utilizing the ecological processes of the site. To put it another way, all because it looks like nature doesn't mean it is necessarily good for nature. Biomimicry, on the other hand, does not just solely focus on aesthetics, but rather on learning from strategies used by species in the natural world to provide solutions – so while biomorphism looks like nature and bio-utilization utilizes nature, biomimicry functions like nature (Biomimicry Institute, n.d.).

In addition to writing a book on the topic called *Biomimicry: Innovation Inspired by Nature*, Janine Benyus co-founded the Biomimicry Institute that established three elements of biomimicry: Emulate, which is the scientific, research-based practice of learning from and then replicating nature's forms, processes, and ecosystems to create more regenerative designs; Ethos, which is the philosophy of understanding how life works and creating designs that continuously support and create conditions conducive to life; and (Re)Connect, which is the concept that we are nature and find value in connecting to our place on the earth as part of life's interconnected systems (Biomimicry Institute, n.d.). Biomimicry doesn't merely utilize nature for a specific benefit but rather functions like nature by emulating and learning from its systems, behaviors, and patterns. For example, the company ECOncrete has developed a bio-enhancing admix that makes concrete more resilient by promoting the biogenic growth of oysters, barnacles, and corals on marine infrastructure such as seawalls, breakwaters, bridge/building foundations, and waterfronts. The company also produces three-dimensional, textured molds with the admix that create ecological niches and provide infrastructure to protect coastlines from rising sea levels (ECOncrete, n.d.). The Sabin Design Lab at Cornell University is also engaged in biomimicry in developing building skins that change a material's opacity and color in response to sunlight – a structural color phenomenon exhibited by the unique cellular behavior of butterflies and hummingbirds.

Building taller: healthy or not?

When faced with horizontal land constraints that are commonplace in dense, urban cities, the traditional solution has been to build taller. According to the International Building Code, a high rise is defined as a building with an occupied floor located more than 75 feet (22.9 meters) above the lowest level of fire department vehicle access (ICC, 2021). There are certainly pros and cons to living and/or working in a high-rise building, many of which are subjective. Those who advocate for high rises cite that they make efficient use of land, offer better views, have less street noise, are safer from crime because access can be controlled, have less air pollutants from traffic, and offer more on-site amenities than their horizontal counterparts. Those against high rises cite the cost and complexity of construction and maintenance, higher energy consumption, evacuation issues when faced with a disaster, wind-induced motion, displacement of low- and median-income residents and businesses, and livability issues due to isolation from both nature and street-level activities. Some argue high-rise construction is "vertical sprawl" with poor-quality buildings being built too close together that block views, block sunlight at the street level, and are out of scale with human mobility, senses, sightlines, and socialization patterns. Researchers at University College London found electricity use is close to two and a half times greater per square meter of floor area for high-rise office buildings 20 or more stories compared to low-rise buildings of six stories or less, gas use increased with height in residential buildings, and carbon emissions were higher as building height increased. More specifically, carbon emissions were found to be 60 percent higher in high-rise, air-conditioned buildings with no natural or mechanical ventilation (UCL, 2017).

There is also the question of human comfort when working and living at heights. A colleague of mine in Buffalo, New York, once told me he had an interview on the 87th floor of the World

Trade Center in the 1990s. The height was so extreme that he became motion sick and dizzy and had a panic attack that forced him to blunder the interview. Although there is a large spectrum regarding acrophobia (fear of heights) and subjective and object fears or dangers of falling, my colleague may not be alone. In evaluating responses to height exposure – including a physiological height imbalance that results from an impaired visual control of balance, visual height intolerance, and severe acrophobia – Huppert et al. (2020) found about one-third of the general population suffers from susceptibility to acrophobia and visual height intolerance. As far as people's satisfaction or comfort levels living in high rises, the data is mixed and somewhat skewed. Obviously, those who *choose* to live in a high rise most likely would not be in the subset of the population that has a fear of heights. A more randomized experiment with a range of building heights and a wide spectrum of height intolerances would have to be examined. In lower socio-economic groups, some people may have limited choices of where they can afford to live and may be forced to live in a high rise despite their comfort level of heights, which might offer a more distributed sample compared with people who have the financial means to choose to live in a luxurious high rise. Another option is to survey people staying in hotels who may have had limited choices in selecting a preferable floor and/or surveying people who have no choice but to work in a high rise. However, as in the case with my colleague from Buffalo, some may have such an aversion to heights that they would refuse a high-paying or high-satisfaction job on an upper floor and opt for a lesser-satisfying or lower-paying job closer to the ground, which once again could skew the data.

After living through the tragedies that occurred on September 11, 2001, the emotional toll on many who experienced the attacks first-hand or others who observed from their television sets remains prevalent. Since the attacks, new structural and safety measures have been implemented in high-rise construction. Shortly after the attacks, Sally Regenhard founded the Skyscraper Safety Campaign in memory of her son Christian who was a firefighter killed in the attacks on the World Trade Center (WTC). The goals of the campaign were to investigate the collapse of the WTC (which has been completed); encourage better compliance with the building and fire codes in New York City and nationwide; allow the fire fighters to have more input in writing building codes; and ensure that all future WTC development by the Port Authority of New York and New Jersey is characterized by quality, safety, security, and codes compliance (SSC, 2003). The terracing strategies discussed earlier – including the Mountain in Copenhagen – provide both a connection to the outdoors and a means of egress in the event of an emergency. Although the horizontal footprint of such designs might be somewhat larger, such strategies may prove to have better livability, safety, and comfort ratings than the tall, confined skyscrapers that are more ubiquitous in cities.

Although the high rise may still be a solution to meeting housing demands in large, dense cities, a better understanding of human comfort, demographics, and impacts to adjacent land is needed. Due to the high costs of construction, some high rises may increase adjacent land values and displace businesses and residents in lower- or middle-income brackets. Given the current national office vacancy rate hovering around 20 percent in the United States in the fourth quarter of 2023 (Lasalvia et al., 2024) combined with an increase in hybrid work schedules, the need for more office space may decrease, which further clouds the relevance of the skyscraper in the near future. Similar to providing a financial range of different housing types to foster diversity and equity in neighborhoods, a mix of high- and low-rise housing that responds to a particular community's preferences and needs can provide choices and ultimately yield higher satisfaction, comfort, and wellness ratings.

Not building as a strategy

I close this chapter on buildings and rooftops with a controversial strategy: not building. That's right, you read it correctly. Although buildings are necessary to meet the spatial demands of society, how many buildings are being built that are not needed, not wanted, or completely unnecessary? How many unnecessary buildings are justified with a GBRS certification? How many vacant buildings can be feasibly reused as an alternative to new construction? In *Indiana Jones: The Last Crusade*, Kazim – the leader of the Brotherhood of the Cruciform Sword who swore to keep the Holy Grail safe and hidden – asks Indiana this question: "Ask yourself, why do you seek the cup of Christ? Is it for his glory, or for yours?" Although it is not my intent to compare the holy grail to a building (nor my intent to discuss religion), the question is relevant and applicable to the discipline of architecture, just worded somewhat differently: "Ask yourself, why do you seek to design this new building? Is it for society or the earth's benefit, or for yours?" In the field of medicine, an "ethical" doctor will not perform a procedure on you that you do not need; in the field of architecture, architects need to start practicing with the same integrity. This often poses the question, "Then what are architects supposed to do?" After reading this book, hopefully, the answer to this will become clearer. I am in no way implying that architects need to abandon new building design in their practice; however, I am advocating for redefining the vocation so that architects are not merely focused on form-making, but rather on developing solutions to specific problems that may or may not require a new building or structure. It's a tough pill to swallow for zealous designers eager to leave their mark on history, but in the long term, it may be the most sustainable solution for both the survival of the profession and the health of the planet.

References

Assefa, S., Lee, H.-Y., & Shiue, F.-J. (2022). Sustainability performance of green building rating systems (GBRSs) in an Integration Model. *Buildings*, *12*(2), 208. MDPI AG. http://doi.org/10.3390/buildings12020208

Association pour la Haute Qualité Environnementale (ASSOHQE). (n.d.). *Haute Qualité Environnementale (HQE)*. https://www.hqegbc.org/

Baldwin, E. (2019, October 7). CopenHill: The story of BIG's iconic waste-to-energy plant. *ArchDaily*. www.archdaily.com/925966/copenhill-the-story-of-bigs-iconic-waste-to-energy-plant

Biomimicry Institute. (n.d.). *What is biomimicry*. https://biomimicry.org/what-is-biomimicry/?gclid=Cj0KCQjw0IGnBhDUARIsAMwFDLljBDcBg5O8DoRbDqqwHMhJMORnzyS-EGx8T0bjTrWH3IxBqS%E2%80%A6

BRE Group (BRE). (n.d.a). *Building Research Establishment Environmental Assessment Method (BREEAM)*. https://bregroup.com/products/breeam/

BRE Group (BRE). (n.d.b). *All about BREEAM*. https://bregroup.com/buzz/all-about-breeam/#:~:text=BREEAM%20is%20an%20internationally%20recognised,further%202%2C217%2C000%20registered%20for%20certification

Building and Construction Authority of Singapore (BCA). (n.d.). *Green Mark certification scheme*. www1.bca.gov.sg/buildsg/sustainability/green-mark-certification-scheme

California Academy of Sciences (CAS). (n.d.). *Green building and operations*. www.calacademy.org/about-us/sustainability-in-action/green-building-operations

Dakheel, J. A., Del Paro, C., Aste, N., & Leonforte, F. (2020). Smart buildings features and key performance indicators: A review. *Architecture, Built Environment and Construction Engineering A.B.C., Politecnico di Milano, Via Bonardi*, *9*, 20133.

Development Bank of Japan (DBJ). (n.d.). *Development Bank of Japan (DBJ)*. http://igb.jp/en/index.html

Earth Pledge. (2005). *Green roofs: Ecological design and construction*. Schiffer Publishing.

ECOncrete. (n.d.). *How ECOncrete works*. https://econcretetech.com/econcrete-technology/

Fitwel. (n.d.a). *Fitwel v3*. www.fitwel.org/

Fitwel. (n.d.b). *What is the Fitwel standard*? www.fitwel.org/fitwel-solutions
Fox, M., Morewood, J., Murphy, T., Lunt, P., & Goodhew, S. (2022, January). Living wall systems for improved thermal performance of existing buildings. *Building and Environment*, *207*, 108491. https://doi.org/10.1016/j.buildenv.2021.108491
German Sustainable Building Council (GSBC). (n.d.). *German Sustainable Building Council (DGNB)*. https://www.dgnb.de/en
Green Building Council of Australia (GBCA). (n.d.). *Green Star*. https://new.gbca.org.au/green-star/exploring-green-star/
Green Building Index (GBI). (n.d.). *Green Building Index*. www.greenbuildingindex.org/
Green Building Initiative (GG). (n.d.). *Green Globes*. https://thegbi.org/why-green-globes/?gclid=CjwKCAiAx_GqBhBQEiwAlDNAZr5fzjPgNwUc0zHzNMNrS02KzDbG2let0saHSQ_t6PakG6ywI3jDz-BoC0RkQAvD_BwE
Greenroofs.com. (n.d.). *California Academy of Sciences living roof*. www.greenroofs.com/projects/california-academy-of-sciences-cas-living-roof/
Gulf Organization for Research and Development (GORD). (n.d.). *Global Sustainability Assessment System*. https://gsas.gord.qa/gsas-overview/
Hadba, L., Silva, L., & Mendonça, P. (2017). *Green walls: An efficient solution for hygrothermal, noise and air pollution control in the buildings*. Conference: Architecture Media Politics and Society Conference – Living and Sustainability: An Environmental Critique of Design and Building Practices, Locally and Globally. London, England. Volume: 1
Harnik, P. (2010). *Urban green: Innovative parks for resurgent cities* (pp. 116–117). Island Press.
He, Y., Kvan, T., Liu, M., & Li, B. (2018, April). How green building rating systems affect designing green. *Building and Environment*, *133*, 19–31. https://doi.org/10.1016/j.buildenv.2018.02.007
Heritage Aviation. (n.d.). *Environmental sustainability*. https://flyheritage.com/galleries/solar-wind-green-roof/#green-roof
Hong Kong Green Building Council (HKGBC). (n.d.). *BEAM Plus*. www.hkgbc.org.hk/eng/main/index
Huppert, D., Wuehr, M., & Brandt, T. (2020). Acrophobia and visual height intolerance: advances in epidemiology and mechanisms. *Journal of Neurology*, *267*(Suppl 1), 231–240. http://doi.org/10.1007/s00415-020-09805-4
Indian Green Building Council (IGBC). (n.d.). *Indian Green Building Council*. https://igbc.in/index.php
Ingels, B. (2019). Quoted in: CopenHill: The story of BIG's iconic waste-to-energy plant. *ArchDaily*. www.archdaily.com/925966/copenhill-the-story-of-bigs-iconic-waste-to-energy-plant
Institute for Building Environment and Energy Conservation (IBEC). (n.d.). *Comprehensive Assessment System for Built Environment Efficiency (CASBEE)*. www.ibec.or.jp/CASBEE/english/index.htm
International Code Council (ICC). (n.d.). *International Green Construction Code (IGCC)*. https://www.iccsafe.org/products-and-services/i-codes/2018-i-codes/igcc/
International Code Council (ICC). (2021). *2021 International Building Code, Section 508.3.1.1: High rise buildings*. https://codes.iccsafe.org/content/IBC2021P1/chapter-5-general-building-heights-and-areas
International Energy Agency (IEA). (n.d.). *Green Building Evaluation Standard*. www.iea.org/policies/7915-gbt-50378-2014-green-building-evaluation-standard
International Living Future Institute (ILFI). (n.d.). *Living Building Challenge*. https://living-future.org/lbc/
International Well Being Institute (IWBI). (n.d.). *WELL standard*. https://v2.wellcertified.com/en/wellv2/overview/
Kiley, R. (2014). *Cutting-edge technology deeply rooted in GIS*. Rochester Institute of Technology. www.rit.edu/news/cutting-edge-technology-deeply-rooted-gis
Lasalvia, T., Chen, L., and Luettke, N. (2024). *Q4 2023 preliminary trend announcement*. Moody's Analytics. https://cre.moodysanalytics.com/insights/cre-trends/q4-2023-preliminary-trend-announcement/
Liu, K. L., & Baskaran, B. (2003). Thermal performance of green roofs through field evaluation. Presented at *Greening Rooftops for Sustainable Communities, the First North American Green Roofs Infrastructure Conference, Awards, and Trade Show*. National Research Council, Institute for Research in Construction.
Maclay, W. (2014). *The new net zero: Leading-edge design and construction of homes and buildings for a renewable energy future* (p. x). Chelsea Green Publishing.
Marchi, L., Antonini, E., & Politi, S. (2021). Green building rating systems (GBRSs). *Encyclopedia*, *1*(4), 998–1009. MDPI AG. http://doi.org/10.3390/encyclopedia1040076
Mattoni, B., Guattari, C., Evangelisti, L., Bisegna, F., Gori, P., & Asdrubali, F. (2018). Critical review and methodological approach to evaluate the differences among international green building rating

tools. *Renewable and Sustainable Energy Reviews, 82*(Part 1), 950–996. https://doi.org/10.1016/j.rser.2017.09.105
Millard, B. (2021, September 7). New York's proliferation of green building rating systems raises the question of what's being measured. *Archpaper*. www.archpaper.com/2021/09/nyc-proliferation-green-building-rating-systems-questions-whats-being-measured/
Minergie. (n.d.). *Standards*. www.minergie.ch/de/standards/neubau/minergie/
Ministry of Housing and Urban-Rural Development of People's Republic of China (MOHURD). (n.d.). *Evaluation Standards for Green Buildings (GB/T)*. https://neec.no/uploads/Article,%20China%20green%20building%20standard.pdf
Minke, G., & Witter, G. (1982). *Haeuser Mit Gruenem Pelz: Ein Handbuch zur*. Frankfurt/Main. ISBN 10: 3881840621.
Monty, L. (2014, September 25). Green roofs grow in popularity. *Burlington Free Press*. www.burlingtonfreepress.com/story/life/green-mountain/2014/09/25/green-roofs-popularity/16207407/
National Association of Home Builders (NAHB). (n.d.). *National Green Building Standard (NGBS)*. https://www.nahb.org/advocacy/industry-issues/sustainability-and-green-building/national-green-building-standard-certification
Sharma, A., Conry, P., Fernando, H. J. S., Hamlet, A. F., Hellmann, J. J., & Chen, F. (2016). Green and cool roofs to mitigate urban heat island effects in the Chicago metropolitan area: Evaluation with a regional climate model. *Environmental Research Letters, 11*(6). https://doi.org/10.1088/1748-9326/11/6/064004
Skyscraper Safety Campaign (SSC). (2003). *Skyscraper safety campaign statement to the NIST federal advisory committee of the national construction safety team*. www.nist.gov/system/files/documents/2017/05/09/MG-NIST-statement4-29-03.pdf
Song, Y., Lau, S.-K., Lau, S. S. Y., & Song, D. (2023). A comparative study on architectural design-related requirements of green building rating systems for new buildings. *Buildings, 13*(1), 124. http://doi.org/10.3390/buildings13010124
Sproul, J., Wan, M. P., Mandel, B. H., & Rosenfeld, A. H. (2014). Economic comparison of white, green, and black flat roofs in the United States. *Energy and Buildings, 71*, 20–27. https://doi.org/10.1016/j.enbuild.2013.11.058
Stanley, S. (2018). *U.S. Green Building Council releases annual top 10 countries and regions for LEED*. USGBC. www.usgbc.org/articles/us-green-building-council-releases-annual-top-10-countries-and-regions-leed
Sweden Green Building Council (SGBC). (n.d.). *Miljobyggnad*. www.sgbc.se/certifiering/miljobyggnad/
Tang, K. H. D., Foo, C. Y. H., & Tan, I. S. (2020). A review of the green building rating systems. *IOP conference series: Materials, science, and engineering 943 012060*. http://doi.org/10.1088/1757-899X/943/1/012060
Top Leaf Farms. (n.d.). *Building a resilient food system: Farm integration*. https://topleaffarms.com/about.html
University College London (UCL). (2017). *High rise buildings much more energy-intensive than low-rise*. www.ucl.ac.uk/news/2017/jun/high-rise-buildings-much-more-energy-intensive-low-rise
University of Vermont (UVM). (n.d.a). *George D. Aiken center: A green, renovated building*. www.uvm.edu/rsenr/george_d_aiken_center_green_renovated_building
University of Vermont. (n.d.b). *Dudley H. Davis Center: Green building and environmental stewardship*. www.uvm.edu/daviscenter/green-building-environmental-stewardship
U.S. Energy Information Administration (EIA). (2022). *How much energy is consumed in U.S. buildings*? www.eia.gov/tools/faqs/faq.php?id=86&t=1
U.S. Environmental Protection Agency (EPA). (2018). *Estimating the environmental effects of green roofs: A case Study in Kansas City, Missouri*. www.epa.gov/sites/default/files/2018-09/documents/green-roofs_casestudy_kansascity.pdf.
U.S. Environmental Protection Agency (EPA). (n.d.a). *Green building*. https://archive.epa.gov/greenbuilding/web/html/about.html#:~:text=Green%20Building%20History%20in%20the%20U.S.,-Some%20practices%2C%20such&text=The%20oil%20price%20increases%20of,experiments%20with%20contemporary%20green%20building
U.S. Environmental Protection Agency (EPA). (n.d.b). *Water: Long term 2 enhanced surface water treatment rule: Basic information*. www.epa.gov/dwreginfo/long-term-2-enhanced-surface-water-treatment-rule-documents#two
U.S. Green Building Council (USGBC). (2023). *LEED v4 energy update FAQ*. www.usgbc.org/resources/leed-v4-energy-update-faq

U.S. Green Building Council (USGBC). (n.d.). *LEED home page*. www.usgbc.org/

Verdinez, D. (2023). *The top 10 countries for LEED demonstrate that green building is a truly global movement*. USGBC. www.usgbc.org/articles/top-10-countries-leed-demonstrate-green-building-truly-global-movement

Vinoly, R. (2014). *Howard Hughes Medical Institute, Janelia research campus*. https://vinoly.com/works/howard-hughes-medical-institute-janelia-research-campus/

White, A. (2023, October 3). *George D. Aiken center becomes UVM's first net-zero energy building*. University of Vermont. www.uvm.edu/news/story/george-d-aiken-center-becomes-uvms-first-net-zero-energy-building#:~:text=Aiken%20Center%2C%20home%20of%20the,uses%20on%20an%20annual%20basis

World Green Building Council (WGBC). (2019). *Bringing embodied carbon upfront*. https://worldgbc.org/advancing-net-zero/embodied-carbon/#:~:text=Buildings%20are%20currently%20responsible%20for,11%25%20from%20materials%20and%20construction

10

ADAPTIVE REUSE

Since the latter half of the twentieth century, adaptive reuse projects have been conducted across various urban and architectural scales, ranging from entire districts to single living spaces. Some prominent adaptive reuse projects include the transformation of the Bankside Power Plant into the Tate Museum of Modern Art in London (Tate Modern), the reclamation of the Emscher River in the Ruhr Valley of Germany, the High Line in New York City, and the Centre National de la Danse in Paris. And then, of course, there are numerous obscure adaptive reuse projects that could fill an entire volume, such as converting old residential houses or industrial buildings into offices, old train stations into restaurants or retail, shopping centers into multi-use villages (see the following text), and the conversion of vacant retail and office space into housing. In addition to preserving history, adaptive reuse of existing buildings and infrastructure presents another strategy that can provide spatial needs without having to construct new projects on undeveloped land. Existing buildings can be preserved if essential elements and materials are intact, restored if elements were modified from the original state, or rehabilitated if there is extensive damage. In many cases, uses can be repurposed without altering their historic character. In cases where a building is not capable of being reused due to structural issues and/or costs associated with making it safe, building materials can be recycled and integrated into a new project to preserve original design elements.

Policies and incentives

Given the amount of vacant building stock in the world, it is difficult to comprehend why we continue to develop open land with new construction. According to the U.S. Census Bureau (2022), there were approximately 15,078,000 vacant homes in the United States in 2022, which is 10.5 percent of the total housing inventory. As a result of the COVID-19 pandemic, roughly 5,000 retail stores have closed and nearly 20 percent of office spaces are vacant in the United States (Jones, 2022; Rezvani, 2023). According to an analysis conducted by the U.S. Government Accountability Office (GAO, 2023), the federal government owns 1,685 buildings with over 500 million square feet of office space and spends roughly $2 billion to operate and maintain them and another $5 billion or so leasing space in 6,712 buildings they do not own. GAO's

DOI: 10.4324/9781032668666-10

analysis showed 17 of 24 federal agencies to have an average capacity rate of around 25 percent or less in selected weeks early in 2023. A 2015 survey by Picture the Homeless found 1,723 abandoned buildings containing 11,170 apartments and 505 vacant lots in Manhattan that, if properly developed, could provide 24,000 housing units that could house more than 9,000 families and 7,000 single adults living in city shelters (Picture the Homeless, 2015). Office vacancy rates showed 22.1 percent vacancy for quarter 3 of 2023 in Manhattan, 34 percent for quarter 3 of 2023 in San Francisco, and 9.4 percent for quarter 2 in London (Cushman & Wakefield, 2023; CBRE, 2023; JLL, 2023).

Many building owners and developers often base their decisions on what to do with an existing building on the cost of renovation versus the cost of constructing a new building. Often, it is simply easier, safer, and less expensive to tear them down. If the lot is reused and a new, more energy-efficient, and green building is constructed in its place, such a decision may yield positive environmental benefits. However, it is obviously less sustainable if the building is left vacant and a new building is constructed on undeveloped land. Some of the large retailers and grocery stores often use this strategy. They close one of their stores, build a new one down the road, but leave the vacant building in place because it may be more profitable to keep it rather than sell the property/building to another business that could compete with their sales.

One solution to this issue is to offer incentives for rehabilitation projects that are equal to or better than new construction projects. Prior to the Tax Reform Act of 1976, the federal taxation policy contained accelerated depreciation rules that encouraged property owners and developers to tear down old buildings in favor of new construction to recoup investment dollars quickly. The Tax Reform Act launched what is known as the Federal Historic Preservation Tax Incentives Program. This is administered by the National Park Service and the Internal Revenue Service in partnership with state historic preservation offices, and since its creation in 1976, it has leveraged over $116.34 billion in private investment to preserve more than 47,000 historic properties (NPS, n.d.a). The current program offers two tax credits: a 20 percent income tax credit for the rehabilitation of historic, income-producing buildings that are certified by the Secretary of the Interior and the National Park Service; and a 10 percent tax credit for the rehabilitation of non-historic, non-residential buildings built before 1936. To qualify for the 20 percent tax credit, the building must be depreciable, expenditures must be greater than $5,000, the property must be returned to use, and it must be a certified historic structure when it is placed in service. To qualify for the 10 percent tax credit, the building must be rehabilitated for non-residential use and must meet three criteria: at least 50 percent of the existing external walls must remain in place as external walls; at least 75 percent of the existing external walls must remain in place as either external or internal walls; and at least 75 percent of the internal structural framework must remain in place (NPS, n.d.b). There are also tax benefits for historic preservation easements where a property owner can protect properties by restricting development on land that has historic, architectural, or archaeological significance. The owner can then transfer the stipulations/restrictions to a conservation/preservation organization or donate the easement in exchange for a federal income tax deduction.

Since many historic buildings pre-date current zoning and building permit regulations, rehabilitating a building may prove to be both financially and physically challenging. Zoning may also restrict certain uses, limiting the reuse options. Modifying the regulations to allow certain physical features of the building that may be difficult to change to be grandfathered and allowing for a mix of uses can increase interest in adaptive reuse. For example, in 1999, Los Angeles adopted an Adaptive Reuse Ordinance that encourages the conversion of existing buildings to residential uses, such as apartments, condominiums, live/work units, or hotels. The ordinance expedites the approval process; waives density restrictions by not limiting the number of housing units in a project, grandfathers in non-conforming floor areas, setbacks, and heights; allows mezzanine spaces to be added if they do not exceed one-third of the size of the floor below; and does not require new loading areas or parking spaces (LA City, n.d.). Prior to the ordinance's adoption in 1999, downtown Los Angeles had 18,000 inhabitants and 11,626 residential units; in 2011, it had 29,429 apartments and condominiums and 46,400 residents (Guzman, 2012). In 2023, the city of Los Angeles proposed an expansion of the ordinance as one of six strategies in the Citywide Housing Incentive Program. The new ordinance encourages converting underutilized buildings into housing, expedites the approval process, and shortens the minimum age criteria of buildings to 15 years (as opposed to the existing system which only allowed incentives for buildings built before 1974) (LA City, n.d.).

Other communities have adopted form-based zoning codes (as discussed in Chapter 4) that use physical form rather than separation of uses as the organizing principle for the code. As opposed to traditional zoning (Euclidean zoning) that often zones sections/blocks for single uses, form-based zoning could allow a mix of uses if the form of the building is consistent with the transect in which it is located. In theory, a form-based code should increase interest in adaptive reuse since there is more flexibility in the allowed uses. In 2010, Nashville, Tennessee, adopted a form-based zoning code as part of its Downtown Community Plan to strengthen the area's character and pedestrian experience. The Downtown Code (DTC) regulates the physical form of buildings to ensure each makes a positive contribution to the public realm and that the interaction between the building and the street is inviting, safe, and comfortable for pedestrians (Nashville, n.d.). The DTC created several historic zoning overlays to incentivize the adaptive reuse of historic buildings. For example, the preservation of historic structures is one of nine options for projects to earn bonus height in the DTC Bonus Height Program.

The adaptive reuse of farm structures

Given our agrarian roots, it is common to still see farm structures, such as barns, silos, carriage barns, milking parlors, stables, stalls, Quonset huts, sheds, and pavilions speckled throughout the landscape. Given their resilient construction, many of these structures have withstood the test of time and can present distinctive spaces to accommodate a variety of functions. The circular design of silo structures, for example, bestows some unique opportunities. Agricultural silos are constructed to house grain or fermented feed (silage). They range from the massive concrete grain elevator complexes used for municipal storage of surplus farm materials to the more common tower silos that are found on many farms throughout the countryside. The tower silo, invented by Franklin Hiram King, can be constructed from wood, steel, rolled steel shells, concrete/stone/brick block, or concrete staves. Sizes can range from 10 to 100 feet (3–30 meters) in diameter and 20–300 feet (6–91 meters) in height. In the late nineteenth and early twentieth centuries, concrete stave silos were the most common form due to ease of construction and concrete's strong compressive strength, which makes it suitable for storing tightly packed material. The staves are small concrete blocks pieced together by tongue and groove joints and tightly wound by a steel hoop system to compress the silo and prevent structural failure. Wood silos were also often wound by steel hoops. Since the compressive forces are strongest toward the bottom, spacing of the hoops is often tighter at the base than at the top.

The circular design of silos makes for an interesting space that can be adapted to a variety of functions. The simple construction also allows for relatively rapid disassembly and reassembly, which enables silos to be moved to and from different sites – an attribute that benefits both agricultural and adaptive reuse needs. For example, the Silo Restaurant in West Dover, Vermont, was constructed in Orwell, Vermont, in 1890, but it was eventually taken apart, moved to its current location, and reassembled. The restaurant also recycled myriad materials, such as lumber, stone, equipment, tools, weathervanes, sleds, skis, and toboggans, from various farms throughout Vermont for the interior decor. Another disassembly/reassembly silo project was completed just outside of Stowe, Vermont, in the town of Morrisville. The top third of a tall wooden silo was "popped" and moved from a farm a couple of miles down the road to a new location off Route 100 – a sort of "parent–child" relationship. Both silos were then refurbished and transformed into living spaces.

The Quaker Smith Point silo in Shelburne, Vermont, was restored by disassembling and reusing materials from the original silo. Quaker Smith Point is a teardrop-shaped peninsula that extends westward into Lake Champlain from the southwest shoreline of Shelburne Farms. The

FIGURE 10.1 Silo adaptive reuse. Top left: original ("parent") portion of the silo, Morrisville, Vermont; bottom left: "child" portion of the silo, Route 100; top right: exterior of the Silo Restaurant, West Dover, Vermont; bottom right: Silo Restaurant interior, using parts of the original silo for decoration.

point was first settled by blacksmith William "Quaker" Smith and his wife Elizabeth in 1783. The Smiths lived and farmed off the land at Quaker Smith Point, as it came to be known, and constructed a family home as well as barns for animals and agricultural storage. The property was eventually handed down to William and Elizabeth's grandson Isaac around 1792. After Isaac passed away, his wife Lucinda sold the property to Seward and Lila Webb in 1886. Seward and Lila's son Watson was deeded the property in 1913. Watson constructed the silo for water storage, a new sheepfold, extensions to barns, and houses for employees and guests. In 1949, Watson and his wife Electra eventually deeded the property to their eldest son Sam (Donnis, 2010).

After Sam passed away in 1988, the property was split between his two children, Holly Webb Darling Froud and Sam Webb Junior. Holly and Sam eventually made the difficult decision to divide and sell the property. Jon Fishman, the drummer of the band Phish, purchased the 18-acre northern parcel of the point in the spring of 2004 and lived in what was once the employee/guest house, constructed by Watson Webb. Michael and Karen Stone purchased the southern parcel, comprising 15.7 acres, also in 2004, and then Fishman's parcel the following year. The Stones have since made some retrofits and additions to the buildings and landscape of the property that have upgraded the site while preserving the original colonial aesthetic imprinted by the Smiths and the Webbs. In 2004, the Stones hired Vermont-based builder Peter Close to disassemble and reconstruct the water silo. Close reused much of the original exterior cedar wood to finish the interior walls and recycled other lumber for the spiraling staircase. Construction alignment lasers were used to ensure the accuracy of the circular geometry. The renovated silo is about 22 feet (6.7 meters) in diameter and 30 feet (9 meters) in height and contains about 760 square feet of interior space (380 square feet on each of the two levels). The new construction features a stronger light wood frame, red hand-split cedar shake siding, double-hung windows, and a metal conical roof. The water cistern now sits below the silo and still distributes water via the well across the property. A yoga studio and a meditation room occupy the ground and upper levels, respectively. Windows were strategically placed around the circumference of the silo to maximize natural light and provide breathtaking 360-degree vistas that showcase the rolling countryside and the Green Mountains to the east and Lake Champlain and the Adirondack Mountains to the west. The silo can be heated in the winter by propane and is naturally cooled by the breezes off Lake Champlain in the summer.

Carriage barns are also still prominent both in rural areas and even in some urban areas. I lived in the carriage house of the Tracy House complex in Shelburne, Vermont, for nearly 20 years. This complex, at 5224/5226 Shelburne Road, was built in 1878 and consists of a 2.5-story Italianate/Gothic Revival brick house and a carriage house complex to the west. The construction was commissioned by Charles Lee Tracy and designed by the Reverend Edgerton. The main house was built with bricks recycled from the Methodist church, also designed by Reverend Edgerton, that once stood on land that is currently occupied by the church of St. Catherine of Siena, at 72 Church Street (Webster, 1994). Lee Tracy was the grandson of Hezekiah Tracy, who settled on what came to be known as Southern Acres at Shelburne Farms in the 1780s. Lee finally moved into the brick house around 1882, after selling Southern Acres to Seward and Lila Webb. Lee lived in the main house with his wife Adeline Amanda Peckham. After Lee passed away in 1897, the property was deeded to his eldest son, Julius. Lee's other sons Henry and Charles also lived in the house. In 1933, Charles built a music studio adjacent to the first ell on the west side of the house. During this time, the eastern portion of the house was occupied by Julius's daughter Hannah and her husband Harris. After Charles passed away in 1933, the house

FIGURE 10.2 Quaker Smith Point water silo. Top left: original silo with water cistern inside; top right: cistern being moved; bottom left: new structural framing; bottom right: view of the entrance on south facade of the completed silo.

Source: Images courtesy of Peter Close (top left and right, and bottom left).

was deeded to Henry and his wife Charlotte, who then rented the house to James Norton and his wife for 14 years, but then moved back into the house in 1959.

At the age of 98, Charlotte moved out of the brick house, and the property was sold to architect Frederick (Fritz) Horton in 1983. Fritz and his firm, Horton–Ptaszynski Architects, simultaneously restored the historic elements of the property and modernized the infrastructure to accommodate residential and commercial activities. In the roughly 8,000-square-foot main house, Horton and Ptaszynski constructed offices on the ground and second floors, a large studio in the attic for their firm, and two apartments in the two western ells. The carriage barn – which also contained a farm barn and a storage shed – was converted into 2,000 square feet of finished space for three apartments: an upstairs and downstairs apartment in the carriage house portion, and a studio with a loft in the farm barn portion. The shed portion is still used for storage. All the structures have purple-slate gable roofs and green clapboard siding with light green eaves, baseboards, and fascia boards. Two large bays of the carriage house are still evident on the east facade, with the southernmost bay now occupied by two entrance doors and the northern bay occupied by a pair of six/six sash windows. The hayloft door on the east gable is also now a window for the upstairs middle apartment. A cupola, originally intended for ventilation, remains in the center, along the ridge of the carriage house roof. A window was placed on the south facade of the cupola to provide light into the middle apartment. The original shutters on the other facades of the cupola are still in place and provide an ideal roosting place for bats, which have most likely used the barn as a habitat since it was built. Structurally, the carriage house employs heavy timber, post and beam construction with a modified queen post/room-in-attic truss system for the roof portion. The new light wood frame structure was integrated between the heavy timbers.

Charlotte Tracy was guest of honor at the ceremony to mark the completion of the renovations in June 1983. She was extremely pleased with the preservation of the historical aesthetic and marveled at the innovations. With respect to the carriage house, Fritz Horton recalls Charlotte commenting: "That was a barn . . . we didn't have any carriages, just a cow, a couple of riding horses, and some pigs . . . that's who lived out there" (Tracy, 1983).

The Tracy House complex was placed on the National Register of Historic Places in 1983. Gillian (Jinx) Senior purchased the property in 1991 and maintained the mix of commercial and residential space. She added a glass sunroom to the north facade of the carriage house in 2004. Gillian sold the property to Ewa Marais in 2017, and Ewa sold the property to Charles Davis in 2020.

Although the retrofit of the Tracy House carriage barn was ahead of its time in 1983, the structure has not been without its challenges. The property was sold in 2017 for $1,050,000 and again in 2021 for $1,200,000. Over the years, the property has been in a steady state of decline. The various owners have claimed the requirements stipulated by the Vermont Department of Historic Preservation (VTDHP) make it too difficult to do repairs that are cost-effective. The slate roof, for example, is in desperate need of repair. It has several broken and/or missing pieces and has been chronically leaking for the past 20 years. There is a shortage of expertise to repair slate roofs and those who do possess the skills charge exorbitant amounts of money for their service. VTDHP does not allow replacement of the roof, so the owners (and tenants) are basically stuck with water damage, mold, and aesthetic decline. The wood sash windows are also completely rotten in areas, with several windowpanes missing or on the verge of falling out with the slightest amount of force. Preservation laws once again do not allow replacement with more energy-efficient, one-piece glazing with "faux" sashes; window replacement must mimic the original design with individual panes of glass, which once again has significant costs.

Indeed, preservation laws can often conflict with retrofits and/or green building strategies. Retrofits that employ solar panels, for example, are often rejected by preservation laws because of the alteration to the original design; similarly, replacing original wood columns with steel columns encased by wood could also constitute a violation, which is ludicrous given steel columns are stronger, last longer, and have better fire resistance. While historic preservation is indeed important to preserve culture and history, it seems many of the laws and regulations are dated and in need of revision. The health, safety, and well-being of the occupants and the impact of the structure on the environment need to be integrated with preservation ordinances.

Reinventing the shopping mall

Like many other fellow Generation X citizens, I have a complex, ambivalent relationship with the mall. Growing up in Rochester, New York, during the 1980s, I have memories of going to Midtown Plaza in the downtown area during the holidays and riding the suspended monorail. As a teen, I used to hang out at Longridge Mall (constructed in 1969) and Greece Towne

FIGURE 10.3 The Tracy House complex. Top left: carriage house in 1983, prior to renovations; top right: carriage house after renovations; middle left: interior of the middle apartment; middle right: glass solarium addition on north facade; bottom: rendering showing original heavy-timber framing.

Source: Top left and top right images courtesy of F.W. Horton & Associates

Mall (constructed in 1967) in Greece, New York. Longridge Mall was anchored by McCurdy's, Woolworth's, J.C. Penny, Sears, and Forman's and had several trippy displays and "contemplation" areas down the center, including a "visual music" area, a video art area, the large, globular Landsman Sculpture, the Electric Ball Circus, and several play structures where children (and sometimes adults) could jump on oversized foam blocks. The visual music area was sunk-in and contained several lounge chairs where the heads used to congregate. The smell of soft pretzels from Hot Sam's and cinnamon from Cinna-buns still ignites a bit of nostalgia. In 1994, the two malls were connected, creating a 1.7 million square feet conglomerate renamed, *The Mall at Greece Ridge*.

I grew to dislike the mall during my college years. They were crowded and noisy and enabled patterns of over-indulgence and consumerism that many rebellious 20-year-olds who think they have the world figured out would balk at. In graduate school, many of my fellow architecture and planning students viewed both indoor malls and outdoor strip plazas as everything that was wrong about design, or more specifically, large, square, lifeless buildings constructed as cheap as possible surrounded by a sea of impervious parking space that was the nucleus of the ever-sprawling suburban hell.

Fast forward to the internet and fast even more forward to the COVID-19 pandemic where e-commerce has become mainstream and threatens all brick-and-mortar retail stores. Although I still have issues with the design of outdoor strip malls, my view of indoor malls has suddenly changed. Call it nostalgia or a personality trait that always seems to favor the under-dog, but I am starting to almost feel sorry for malls given their losses to the pandemic and e-commerce. And I am a total culprit – I order just about everything online. Although e-commerce is here to stay and resurrecting malls as shopping meccas may be unrealistic at this point in history, the indoor mall has become a significant cultural icon that is worth preserving. With its open courtyards and its linear, T-shaped, Z-shaped, H-shaped, or donut-shaped designs, the mall contains ample and unique spaces that can be adapted for a variety of uses other than shopping. The large parcels that malls occupy also present opportunities to create villages or "second cities" in suburbs that never contained dense centers in the first place. And contrary to their original intent of being drive-up experiences, creating a mix of uses at malls can actually cluster development and prevent further sprawl in communities.

Brief history of shopping malls

Although open-air markets and public forums have most likely been around since humans first began trading, one of the first marketplaces that influenced modern enclosed malls was Trajan's Market in ancient Rome constructed between 100 and 110 AD. Trajan's Market utilized some of the multi-use strategies that are being studied to resurrect malls today. The structure was multi-storied, built into the hillside, and included a combination of ground-level shops facing the street, enclosed arcades, refreshment areas, offices, and most likely residential spaces on the upper floors (Mandal, 2017). Other prominent forerunners to the modern enclosed shopping mall include the agora in ancient Greece, the bazaars in the Middle East and Asia, including the Grand Bazaar in Istanbul, Turkey constructed in the mid-fifteenth century that contains 61 covered streets and over 3,000 vendors, covered arcades including the Passage du Caire covered arcade constructed in 1798 in Paris, France, the Arcade Providence (Westminster Arcade) built in 1828 in Providence, Rhode Island, the four-story, double-arcade Galleria Vittorio Emanuele II built in 1877 in Milan, Italy, and the Arcade Cleveland built in 1890 in Cleveland, Ohio (Barnes, 2016).

Open-air shopping villages also emerged in America in the 1920s and 1930s in Missouri (Country Club Plaza), Ohio (Grand Avenue Shopping Center), Dallas, Texas (Highland Park Village), and Illinois (Park Forest Plaza), the last of these utilizing a pedestrian-only shopping center. The automobile-free, pedestrian experience is what later influenced Victor Gruen in his design of Northgate in Seattle, Washington in 1950 and in Southdale Center in Edina, Minnesota in 1956 (Dunham-Jones & Williamson, 2011). At Southdale Center, Gruen's concept of an enclosed version of a village street came to fruition. As stated by Alexandra Lange (2022) in her book, *Meet Me by the Fountain: An Inside History of the Mall*,

> The model that Southdale provided was of an exurban island with as many civic functions as possible under one roof. If downtown could not be made to hitch up to the mall, the mall would become downtown, with its own plazas and lanes, post offices and cafes, play spaces, and of course, shops. Once you had parked in the lot, you could move freely through a variety of environments *(sic)* 365 days per year.

Indeed, since much of suburbia lacks well-defined village centers, the faux city that enclosed malls created provided a place for the suburban community to congregate.

Interestingly, Gruen and Associates (including a young Frank Gehry) designed the urban mall in Rochester, New York, I used to visit during the holidays as a kid. Midtown Plaza was opened in 1962 and was the nation's first enclosed *urban* shopping mall (Lange, 2022). The open, two-level mall was positioned below the 17-story Midtown Tower that included offices, a restaurant, and a hotel, and the entire complex was linked to adjacent buildings and parking garages through a system of skyways and tunnels. Gruen's innovative design attracted a visit from Walt Disney and was an influence in the development of Disney World, Epcot Theme Park, and the Celebration planned community in Florida (Astor, 2016). As I mentioned at the beginning of this section, at Christmastime, Midtown used to set up Santa's Magic Mountain and a suspended monorail that ran along the perimeter of the mall. Although there was an exodus of people from Rochester's downtown center in the second half of the twentieth century, Midtown Plaza used to bring people back to the city – and not just at Christmastime. Before the suburban malls were built in the area, Midtown had cutting-edge fashion and products. However, with the construction of the Greece Towne Mall in 1967, Longridge Mall in 1969, Eastview Mall in Victor in 1971, Marketplace Mall in Henrietta in 1982, and the Irondequoit Mall in 1990, Midtown could not compete. Despite several marketing attempts to keep it open, Midtown officially closed in 2008 and was demolished in 2010.

Suburban malls continued to gain popularity in the 1970s and 1980s and seemed to peak around 1990 when 19 new malls opened across America (Geiling, 2014). The International Council of Shopping Centers (ICSC) estimated there were about 1,000 enclosed malls in 2007 (Dunham-Jones & Williamson, 2011). The last enclosed mall was built in 2006 and between 2007 and 2009, four hundred of America's largest 2,000 malls closed (Geiling, 2014). And then the pandemic hit. According to Moody's Analytics, the vacancy rate for regional malls in the United States hit a record high of 11.4 percent in the first quarter of 2021 from 10.5 percent in the fourth quarter of 2020 (Thomas/CNBC, 2021). But even before the pandemic, e-commerce and an over-saturation of retail space were hurting sales. In 2005, the amount of retail area per person was 20.2, which was six times greater than Sweden, which leads Europe in the same category (Dunham-Jones &Williamson, 2011, p. 116).

Mall parcels are significant in their size and present the design and planning community with several opportunities for redevelopment. The International Council of Shopping Centers

(ICSC) classifies U.S. shopping centers as General Purpose Centers, Specialized Purpose Centers, and Limited Purpose Property. General Purpose Centers include Super Regional Malls that have 800,000 or more square feet, occupy 60–120 acres, and have a trade area of 5–25 miles; Regional Malls that have 400,000–800,000 square feet, occupy 40–100 acres, and have a trade area of 5–15 miles; Community Centers that are usually strip malls that have 125,000–400,000 square feet, occupy 10–40 acres, and have a trade area of 3–6 miles; Neighborhood Centers that have 30,000–125,000 square feet, occupy 3–5 acres, and have a trade area of three miles; and Strip/Convenience that have less than 30,000 square feet, occupy less than 3 acres, and have a trade area less than one mile. Specialized Purpose Centers include Power Centers, Lifestyle, Factory Outlets, and Theme/Festival areas, and Limited Purpose Properties include retail stores within a commercial airport (ICSC, 2017). For the purposes of this section, the focus is on the inward-facing, enclosed malls that generally fall under Super Regional Malls or Regional Malls.

Retrofitting strategies

Several strategies have been utilized to resurrect dead malls, including creating destination centers with amenities such as amusement parks, water parks, fitness centers, indoor ski/snowboard hills (such as New Jersey's American Dream), kayaking canals, running/walking tracks, ice skating, and climbing walls, to name a few; demolishing the entire structure and regreening the site with parks and/or green stormwater infrastructure to restore wetland function; demolishing portions or the entire structure and redeveloping the site as a new village center; or keeping the structure in place but in addition to retail, reprogramming it with a mix of new uses, including health clinics, community centers, housing, restaurants, fitness centers, transit stations, office space, libraries, classroom space, or other civic uses. Some strategies use a combination of these, such as keeping the structure in place but still designing it as a village center by extending the wings to the street front and/or creating liner buildings around the perimeter that connect the street to the enclosed portion of the mall.

The decision whether to knock down and regreen, knock down and completely redevelop, or preserve the entire structure or portions of the original inward-facing, enclosed portion of the mall is influenced by geography/climate, socioeconomics of the community, the juxtaposition of the mall relative to the surrounding community, local zoning codes, budgetary and political constraints/forces, and the intentions of the those who own the property. For the last of these, malls are often owned by Real Estate Investment Trusts (REITs) that must present the projects as a form of a tradable commodity to their Wall Street investors (Dunham-Jones & Williamson, 2011, p. 114). Often, investors view many of the mixed-use, new urbanist concepts proposed by designers as risks that will not generate profit. Some projects – such as Mizner Park in Boca Raton, Florida, and Laguna Hills Mall in California – demolished the original structure and redeveloped the sites into mixed-use, village centers. Other projects, such as Centurian American in Plano, Texas, and the proposal for Main Place Mall in Santa Anna, California, retain the core mall structure and infill development around the site. Centurion American retained 325,000 square feet of retail but tore down the big boxes and replaced them with 2,300 units of multifamily housing, 500 single-family residences, and 1.5 million feet of office space; the Main Place Mall proposal densifies the perimeter of the site with "liner" buildings that contain entertainment, office space, and housing (Kiger, 2023).

As mentioned earlier, the Irondequoit Mall near Rochester, New York, opened in 1990 with over 100 stores in an open, two-story skylight-lined arcade. After lackluster profit margins and

FIGURE 10.4 Top: the Mall of America in Syracuse, New York, an example of a mall destination center; bottom: University of Rochester's Orthopedic and Physical Performance Center at the former Sears site in Marketplace Mall, Henrietta, New York.

Source: Top image courtesy of Kimberly A. Richards

the closing of several anchors, the original owners (Wilmorite Incorporated) sold the mall to an insurance company in 2001. The mall was again sold in 2005 to developer Adam Bersin who renamed the complex to the Medley Center. Bersin sold the mall two years later to Scott Congel who announced plans to convert the mall into a mixed-use complex. However, Congel eventually foreclosed on the property and it was then sold in an auction to Angelo Ingrassia in 2014. In 2021, the Medley Center was once again renamed to Skyview on the Ridge and opened a 41,339 square foot community center that includes a fitness center, gymnasium, dance studio, walking track, meeting rooms, and lounges for seniors and teens. The former Sears building was converted into 73 apartment units and connects to a new four-story building with 84 apartments. The apartment complex – called Skyview Apartments – is for people 55 years of age and older earning at or below 60 percent of the median income (Rudd, 2022). Seventy-eight apartments are reserved for households at risk of homelessness and receive on-site support through Rochester General Hospital's ElderONE program (Driffill, 2022). In addition, the space formerly occupied by Macy's now houses Rochester Regional Health's Isabella Graham Hart School of Nursing. The cost of the project thus far is $44.3 million, with funding provided by permanent tax-exempt bonds, the New York State Energy Research and Development Authority, the town of Irondequoit, Monroe County, and low-income housing credits from the New York State Homes and Community Renewal program (Rudd, 2022).

Wilmorite Incorporated also developed the Marketplace Mall in Henrietta, New York, in 1982. Although the mall has not suffered the same fate as the Irondequoit Mall, the space formerly occupied by Sears has been converted into the University of Rochester's new Orthopedic and

FIGURE 10.5 Skyview on the Ridge. Top: Community Center walking loop adjacent to the interior of the former Irondequoit Mall; middle: Multi-purpose athletic court taking advantage of the mall's skylights; bottom: Skyview Apartments attached to the former mall.

FIGURE 10.5 (Continued)

Physical Performance Center. The center contains 330,000 square feet and includes 14 operating rooms, four procedure rooms, orthopedic urgent care, ambulatory surgery, two therapy pools, a motion analysis lab, clinical trial space, and radiological imaging (CT scan, MRI, Pet Scan, X-ray). Seeing humans will never stop getting sick, anchoring a mall with a medical clinic may provide a long-term tenant that can rejuvenate the mall. A person can get an MRI, go to lunch at Dave and Buster's (also an anchor at Marketplace), and go shop at close to 100 different stores.

Figure 10.6 shows a proposed redesign of The Mall at Greece Ridge introduced at the beginning of this section with a combination of some of the aforementioned strategies. Since Greece, New York, is a suburb that lacks a clearly defined village center, the redeveloped mall site has the potential to create the "downtown" that never existed in the first place. The design extends two sections of the core of the mall to the north of Ridge Road and to the west of Long Pond

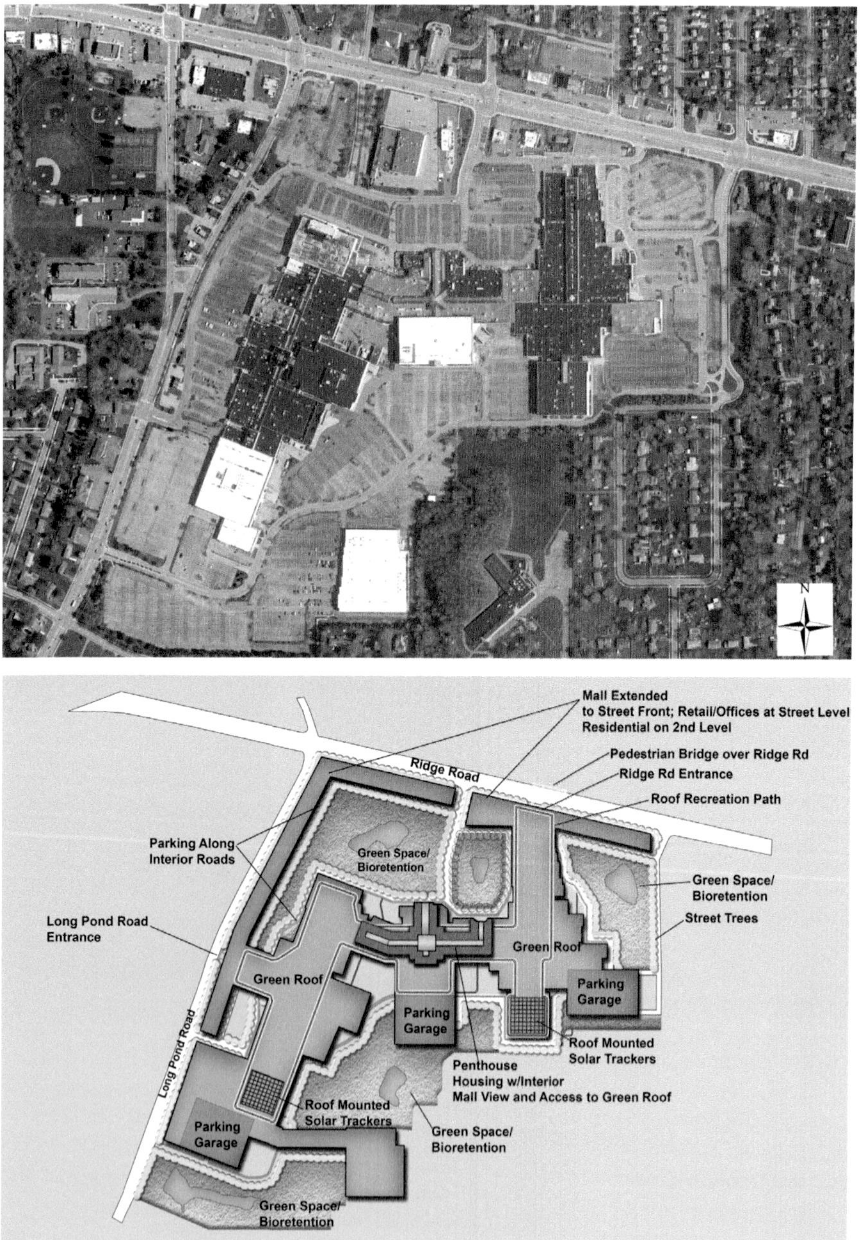

FIGURE 10.6 Retrofit of Mall at Greece Ridge. World Imagery courtesy of Esri, DigitalGlobe, GeoEye, i-cubed, USDA FSA, USGS, AEX, Getmapping, Aerogrid, IGN, IGP, swisstopo, and the GIS User Community.

Road, providing pedestrian access from the street. A pedestrian bridge is placed over Ridge Road, facilitating an uninterrupted pedestrian connection from the north side of Ridge Road to the mall. Liner buildings are also placed around the perimeter to address the street fronts with retail and office space at the ground level and residences on the second level. Three parking structures and on-street parking along the interior roads facilitate the removal of the large surface parking. Green space and bioretention ponds replace the surface parking and trees are planted along Ridge Road, Long Pond Road, and the interior mall roads. A green roof is placed on top of all buildings and contains a recreation path, solar trackers, and several gathering spaces. Penthouse housing is placed in the center of the mall with interior windows/balconies facing the interior of the mall and windows and patios connecting to the green roof.

References

Astor, W. (2016, November 18). Original tower was jewel of national wonder. *Rochester Business Journal.* https://rbj.net/2016/11/18/original-tower-was-jewel-of-national-wonder/

Barnes, S. (2016, January 14). *History's oldest shopping mall remains one of the world's most visited attractions.* My Modern Met. https://mymodernmet.com/istanbul-grand-bazaar/

CBRE. (2023). *San Francisco office snapshot: Q3 2023.* www.cbre.com/insights/figures/san-francisco-office-snapshot-q3-2023

City of Nashville (Nashville). (n.d.). *Downtown community plan: Downtown code (DTC).* www.nashville.gov/sites/default/files/2022-12/Downtown-Code-updated-2022-12-12-opt.pdf?ct=1671135144

Cushman & Wakefield. (2023). *Market beat Manhattan: Office Q3 2023.* https://cw-gbl-gws-prod.azureedge.net/-/media/cw/marketbeat-pdfs/2023/q3/us-reports/office/manhattan_americas_marketbeat_office_q3-2023.pdf?rev=446b552b8ecf4882a2179ce6dad341b4

Donnis, E. H. (2010). *The history of Shelburne Farms: A changing landscape, an evolving vision* (pp. 195–198). Vermont Historical Society.

Driffill, M. (2022, May 4). *$44M affordable housing development completed at Skyview on the Ridge Irondequoit.* Rochester First. www.rochesterfirst.com/development/44m-affordable-housing-development-completed-at-skyview-on-the-ridge-irondequoit/

Dunham-Jones, E., & Williamson, J. (2011). *Retrofitting suburbia: Urban design solutions for redesigning suburbs.* John Wiley & Sons.

Geiling, N. (2014, November 15). The death and rebirth of the American mall. *Smithsonian Magazine.* www.smithsonianmag.com/arts-culture/death-and-rebirth-american-mall-180953444/

Guzman, Richard. (2012, Jan 31). *Adapting the adaptive reuse ordinance.* DT News. https://www.ladowntownnews.com/news/adapting-the-adaptive-reuse-ordinance/article_f9bf41da-493a-11e1-b6c4-0019bb2963f4.html

International Council of Shopping Centers (ICSC). (2017). *U.S. shopping center classification and typical characteristics.* www.icsc.com/uploads/research/general/`_CENTER_CLASSIFICATION.pdf

JLL. (2023). *Q2 2023 Central London office market report.* www.jll.co.uk/en/trends-and-insights/research/q2-2023-central-london-office-market-report

Jones, S. (2022). *How adaptive reuse gives defunct buildings new leases on life.* Autodesk. https://www.autodesk.com/design-make/articles/adaptive-reuse

Kiger, P. (2023, February 27). *Turning malls into neighborhoods.* Urban Land. https://urbanland.uli.org/public/turning-malls-into-neighborhoods/

Lange, A. (2022). *Meet me by the fountain: An inside history of the mall.* Bloomsbury Publishing.

Los Angeles City Planning. (LA City Planning). (n.d.). *City wide adaptive reuse ordinance.* https://planning.lacity.org/odocument/6725f347-7fdb-42fa-aa6e-44c37f8fa999/Fact_Sheet_-_Adaptive_Reuse_Ordinance.pdf

Mandal, D. (2017, January 3). *Trajan's Market: Possibly the world's oldest known public shopping center.* Realm of History. www.realmofhistory.com/2017/01/03/trajans-market-oldest-shopping-center/

National Park Service (NPS). (n.d.a). *Tax incentives for preserving historic properties.* www.nps.gov/subjects/taxincentives/index.htm

National Park Service (NPS). (n.d.b). *About the incentives.* www.nps.gov/subjects/taxincentives/about.htm

Picture the Homeless. (2015). *Homeless people count: Vacant properties in Manhattan.* www.picturethehomeless.org/wp-content/uploads/2015/12/Homeless_People_Count2.pdf

Rezvani, A. (2023). *A lot of offices are still empty – And it's becoming a major risk for the economy*. NPR. https://www.npr.org/2023/05/16/1174938708/commercial-real-estate-property-offices-work-from-home-remote-work

Rudd, N. (2022, May 3). *Transformation of former Irondequoit Mall unveiled*. WHEC News 10. www.whec.com/archive/transformation-of-former-irondequoit-mall-unveiled/

Thomas, L. (2021, April 7). *Mall vacancies jump at fastest pace on record, hitting new high, as retailers cull store counts*. CNBC. www.cnbc.com/2021/04/07/us-mall-vacancies-jump-at-fastest-pace-on-record-hit-high-moodys.html

Tracy, C. (1983). Quoted from interview. Source unknown.

U.S. Census Bureau. (2022). *Housing vacancies and homeownership*. www.census.gov/housing/hvs/data/prevann.html

U.S. Government Accountability Office (GAO). (2023). *Federal real property: Preliminary results show federal buildings remain underutilized due to longstanding challenges and increases telework*. https://www.gao.gov/assets/gao-23-106200.pdf

Webster, T. (1994). *Shelburne: Pieces of history*. Shelburne Historic Sites Committee.

11

INTERIORS

As a result of the digital revolution and the subsequent proliferation of smartphones/tablets, streaming services, e-commerce, artificial intelligence, and virtual meeting platforms, it is not surprising that we are spending more time indoors than at any other time in history. Even before the explosion of virtual technology, data collected for the National Human Activity Pattern Survey (NHAPS) between 1992 and 1994 showed about 87 percent of Americans spent their time in enclosed buildings and 6 percent of their time in enclosed vehicles (Klepeis et al., 2001). Another study examining seven regions of Europe found that the average person spends 22 hours a day indoors, amounting to 90 percent of the day (Schweizer et al., 2007). At the height of the COVID-19 pandemic in 2020–2021, data from the Pew Research Center (2022) showed that 71 percent of people with jobs that could be done from home were working from home all or most of the time in October 2020 compared to 59 percent in February 2022. Although the numbers are decreasing, data shows that 64 percent who are still working from home some of the time say it is easier to balance work with their personal life, 60 percent say remote work has made them feel less connected to co-workers, and 44 percent say working from home has made it easier for them to get their work done and meet deadlines (Parker et al., 2022). Although many companies are requiring workers to return and/or allowing hybrid schedules, the Pew Research data also shows that, given a choice, 60 percent would still prefer to work from home all or most of the time when the coronavirus is over.

How this will unfold and impact the amount of time people spend indoors is difficult to say, but one thing is for certain: regardless of where we are working, we are not getting outside enough. A study by Forrest & Stuhldreher (2011) showed the overall prevalence rate of vitamin D deficiency to be 41.6 percent. A global survey of 81 countries also found 15.7 percent had vitamin D levels less than 30 nmol/l, 47.9 percent less than 50 nmol/l, and 76.6 percent less than 75 nmol/l (Cui et al., 2023). Studies show that the risk factors that are responsible for vitamin D deficiency include air pollution, latitude, season, the use of sunscreen, sedentary jobs, and diet (Arabi et al., 2010). Whether we are forced to be indoors due to pandemics or extreme weather events or whether we choose to stay home because we are introverted, germophobic, or agoraphobic, the quality of our interior environments is a common denominator that directly impacts health, comfort, and wellness.

DOI: 10.4324/9781032668666-11

Environmental psychology

While many designers and sociologists assume specific metrics as to what most people consider "pleasant" or "comfortable," understanding human relations to and perceptions of space along with understanding different physical and mental abilities is pivotal to truly making interior space equitable, comfortable, and appealing. While accessibility has specific parameters defined by the Americans with Disabilities Act (ADA) or the International Code Council (ICC), there is less design guidance surrounding the relationship between human beings and their surroundings. Although the field of environmental psychology saw a surge in the 1970s and 1980s, it seems to have lost momentum in recent times. One of the challenges in the field has been how to translate human behavior and perception into specific design recommendations, and given the complexity of the human brain, this could be a daunting task, to say the least. In addition, the theory of whether form dictates function or whether human behavior will adapt and define its own behavior settings regardless of the form of the space is still an ongoing debate.

However, it is still important to be aware of behavior, psychology, or sociologic patterns when programming space. For example, a person who is claustrophobic will feel uncomfortable in confined spaces, a person who is agoraphobic may feel uncomfortable in open or crowded spaces where escape is difficult, a person with acrophobia will be uncomfortable with heights, a person with nyctophobia will be afraid of dark spaces, a person who is hydrophobic will feel uncomfortable near water, a person who is mysophobic will be afraid of dirt and germs, and although it is a wide spectrum, people with autism may experience sensory overload given crowding, lighting, acoustics/noise, temperature, colors, patterns, or visual stimulus, to name a few. There is also a large spectrum in all psychological conditions, so although someone may not be officially diagnosed with a specific phobia or mental illness, they still may feel occasional discomfort given a specific environmental condition or stimulus.

Restaurants are a good example. While certain extroverted personality types might get energized from being in a crowded place with ample socialization opportunities, more introverted types may simply feel it is too loud. But regardless of a person's Meyers-Briggs personality type or preferred personal space dimensions, commotion and people moving around you or bumping into you while you eat may spawn irritability for many people. Many domestic dogs often engage in "food caching" where they take their food to a safer location to eat; this behavior may be preferred by some humans as well. Many restaurant owners want to get as many people in the space as possible to maximize their profits, ultimately neglecting any notion of human comfort or safety. In many old Italian restaurants, all the tables were designed to face outward, putting the person's back against the wall so they could see who was approaching them. Although this was to prevent an ambush in the heyday of organized crime, it seems many non-gangster types would also prefer not to have their back in the room.

Some restaurants, cafeterias, and dining halls are also designed like military dining barracks with one large, flat rectangular space filled with tables in a grid-like fashion. The result is a noisy, lifeless space devoid of any intricacy, hierarchy, character, or behavior settings that foster different types of activities. The high school cafeteria may come to mind for many reading this section. Long, rectangular tables may provide more equity than groups of circular tables since everyone sits in a line with a limited hierarchy; however, although circular tables may provide more intimacy, they also can foster the congregation of social cliques that leave students who don't identify with a specific social group limited places to sit. A better design for a cafeteria would be to provide a variety of seating and space types to suit different personality types.

FIGURE 11.1 Incorporating spatial hierarchy at the Corning Museum of Glass using ramps, bridges, textures, colors, lighting, and varying ceiling and floor heights.

Perimeter, bar-style seating along the windows would give options to people that are not part of a specific clique and/or who prefer to eat alone. Mixing different sized tables in different zones of the cafeteria also provides options that can match a variety of different socialization preferences. A less formal lounge space away from the action may also be preferred by some during or after eating. Another more equitable option is the donut arrangement where tables are placed in a rectangle or circle around the perimeter with everyone facing the center of the room.

Indeed, spatial hierarchy is something that is often neglected in spaces. Altering ceiling heights, floor heights, textures, colors, geometries, materials, and lighting styles establishes unique behavior settings and breaks the monotony of bland, open spaces. In addition to providing variety, spatial hierarchy can make spaces more intuitive to navigate and can better delineate spaces for large group gatherings, small group gatherings, circulation, production, or back-of-the-house operations.

The design of lavatories is also something that can be frustrating. During my college years, there was a bar on the Main Street strip in Buffalo that we would all try to avoid because of its horrible bathroom. The men's room consisted of a long, steel trough down the side of the wall with little space between the wall and the trough and with no doors into the bathroom and no partitions on the trough. I think this is where many of us first learned what was meant by "stage fright" when going to the bathroom. A worse bathroom design was at Highmark Stadium (formerly Rich Stadium) where the Buffalo Bills football team play. This trough outdid the one at the college bar because it was shaped as a semi-circle, so in addition to looking directly at another person while they were urinating, you had to worry about someone intentionally or unintentionally spraying you with their stream. It is vexing that bathroom design is not given a higher priority, because despite the specific activity in a building, one thing that will always remain constant is that people will have to use the bathroom.

Luckily, we are starting to see the error of our ways in restroom design. Gang showers, urinals without partitions, toilets without doors, and troughs are starting to become a thing of the past. As part of the work done by the University of Vermont's (UVM's) Gender Inclusive Restroom Task Force, all single occupancy restrooms were made to be gender inclusive along with providing a

FIGURE 11.2 Privacy, anyone? Gang-style urinals, which are hopefully something of the past.

mix of gender-inclusive multiple occupancy and men's/women's multiple occupancy restrooms. Another recommendation was to rid the urinal once and for all. Sorry, Andrew Rankin, but I am not sure what you were thinking in 1866. As a result of student demand, all the restrooms in the new Central Campus Residence Hall complex at UVM are also gender inclusive. The restrooms contain individual, private suites with floor-to-ceiling walls and locking doors and contain a shower, sink, changing area, and toilet. As a result of green building rating systems, many non-residential buildings are also incorporating private, gender-inclusive shower spaces and indoor and/or covered bicycle parking to foster bicycle ridership.

People above or below the average height and weight range may also struggle with personal space and/or fitting comfortably in a space. Speaking from my own experiences of being 6′5″, I have struggled to fit in cars, showers, patient exam tables at the doctor's office, MRI machines, and showers as well as in the seating on/in public transportation, cars, movie theatres, assembly halls, and classrooms. Someone who is shorter than average may not be able to reach things at the grocery store, see over the person in front of them at theatres, or may have trouble reaching the pedals in a car. Someone who is overweight may experience some of the same spatial difficulties as someone who is tall, just with more ridicule and bullying from people in society who are insensitive to others who don't fit their definition of what a "normal" anatomy should look like.

Although I am not claiming to be an environmental psychologist, this section is meant to make the reader aware of some of the issues that are difficult to quantify or qualify regarding people's relationships with their surroundings. Given the variety of personality types, body types, abilities/disabilities, and mental/physical ailments, it is no longer practical to design exclusively for the norm nor is it feasible to create a one-size-fits-all solution. Difference and diversity should be accepted and included rather than being marginalized or excluded. If everyone looked and felt the same and we designed everything to look the same to match the sameness of said people, the world would be a boring place. Although it can be challenging to translate human perception and emotion into specific design parameters, there have been advances made regarding accessibility, inclusivity, safety, and health/well-being for interior spaces that are discussed in the next section.

Accessibility, universal design, and inclusive design

The disability rights movement started around the same time as the civil rights movement in the 1960s with a mission to secure equal opportunities and rights for all people with disabilities. In 1968, the Architectural Barriers Act was passed that required buildings of facilities that were built or renovated with federal dollars by federal agencies after August 12, 1968, to be accessible. In 1973, passage of Section 504 of the Rehabilitation Act banned discrimination based on disability by recipients of federal funds. After several hearings before Congress and persevering activism by people in the movement, section 504 regulations were issued in 1977 that would form the foundation of the Americans with Disabilities Act (ADA). In 1988, the Fair Housing Amendments prohibited housing discrimination on the basis of race, color, religion, sex, disability, familial status, and national origin in private housing, housing that receives federal financial assistance, and state and local government housing. The first draft of the ADA was presented to Congress in 1988, passed in 1990, and became enforceable in 1992. The ADA has four sections: Title I protects individuals with disabilities in employment; Title II covers state and local government services and public transportation; Title III covers all public covers all public accommodations; and Title IV covers telecommunications. Title II and Title III regulations are augmented by the ADA Standards for Accessible Design, which covers newly constructed buildings and facilities; alterations and renovations; architectural changes in existing state and government buildings; and removing architectural barriers that are easily accomplishable without much difficulty or expense in existing buildings of business (ADA, n.d.).

State and local municipalities can develop their own accessibility standards, but they must meet or exceed what is specified in the ADA. Some municipalities and institutions have also implemented Universal Design and/or inclusive design standards into their codes. As introduced in Chapter 2, Universal Design seeks to "enable and empower a diverse population by improving human performance, health and wellness, and social participation" and inclusive design considers the full range of human diversity with respect to ability, language, culture, gender, age, sexual orientation, ethnicity, and other forms of human difference (Steinfeld & Maisel, 2012; IDRC, n.d.). Both Universal Design and inclusive design can develop products, experiences, and designs. Inclusive design strategies can include interactive signage that interfaces with an app; signage with intuitive graphics equipped with brail and/or audio controls; flexible workspaces that can be reconfigured to accommodate different functions, working styles, volumes, or in response to unknown potential pandemic spacing requirements; providing a range of different study spaces in schools (group, small group, personal) to accommodate different learning/study styles; dedicating spaces for culturally or historically significant events or people; providing a variety of seating types in restaurants, classrooms; providing spaces for public markets to stimulate local merchants; programming space that reflects the cultures of the community; and providing flexible gathering spaces where people can engage in informal placemaking.

Daylighting

During the winter months back in the 1980s, my family used to travel to a Sheraton hotel (now M Hotel Buffalo) in Cheektowaga, New York, that contained an indoor courtyard pool positioned in the "donut hole" of the rectangular building. The roughly 16,000 square feet courtyard was designed in a Hawaiian theme and contained several tropical plants, a water feature, a pool, a hot tub, a sauna, and a grass-shack Tiki bar. The roof contains two barrel-shaped rows of skylights that have recently been upgraded to retract when the weather outside is pleasant. This space offered a much-needed reprieve for people in western New York who experience never-ending winters that can last six to eight months and constant overcast skies from the Great Lakes that exist on a year-round basis.

The central atrium is hardly a new concept. The ancient Greeks employed interior peristyle courts and the ancient Roman domus house contained a central atrium surrounded by columnated courts. Although the importance of daylighting has long been recognized, it is unfortunate it is still a luxury afforded by the wealthy or something many people have to seek out at a resort, a conservatory, or in my family's case, the Sheraton Hotel.

Atriums and solariums can be constructed in the center of a space or attached to a side of a space. Natural daylighting can decrease electrical lighting demand and decrease heating demand in winter months by passively absorbing sunlight. In the warmer months, retractable panes, tinting, smart glass, and fans may be needed to keep interior spaces cool. In addition, atriums provide space for indoor gardens, respite, relaxation, and mood enhancement. In climates with long winters, atriums can provide a means to absorb sunlight that increases vitamin D, stimulates blood circulation, and improves the immune system.

FIGURE 11.3 Atriums on the side (left) and down the middle (right).

The University of Vermont took advantage of an open courtyard of the Given Medical Complex by infilling it with new space for offices, laboratories, and support spaces for the College of Medicine. Designed by Black River Design Architects, the Courtyard at Given is a four-story, "ship in a bottle," 33,000 gross square foot space that contains two pods with a central staircase that are connected to each other and to the surrounding buildings with a series of bridges/catwalks. A barrel dome of skylights caps what was once an opening and provides the space with ample daylight. The building also features an underfloor air distribution system, an enthalpy wheel recovery ventilator, low volume water fixtures, daylight control of atrium light fixtures, low VOC paints, coatings, and finishes, and green label plus carpets. The office cubicles are also easy to reconfigure, providing flexibility and adaptability and giving new space needs. The building was awarded LEED Gold in 2009.

In addition to atriums, skylights, and solariums, effective daylighting can be achieved by strategic window placement, clerestory windows, light-reflecting surfaces, solar tubes, louvers, glazing materials, and smart glass. Window types, placement, and roof overhangs are dependent on the orientation of the building with respect to the sun and the latitude of the region. For example, in northern climates, an overhang can be designed to keep the summer sun from penetrating when the sun is at a higher angle but can allow the sun to enter in winter months when the sun angle is lower. In southern climates closer to the equator, overhangs can be designed to keep sun out year-round since the sun stays at a higher angle. Morning sun angles from the east and afternoon sun angles from the west may have to use deeper overhangs (depending on the desired effect), vegetation, window treatments, or glazing treatments to control direct solar gain. The orientation of the building is also directly proportional to the solar gain, with

FIGURE 11.4 Left: the Courtyard at Given at the University of Vermont; right: daylighting the inner atrium of Sustainability Institute Hall at the Rochester Institute of Technology.

a building at 0 degrees facing directly south getting 100 percent solar gain and a building at a 45-degree angle to the south getting about 70 percent of solar gain. Using Rochester, New York, as an example, Figure 11.5 shows the sun's position at 12:00 PM on the summer solstice, winter solstice, and vernal and autumnal equinoxes using the noon sun angle equation derived from zenith angles (latitude +/− the subsolar point, depending on the subsolar point location). Using a computer-aided drafting system or trigonometry, you can calculate the length of overhang needed to block out the sun at specific times/seasons. In the example, a 2-foot eave would block out the summer sun at high noon, but a 6-foot eave is needed to block the sun on the equinoxes at high noon.

Clerestory windows are generally placed in a row above eye level toward the top of the wall. The concept dates to ancient Egypt and later in Romanesque and Gothic churches to illuminate halls, temples, and churches. Frank Lloyd Wright utilized clerestory windows in several of his designs and the "ribbon window" was part of Le Corbusier's Five Points of a New Architecture. The clerestory window allows light to enter the space while maintaining privacy and wall space at eye level, which is ideal for offices or medical practices with interior cubicles/operatories away from outward-facing windows. To avoid glare, light shelves can be installed below the clerestory window that reflects light toward the ceiling. Transom windows are similar to clerestory windows but are placed above and along the side of doors. In addition to providing light, the transom window allows visibility which can prevent human collisions when the door swings open. Transom windows can also be hinged and opened to allow air flow to pass through corridors.

Adjustable louvers, tinted glazing, and/or smart glass can also be utilized to control the amount of sunlight that enters a space. Louvers can be manually controlled or can adjust automatically according to the sunlight, wind, or rain. Likewise, smart glass can be active or passive, with passive systems utilizing low-E (Emissivity) glass that contains thin metallic coatings such as tin oxide to reflect heat back to its source, photochromic glass that tints the glass in response to light, or thermochromic glass that tints the glass in response to temperature. Active systems require an electrical charge such as polymer dispersed liquid crystal (PDLC), suspended particle device, and electrochromic and can turn glass from fully transparent to fully opaque, making them ideal for situations where privacy is needed. Active systems can also use controllers that allow dimming to have various degrees of opacity, or in the case of PDLC, a variety of patterns or colors.

Another way to bring light into sub-grade levels is by using a light trench around the building. As discussed in Chapter 7, the Cliniques universitaires Saint-Luc Hospital parking garage in Brussels contains three underground levels and two above-ground levels. The three underground levels are encircled by an open trench or "canyon" that allows natural light to enter, alleviating the electrical lighting demand and providing a more aesthetically pleasing environment. When light trenches, atriums, or skylights are not feasible, light can be brought into spaces using heliostats, which are dish-shaped mirrors that focus light onto a mirror and distribute light downward into the space via a light pipe. This concept is utilized for the underground Lowline proposal that is discussed in Chapter 12.

Artificial lighting

The expression "it's all in the lighting" is a valid point. People with photophobia (fear or sensitivity to light), migraines, vestibular disorders, or optic disorders are particularly

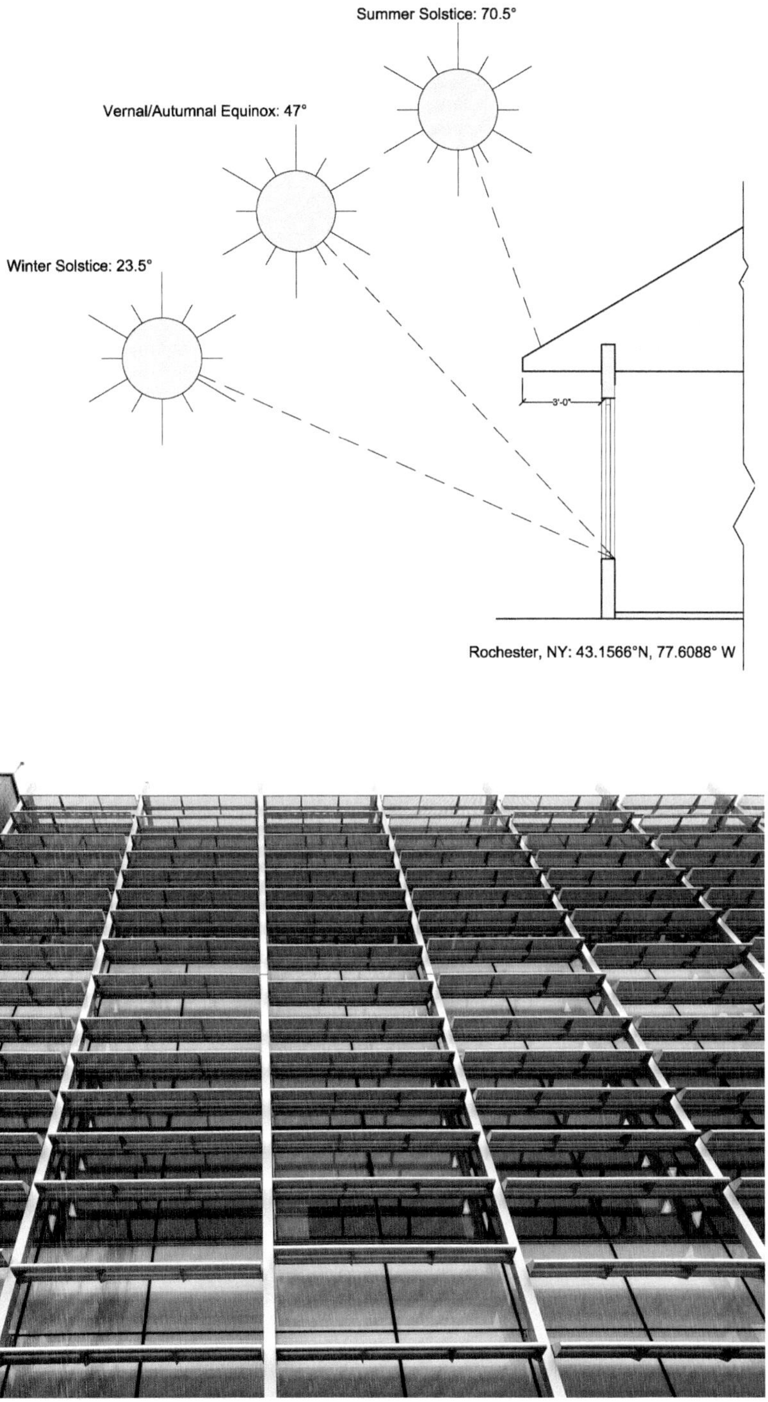

FIGURE 11.5 Top: Passive solar design and roof overhang sizing using Rochester, New York, as an example; bottom: stationary louvers on Sustainability Institute Hall at the Rochester Institute of Technology deflect direct sun exposure and reduce heat gain.

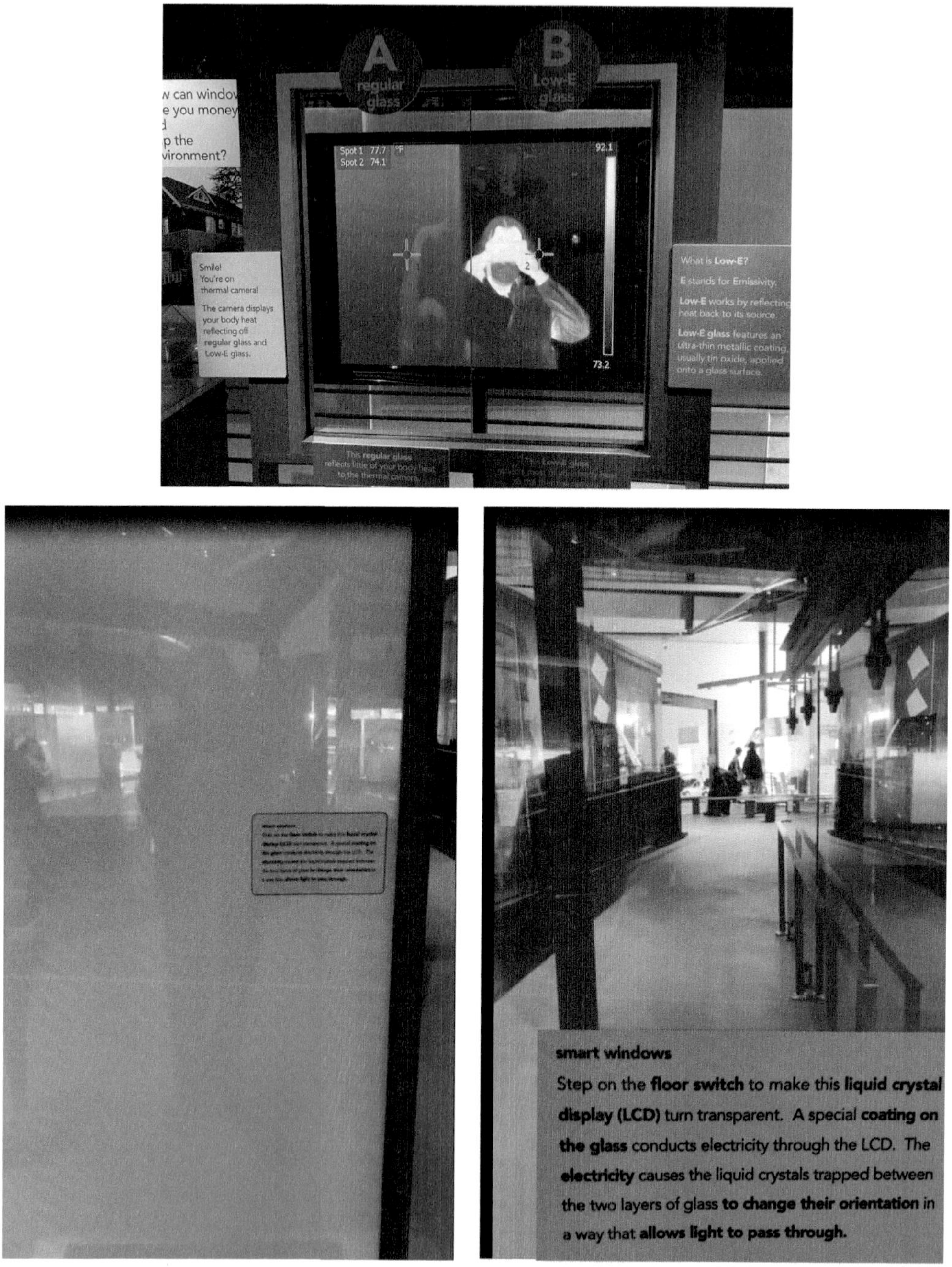

FIGURE 11.6 Top: passive low-E smart glass demonstration at Corning Museum of Glass (Note: this picture requires color to see the heat signatures); bottom: active smart glass demonstration at Corning Museum of Glass using Liquid Crystal Display (LCD). The electricity causes the liquid crystals trapped between the two layers of glass to change their orientation to allow the glass to be opaque or transparent. Bottom left: opaque; bottom right: transparent.

impacted by artificial lighting – combine this with the amount of time we spend on computer screens and smart devices, and the sensitivity can be exacerbated. Although there are ways to turn off the blue light from devices and/or glasses you can wear that filter specific wavelengths of light, the interior design of lighting should prevent the need for such interventions.

Quality illumination should seek to eliminate flicker and eliminate direct and indirect glare. Direct glare can be reduced by decreasing the luminance of the light source, increasing the angle between the glare source and the line of sight, using shielded luminaires, and reducing the brightness ratio between the light source and the background. Indirect or reflected glare can be reduced by using matte background surfaces that are not glossy, ensuring the reflectance of the background does not exceed that of the task, or by adjusting the angles of the light or work surfaces.

One strategy that is often neglected is to provide a mix of lighting types, including ambient, task, and accent lighting. Systems can be direct, where 90–100 percent of light is emitted downward; indirect, where 90–100 percent of light is emitted upward; diffuse, where light is emitted in all directions; semi-direct, where 10–40 percent of light is emitted upward and 90–100 percent is emitted downward; or semi-indirect, where 60–90 percent of light is emitted upward and 10–40 percent is emitted downward (Hoke & Ramsey, 2000).

Mounting, placement, and spacing of different lighting types also impact the aesthetic and efficiency. Lighting can be recessed, adjustable, suspended, coffered, flush, surface mounted, wall-mounted, soffit mounted, cove, cornice, or valance. Often, installers will use a simple formula that divides the ceiling height by two and space lights at equal intervals. While this may provide proper task lighting, it does not address ambiance or accent-specific features. Ideally, different lighting types can be paired with different ceiling heights in a space to produce spatial hierarchies, especially for large, open spaces that lack partitions. For example, recessed lighting can be placed under a soffit that lines the perimeter of a ceiling (such as in a tray ceiling) and can be put on a different switch from the central task lighting to provide variety. Lighting placed in coves around the perimeter of a room can create an indirect glow that illuminates the ceiling.

Providing controls for lighting such as dimmer switches, timers, and/or occupancy sensors provides more options and can cut energy usage. Smart bulbs make it possible to control lights remotely with an app on a smartphone, tablet, computer, or a smart speaker such as Alexa or Google Home. According to the U.S. Department of Energy (U.S. DOE, n.d.), light-emitting diode (LED) bulbs are currently considered to be the most energy-efficient lighting option available, with Energy Star-certified LED bulbs using at least 75 percent less energy and lasting 25 times longer than incandescent lighting. As a comparison, LED bulbs produce 40–90 lumens per watt, followed by compact fluorescent that produce 30–60 lumens per watt, halogen bulbs about 12–15 lumens per watt, and tungsten at 8–10 lumens per watt (Lumen Now, n.d.). When replacing an incandescent with LED, Efficiency Vermont (n.d.) provides this helpful conversion for brightness and color/warmth: LED bulb with a brightness of 800 lumens to replace a 60-watt bulb; LED bulb with 1,100 lumens for a 75-watt bulb; LED bulb with 1,600 lumens for a 100-watt bulb; LED bulb with a color of 2,700–3,000 kelvins for warm-yellow light; LED bulb with a color of 5,000–6,000 kelvins for a cooler, whiter, natural-daylight color. The U.S. Department of Energy estimates that given the current rate of conversions to LED technology, energy savings could reach 569 terra-watts annually by 2035 (U.S. DOE, n.d.).

FIGURE 11.7 Pool at the Mirror Lake Inn in Lake Placid, New York. In addition to a soothing interior water space, indirect cove lighting creates ambiance; varying ceiling heights and ambient lighting over the glass doors and windows create an intricate sitting space; natural wood materials used on the walls and columns and granite floors simulate the aesthetic of a Greek grotto.

Interior vegetation and natural materials

In addition to providing daylighting, atriums, skylights, solariums, greenhouses, and effective window placement provide conditions for interior plants to thrive. A study by Han et al. (2022) showed positive effects of interior plants on people, including a more relaxed physiology, improved cognition, better academic achievement, and lower blood pressure. Although the data is mixed, some studies also do show indoor plants can purify indoor air through absorption, dilution, precipitation, and filtration of airborne molecules and pollutants such as carbon dioxide, volatile organic compounds, carbonyl, particulate matter, organic compounds, nitrates, sulfates, ammonia, calcium, ozone, and carbonate (Kim et al., 2018; Agarwal et al., 2018; Pegas et al., 2012; Soreanu et al., 2013; Abbass et al., 2017; Parseh et al., 2018; Wei et al., 2021).

FIGURE 11.8 Interior vegetation. Left: Doubletree Hotel glassed in courtyard space (formerly the Holi-dome); bottom: Strong Museum of Play Butterfly Conservatory.

Scientists at NASA (Wolverton, 1989) also conducted a study in 1989 and found that plants, particularly in the plant-root-soil zone, can remove trace organic pollutants (formaldehyde, benzene, trichloroethylene (TCE)) that contribute to sick building syndrome. When exposed to high concentrations of 20 parts per million, Mass Cane, Pot Mum, Gerber Daisy, Warneckei, and Ficus removed 70, 61, 50, 50, 47.4 percent of formaldehyde, respectively; 21.4, 53, 67.7, 52, and 30 percent of benzene, respectively; and 12.5, 41.2, 35, 10, and 10.5 percent of TCE, respectively. One criticism of the study is that it tested conditions in a closed environment to simulate conditions on a space station that are significantly different than real-world, interior spaces. A memo from the U.S. Environmental Protection Agency (Girman, 1992) commenting

FIGURE 11.9 Interior gardens at Longwood Gardens in Philadelphia, Pennsylvania (top, bottom left, bottom center) and at the Lamberton Conservatory in Rochester, New York (bottom right).

on the NASA study claimed that a typical house would have to contain 680 plants to achieve the results realized in the NASA test chamber. Some of the atrium designs discussed earlier in the chapter and the green wall systems discussed in the following text could potentially support this quantity and variety.

As discussed in Chapter 9, indoor green wall systems can be constructed with any of the methods used for exterior walls, however, most interior applications seem to utilize modular or living wall systems (see Figure 9.6). In addition to their aesthetical appeal and potential to filter pollutants as described earlier, green interior walls offer the same benefits as exterior green walls, including noise reduction and added thermal insulation that can cool the interior in warmer seasons and warm it during colder seasons.

Interior water features such as fountains, water walls/curtains, waterfalls, aquariums, and indoor pools can create a relaxing atmosphere, ambient "white" noise to block out distracting sounds and can improve air quality by adding humidity (which is good for indoor plants), reducing dust particles, and removing contaminants through the generation of negatively charged ions. Integrating natural materials, textures, and forms using wood, stone, glass, wicker, cork, bamboo, and natural fibers can also create biophilic interiors that create calming and welcoming spaces. Many Green Building Rating Systems (GBRS) also give credits for sourcing local materials and for using low VOC-emitting materials in paints, coatings, adhesives/sealants, flooring, walls, ceilings, insulation, and furniture.

Indoor air quality and thermal comfort

When I was in graduate school at the University at Buffalo, the architecture department was ironically (or maybe suitably) located in two of the oldest buildings on campus. The temperature was smothering in the warmer months and freezing in the cooler months. There was no natural or mechanical ventilation, which yielded an interesting aroma of wood shavings, solder, glue, lead paint, asbestos, and body odor from students who frequently pulled all-nighters to finish their projects on time. One of my classmates used to enter every room and yell, "Someone open a window!" And seeing the windows were painted shut with lead paint, opening them often required a crowbar and a gas mask. Fortunately (but unfortunately for my class), the two buildings (Hayes and Crosby Halls) have recently undergone extensive renovations that have improved lighting, air flow, acoustics, and energy efficiency.

Indoor air quality (IAQ) is an issue in many spaces, regardless of their age. According to the U.S. Environmental Protection Agency (EPA, n.d.), concentrations of pollutants indoors are often 205 times higher than typical outdoor concentrations. Interior pollutant sources can come from appliances, materials, furniture, finishes, cleaning products, improperly maintained heating, ventilation, and air conditioning (HVAC) systems and can permeate from outside air pollution. Sources of pollution include asbestos, lead, carbon monoxide, carbon dioxide, nitrogen dioxide, pesticides, radon, indoor particulate matter, smoke from tobacco or cannabis, ozone, or volatile organic compounds such as formaldehyde, trichloroethylene, vinyl chloride, carbon tetrachloride, toluene, acetone, isopropyl alcohol, hexanal, or carbon disulfide. Then there are, of course, biological contaminants that we all became aware of during the COVID-19 pandemic that forced us to reduce occupancy loads, improve cleaning/disinfecting methods, and improve ventilation systems.

Like Hayes and Crosby Halls before they were renovated, many air quality issues are the result of inadequate natural or mechanical ventilation that results in a low air exchange rate, which is the rate at which outdoor air replaces indoor air. Natural ventilation relies on convection where differences in densities of warmer and cooler air transport air. Passive heating and cooling strategies take advantage of this air buoyancy. Using some of the orientation and shading strategies discussed earlier in this chapter, buildings can be designed to capture sunlight in the winter months and reflect it in the summer months. For passive heating, the sun is entered through south-facing glass and is absorbed by materials with high thermal mass that slowly release the heat through the night. Passive cooling can be accomplished by placing vents low on the windward side of prevailing nighttime breezes and high vents on the leeward side. Thermal chimneys, soffit vents, or vented cupolas on the roofs of buildings can be used where there are minimal prevailing winds. Cool air comes in from lower levels and warm air rises and is pushed out the vents. In cooler months, fans can be installed that can run in reverse to push warm air down into the living space.

The American Society of Heating, Refrigerating, and Air-Conditioning Engineers (ASHRAE) specifies minimum standards for ventilation based on the type of space, the number of occupants, and outdoor air quality (ANSI/ASHRAE 62.1–2022 Ventilation for Acceptable Indoor Air Quality). For example, the recommended ventilation for an office space is 17 cubic feet per minute (cfm) or 8.5 liters per second (l/s) of outside air per person; as a comparison, an auditorium seating area is 5 cfm per person or 2.7 l/s person and a storage room for liquids or gels is 65 cfm per person or 32.5 l/s per person (ASHRAE, 2022). To achieve these standards, natural systems are usually augmented with mechanical systems that heat or cool air with air handler units, ducts, vents, condensers, and heat exchangers. Air can be filtered with mechanical filters, electrostatic filters, or with ultraviolet (UV) light in the case of medical spaces/hospitals. Advances in air sensor technology have enabled the detection of airborne pollutants. The sensors can activate an alarm (such as in a carbon monoxide detector) or activate an air cleansing action or device. Portable air cleaners/purifiers can also supplement filtration on the central HVAC system.

Water

According to the U.S. Environmental Protection Agency (EPA, n.d.), the three largest uses of water in office buildings are restrooms, heating and cooling, and irrigation for landscaping. To reduce water consumption, showerheads, toilets, and faucets can be retrofitted with efficient, Water Sense-certified fixtures. Some of the efficient toilets contain two flushing volumes: one for liquid waste (number one) and one for solid waste (number two). Other measures that can be done to conserve water are using timers for irrigation systems, ensuring there are no leaks in the system, changing our habits, installing smart meters that can monitor water usage in real-time, and reclaiming gray water.

Potable or white water is fresh, treated water coming in, black water is sewage going out, and gray water is unfiltered water that is used in sinks, showers, and washing machines and that has not come into contact with human waste. Gray water can be used for irrigation using gravity-based plumbing systems that use diverters to channel the water to the subsurface instead of flowing into the sewer pipe. The piping terminates at a basin that allows irrigation of the root zone of plants. Depending on what people put down the drain, gray water can contain chemicals, oils, grease, soaps, nitrates, salts, bacteria, and even low levels of pathogens. Although municipalities have different codes as to what types of gray water need to be pre-treated before they are recycled, given the new pandemic era we are living in, it is probably wise to pre-treat all gray water before it is released. This can be done by using first-flush diverters, screening, filters, and disinfection using chlorination, UV light, or ozonation. An alternative to using these components is to divert the water into constructed wetlands that naturally absorbs nutrients and filter particles before the water is reused. Rainwater harvesting can also capture runoff on the outside of the house using a collection area (usually the roof), conveyance (gutters, downspouts, piping), filtering devices, storage (cistern, rain barrels), and/or more complex systems that treat and distribute the collected water using filters and pumps.

Acoustics

Although it may not always be tangible, acoustics are also a major factor that directly influences comfort and wellness in spaces. Noise disturbances can increase stress levels, lower productivity, create distractions, hinder communication, and infringe on privacy. As discussed in Chapter 3, studies show people are less willing to help others, quicker to cast judgment, and in worse moods in loud environments compared to low noise conditions (Dangerous Decibels, n.d.; Gifford, 2002, pp. 278–279). While the sound is not visually tangible, it can disassociate people and/or serve as a barrier just as divisive as visual obtrusions; more simply put, we tend to run away from the noise.

Good hearing criteria seek to silence background noise (unless a certain level of white noise is desired), ensure signal loudness is adequate (in the cases of auditorium or lecture spaces), balance the distribution to achieve acoustic uniformity, and ensure sound is blended with separation to minimize reverberation time and flutter echo. Recommended background noise is around 25 decibels (dB) for sensitive listening spaces such as recording studios or concert halls, 35 dB for general presentation spaces, and 45 dB for public spaces, offices, and restaurants. The sound transmission class (STC) is a number rating of the performance of a material in preventing the transmission of airborne sound. The lower the number, the more sound is transmitted. For example, single-pane glass windows have an STC of 25 while a 10-inch masonry wall has an STC of 50. Recommended STCs between adjacent classrooms are around 40, adjacent hotel rooms around 50, surrounding bathrooms around 56, surrounding music rooms around 55, and between floors with different dwellings around 52 (Hoke & Ramsey, 2000).

Sound can be controlled through materials, geometries, and volume. Hard materials reflect sound while soft, porous materials absorb sound. Porous absorption includes soft materials such as fiber glass, insulation, foam, or carpet; panel absorption uses acoustic ceiling tiles, gypsum, or plywood; cavity absorption uses an enclosed air cavity in the wall layers; heavy construction uses double-stud walls, concrete-masonry, double layered subflooring, floor underlayment, and floating floors; surface treatment includes splayed surfaces and suspended baffles; sound masking uses white or background noise to mask sound transmission.

In terms of geometry, parallel surfaces cause reflection and excessive reverberation when spaces are more than 60 feet apart and flutter echo when spaces are less than 60 feet apart. Concave surfaces focus sound and create hot spots and convex surfaces diffuse sound, which may be desirable in listening areas. In "cubicle-city" open office layouts that do not have floor-to-ceiling partitions, offsetting cubicle openings and desk locations can minimize sound transmission.

References

Abbass, O. A., Sailor, D. J., & Gall, E. T. (2017). Effectiveness of indoor plants for passive removal of indoor ozone. *Building Environ, 119*, 62–70. http://doi.org/10.1016/j.buildenv.2017.04.007

Agarwal, P., Sarkar, M., Chakraborty, B., & Banerjee, T. (2018). *Phytoremediation of air pollutants: Prospects and challenges*. Elsevier.

American Society of Heating, Refrigerating, and Air Conditioning Engineers (ASHRAE). (2022). *Ventilation and acceptable indoor air quality*. ANSI/ASHRAE Standard 62.1–2022.

Americans with Disabilities Act (ADA). (n.d.). *ADA standards for accessible design*. www.ada.gov/law-and-regs/design-standards/

Arabi, A., El Rassi, R., & Fuleihan, E.-H. G. (2010). Hypovitaminosis D in developing countries-prevalence, risk factors and outcomes. *Nature Reviews Endocrinology*, *6*, 550–561. http://doi.org/10.1038/nrendo.2010.146

Cui, A., Zhang, T., Xiao, P., Fan, Z., Wang, H., & Zhuang, Y. (2023). Global and regional prevalence of vitamin D deficiency in population-based studies from 2000 to 2022: A pooled analysis of 7.9 million participants. *Frontiers in Nutrition*, *10*. http://doi.org/10.3389/fnut.2023.1070808. www.frontiersin.org/articles/10.3389/fnut.2023.1070808

Dangerous Decibels. (n.d.). *Noise induced hearing loss*. www.dangerousdecibels.org/education/information-center/noise-induced-hearing-loss/

Efficiency Vermont. (n.d.). *How to choose the best quality LED bulb*. www.efficiencyvermont.com/blog/how-to/how-to-choose-the-best-quality-led-bulb

Forrest, Y. Z., & Stuhldreher, W. L. (2011). Prevalence and correlates of vitamin D deficiency in U.S. adults. *Nutrition Research*, *31*(1), 48–54. https://doi.org/10.1016/j.nutres.2010.12.001

Gifford, R. (2002). *Environmental psychology: Principles and practice* (3rd ed., pp. 278–279). Optimal Books.

Girman, J. (1992). *Comment on the use of plants as a means to control indoor air pollution*. U.S. EPA. https://nepis.epa.gov/Exe/ZyNET.exe/000002IB.txt?ZyActionD=ZyDocument&Client=EPA&Index=1991%20Thru%201994&Docs=&Query=&Time=&EndTime=&SearchMethod=1&TocRestrict=n&Toc=&TocEntry=&QField=&QFieldYear=&QFieldMonth=&QFieldDay=&UseQField=&IntQFieldOp=0&ExtQFieldOp=0&XmlQuery=&File=D%3A%5CZYFILES%5CINDEX%20DATA%5C91THRU94%5CTXT%5C00000002%5C000002IB.txt&User=ANONYMOUS&Password=anonymous&SortMethod=h%7C-&MaximumDocuments=1&FuzzyDegree=0&ImageQuality=r75g8/r75g8/x150y150g16/i425&Display=hpfr&DefSeekPage=x&SearchBack=ZyActionL&Back=ZyActionS&BackDesc=Results%20page&MaximumPages=1&ZyEntry=1

Han, K. T., Ruan, L. W., & Liao, L. S. (2022). Effects of indoor plants on human functions: A systematic review with meta-analyses. *International Journal of Environmental Research and Public Health*, *19*(12), 7454. http://doi.org/10.3390/ijerph19127454

Hoke, J. R., & Ramsey, C. G. (2000). *Architectural graphic standards* (10th ed.). John Wiley & Sons.

Inclusive Design Research Centre (IDRC). (n.d.). *What is inclusive design*? OCAD University. https://legacy.idrc.ocadu.ca/about-the-idrc/49-resources/online-resources/articles-and-papers/443-whatisinclusivedesign

Kim, K. J., Khalekuzzaman, M., Suh, J. N., Kim, H. J., Shagol, C., Kim, H.-H., & Kim, J. K. (2018). Phytoremediation of volatile organic compounds by indoor plants: A review. *Horticulture, Environment, and Biotechnology*, *59*(2), 143–157. http://doi.org/10.1007/s13580-018-0032-0

Klepeis, N. E., Nelson, W. C., Ott, W. R., Robinson, J. P., Tsang, A. M., Switzer, P., Behar, J. V., Hern, S. C., & Engelmann, W. H. (2001, May–June). The national human activity pattern survey (NHAPS): A resource for assessing exposure to environmental pollutants. *Journal of Exposure Analysis and Environmental Epidemiology*, *11*(3), 231–252. http://doi.org/10.1038/sj.jea.7500165

Lumen Now. (n.d.). *4 types of light bulbs: Which is more energy efficient*. https://lumennow.org/choosing-a-bulb/

Parker, K., Horowitz, J. M., & Minkin, R. (2022). Covid-19 pandemic continues to reshape work in America. *Pew Research Center*. www.pewresearch.org/social-trends/2022/02/16/covid-19-pandemic-continues-to-reshape-work-in-america/

Parseh, I., Teiri, H., Hajizadeh, Y., & Ebrahimpour, K. (2018, April). Phytoremediation of benzene vapors from indoor air by *Schefflera arboricola* and *Spathiphyllum wallisii* plants. *Atmospheric Pollution Research*, 0–1. http://doi.org/10.1016/j.apr.2018.04.005

Pegas, P. N., Alves, C. A., Nunes, T., Bate-Epey, E. F., Evtyugina, M., & Pio, C. A. (2012). Could houseplants improve indoor air quality in schools? *Journal of Toxicology and Environmental Health, Part A*, *75*(22–23), 1371–1380. http://doi.org/10.1080/15287394.2012.721169

Schweizer, C., Edwards, R., Bayer-Oglesby, L., Gauderman, W. J., Ilacqua, V., Jantunen, M. J., Lai, H. K., Nieuwenhuijsen, M., & Künzli, N. (2007). Indoor time – microenvironment – activity patterns in seven regions of Europe. *The Journal of Exposure Science and Environmental Epidemiology*, *17*, 170–181. https://doi.org/10.1038/sj.jes.7500490

Soreanu, G., Dixon, M., & Darlington, A. (2013). Botanical biofiltration of indoor gaseous pollutants – a mini-review. *Chemical Engineering Journal*, *229*, 585–594. http://doi.org/10.1016/j.cej.2013.06.074
Steinfeld, E., & Maisel, J. (2012). *What is universal design*? University at Buffalo, School of Architecture and Planning, Center for Inclusive Design and Environmental Access (IDEA). https://idea.ap.buffalo.edu/about/universal-design/
U.S. Department of Energy (DOE). (n.d.). *LED lighting*. www.energy.gov/energysaver/led-lighting
U.S. Environmental Protection Agency (EPA). (n.d.). *Saving water in office buildings*. www.epa.gov/sites/default/files/2017-01/documents/ws-commercial-factsheet-offices.pdf
Wei, Z., Van Le, Q., Peng, W., Yang, Y., Yang, H., Gu, H., Lam, S. S., & Sonne, C. (2021). A review on phytoremediation of contaminants in air, water and soil. *Journal of Hazardous Materials*, *403*(July 2020), 123658. http://doi.org/10.1016/j.jhazmat.2020.123658
Wolverton, B. C. (1989). *Interior landscape plants for indoor air pollution abatement*. NASA. https://ntrs.nasa.gov/api/citations/19930073077/downloads/19930073077.pdf

12

ALTERNATIVE SITES

So what exactly is an alternative site? Alternative to *what*, you might ask? I'll provide a bit of context. I had the privilege of taking Gary Scott Danford's course, *Environmental Behavior* and *Design*, while I was an undergraduate student at the University at Buffalo in the early 1990s. On the first day of class, he walked into the classroom looking quite flustered. He took off his brown leather bomber jacket and put on a white lab coat that he retrieved from his briefcase. He then went on to say,

> Some recent events have drastically altered what we will be working on this semester. This past August, a large meteor struck a sparsely populated area along the former Sino-Soviet border. The impact produced a crater over one quarter mile (0.4 km) in length, 300 feet (91 meters) in width, and 100 feet (30 meters) in depth. The meteor was extremely dense, highly radioactive, and only partially fragmented on impact.
>
> *(Danford, 1994)*

Danford then went on to explain how all teams of scientists sent to investigate the collision were dead within 36 hours of exposure. Over several weeks, all attempts to contain the contamination failed, and based on the spread rates/patterns, epidemiologists were predicting the end of all human life within a 25-year period. Preliminary research also indicated that the contaminant could not penetrate 2 feet (0.6 meters) of solid salt, 10 feet (3 meters) of water, or 500 feet (152 meters) of vacuum. Our project was to design a colony for 12,000 people in one of these three mediums to ensure the survival of humanity with a possible return to the earth if radiation levels decreased to normal levels. (Yes, I realize there is a series called *the 100* about this, but Danford was way ahead of it!) To ensure the security of the project, it was decided that it should be hidden in plain sight by conducting it as an academic, class exercise.

At the end of the class, not a single student left their seat. Granted, this was before the existence of the internet and smartphones, so there was no instant way to verify if this was true or not. One student approached Danford and asked him point-blank, "Is this for real?" Danford had already changed back to his brown leather bomber jacket and replied, "Dr. Saxxon is no longer

DOI: 10.4324/9781032668666-12

here. I'm Dr. Danford. But Saxxon will be back next class. Just look for the white lab coat." The student's jaw dropped in confusion, and he again pleaded with Scott, "Come on, you have to tell us if this is true or not!" Danford simply replied, "It could be!" I remember going home and calling my mom on the public telephone in the hall of my dorm and her telling me she didn't read anything about it or see anything in the news. As the semester went on, I think we all accepted that it could be true, and, even if it wasn't, why not be prepared if something like this actually *does* happen in the near or distant future?

Although Danford assumes the role of the "fictional" character Dr. Scott Saxxon, there were many similarities between the two personas. The "real" Danford received faculty fellowships from NASA and the American Society for Engineering Education where he researched psycho-social and organizational factors and their relationship to human productivity, particularly for the long-term habitation of space and for NASA's then-proposed International Space Station. So maybe it was all true?

Although thinking about cataclysmic threats to our planet and to humanity might seem grim to some, there are other groups – particularly at NASA's Center for Near Earth Objects and organizations like the World Future Society – that have dedicated their careers to preparing for the unfathomable in the near or distant future. Although not all the strategies discussed in this chapter would necessarily sustain life in response to a cataclysmic event, this existential thinking is what spawned the idea of an "alternative site." For example, what happens if the ground becomes contaminated and building on it is not possible? What happens if the earth becomes one giant ball of impervious concrete and there are no sites left to meet the spatial needs of society? What happens if sea levels rise and submerge large portions of our coastlines? What happens if the sun burns out and our solar system plunges to its death into a black hole? What happens if the world destroys itself with war or worse yet, nuclear weapons? And obviously considering the recent COVID-19 pandemic, what happens if an even more virulent strain threatens the health and well-being of society? All these scenarios provoke boundless questions that can spawn serendipitous solutions to make humans and the built environment more resilient to the unknown.

Rising water

As discussed in Chapter 3, climate models indicate that storm surges, flooding, and rising water levels are inevitable, with sea-level rise estimates in the 0.95–6.6 feet (0.3–2 meters) range by the end of this century depending on different greenhouse gas emission scenarios (IPCC, 2018; Lindsey, 2022, April 19). Flooding can be alluvial in the form of flash floods in deserts or mountainous areas, coastal from ocean storms, tides, and surges, riverine from streams and rivers, or shallow flooding that can occur in any location due to a storm event or long period of rain.

In the United States, the Federal Emergency Management Agency (FEMA) manages the National Flood Insurance Program (NFIP). Communities that desire to participate in the program must meet or exceed the NFIP's minimum standards and requirements that use the Floor Insurance Rate Maps and data produced by FEMA to define permit requirements for small and large developments and set standards for development in different flood zones. The maps delineate zones for both 100- and 500-year flood occurrences. The 100-year flood or "base floodplain" is an area that has a 1 percent chance of flooding in any given year, and the 500-year flood has a 0.2 percent chance of flooding in a given year.

NFIP flood zones include Special Flood Hazard Areas (SFHAs) within the 100-year flood, Moderate Flood Hazard Areas that fall between the 100- and 500-year flood, and Minimal Flood Hazard Areas outside of the 500-year flood.

In general, there are three broad design strategies for flooding: relocation/avoidance, resistant design, and resilient design (see Chapter 2 for resistance vs. resilience). The NFIP (FEMA, 2005, 1.31) also established four general floodplain management strategies:

- *Modify human susceptibility to flood damage.* This includes using zoning codes to prohibit development in hazardous or high-risk flooding areas; establishing development policies on the design and location of utilities, public services, and critical facilities; acquiring land in floodplains to preserve open space and relocate buildings; elevating or floodproofing new and existing buildings; more effective forecasting, warning systems, and emergency plans; and restoring and preserving natural resources and functions of floodplains.
- *Modify the impact of flooding.* This includes providing information and education; following flood emergency protocols; providing disaster assistance through flood insurance and tax adjustments; and preparing post-flood recovery plans to help people rebuild.
- *Modify flooding itself.* This includes building dams and reservoirs that store excess water upstream from developed areas; building dikes, levees, and floodwalls; making channels more efficient; diverting high flows around developed areas; utilizing green stormwater infrastructure to increase water infiltration and reduce runoff; on-site stormwater detention; shoreline protection measures that account for natural movement of shoreline features; and controlling runoff from areas outside of the floodplain.
- *Preserve and restore natural resources.* This includes land use regulations to steer development away from floodplains, wetlands, and coastal areas; land acquisition, open space preservation, and floodplain and wetland restoration; beach nourishment and dune building; tax incentives to preserve and restore lands to their natural state; and information and education to increase awareness, knowledge, and protection of natural floodplain resources.

At the site level for coastal communities, flooding design strategies may include relocating existing buildings to have a greater setback from the shorelines/riverbanks and/or avoiding any new development in flood-prone areas. Shoreline stabilization can be implemented by using seawalls, revetments, gabions, ripraps, breakwaters, groins, or jetties. In addition, ecologically restoring shorelines using riparian buffers, tidal wetlands, and salt ponds can naturally absorb and filter water from storm events. At the building level, resistant flood design measures include using wet-floodproofing that allows floodwaters to enter portions of the building and dry-floodproofing to exclude water from entering; elevating critical equipment and protecting utilities; installing backwater valves and sump-pumps; anchoring the structure to the foundation to prevent drift or collapse; and raising the lowest level above the base flood elevation by using fill, piles, posts, piers, columns, walls, or crawlspaces. To prevent damage to walls from hydrostatic/hydrodynamic pressure, walls can be fitted with openings or can be run parallel to the direction of the water flow to allow water to pass through. In addition, the NFIP (FEMA, 2005, 5.43) recommends the following flood-resistant materials:

- Concrete, concrete block, or glazed brick
- Clay, concrete, or ceramic tile

- Galvanized or stainless-steel nails, hurricane clips, and connectors (in areas subject to salt-water flooding)
- Indoor–outdoor carpeting with synthetic backing (do not fasten down)
- Vinyl, terrazzo, rubber, or vinyl floor covering with waterproof adhesives.
- Metal doors and window frames
- Polyester-epoxy paint (do not use mildew-resistant paint indoors, especially on cribs, play-pens, or toys because it contains an ingredient that is toxic)
- Stone, slate, or cast stone (with waterproof mortar)
- Mastic, silicone, or polyurethane formed-in-place flooring
- Styrofoam insulation
- Water-resistant glue
- Pressure-treated (.40 CCA minimum) or naturally decay-resistant lumber, marine-grade plywood

Levees, dikes, and dams: Venice and the Netherlands

Venice, Italy, represents an interesting case study of both resistance and resilience to water. The network of islands and canals that sit within a marshy lagoon along the Adriatic Sea has been subject to flooding throughout history, with the largest flood on record occurring in 1966 when water rose 6.4 feet (~2 meters) above sea level and the second-largest event occurring in 2019 where the tide reached a peak height of 6.1 feet (~1.9 meters) above sea level (Phelan, 2022). According to a report by the European Geosciences Union (Zanchettin et al., 2021), relative sea level rose 0.09 inches (2.5 mm) per year in the past 150 years due to both subsidence and sea-level rise, and sea levels are expected to rise 13 inches (32 cm) by 2100 given global temperatures are kept below a 2 degrees Celsius rise above pre-industrial levels. The original buildings in Venice were built with long, wooden piles of oak and larch that were driven about 13 feet (4 meters) through the soft layers of silt and dirt down to a layer of hard clay. The lack of oxygen in the mud and the petrification from absorbing silt and soil prevented the wood from rotting. Spaces between the piles were filled with rocks and stone. On top of the piles, two layers of wood were added, followed by a limestone base layer and brick construction for the building (Venezia Autentica, n.d.). To protect the city, Project Mose (Modulo Sperimentale Elettromeccanio or Experimental Electromechanical Module) was started in 2003 and consists of 78 mobile hydraulic gates that create a coastal cordon that rises when waters rise to 3.6 feet (1.09 meters). The Mose system has had several challenges, including not providing protection against smaller floods that do not trigger the gates to rise and the reduction of sediment into the Venetian Lagoon (Tognin et al., 2022). The song *When the Levee Breaks* – originally written by Memphis Minnie and Kansas Joe McCoy in response to the Great Mississippi Flood of 1927 and later re-worked by Led Zeppelin – warns us that if it keeps on raining, the levee's going to break. Some experts fear this will be the case with the Mose.

If/when Mose reaches the end of its lifespan, engineers may have to model a new system after the flood control measures used in the Netherlands. After the North Sea Flood in 1953, the Delta Works project in the Netherlands was launched that built 13 dams – including barriers, sluices, locks, dikes, and levees – spanning 18.5 miles (29.8 km) that close off the mouths of the Rhine, Meuse, and Schelde estuaries. Several sections, including the 1.75-mile (2.8 km) Oosterschelde dam, the 5.6-mile (9 km) Eastern Scheldt Storm Surge Barrier, and the Maeslant Storm Surge

Barrier, have gates that are left open during normal weather to preserve the marine tide nature but close to create barriers during storm events.

In response to Hurricane Sandy's devastation of the northeastern United States in 2013, the Hurricane Sandy Rebuilding Task Force of the U.S. Housing and Urban Development launched Rebuild by Design – a multi-stage, regional planning and design competition to foster resilience and innovative ideas for the region. The "Big U" was one of the winning entries designed by the Bjarke Ingels Group for 10 miles (16 km) of coastline in lower Manhattan. The design uses a series of floodwalls, movable flood gates, raised parklands, and terraced landscaping that mitigate flooding and provide social and recreational opportunities within three major compartments, including East River Park, Two Bridges and Chinatown, and Brooklyn Bridge to the Battery. The competition promoted interdisciplinary collaboration between scientists, architects, planners, and engineers who developed "regionally scalable but locally contextual" solutions for the region. The success of Rebuild by Design influenced the creation of the National Disaster Resilience Competition in 2014 and the Global Resilience Design Challenge, launched by the Rockefeller Foundation and the U.S. Agency for International Development to help countries in Africa and Asia improve their resiliency to climate change.

Floating buildings/cities and artificial islands

Given intense storm surges are becoming more commonplace and that rising sea levels are inevitable, building *with* water may be a more viable alternative to making buildings resilient to water. Structures and/or entire communities can be built on the floor of a water body, can cascade up the cliffs of a shoreline, or can even float on water. Floating buildings can be buoyant, sit atop floating pontoons, be constructed on fixed platforms/columns anchored to the seabed, or be fitted with hydraulic/telescoping columns that can raise and lower the building depending on water height. In response to Hurricane Katrina, the Make it Right Foundation was established in 2009 to foster new ideas in the rebuilding of New Orleans. As part of the effort, Morphosis Architects developed the FLOAT House for New Orleans's Lower Ninth Ward. The modular house is designed for efficiency and affordability and is prefabricated as a single unit that is constructed with expanded polystyrene foam coated in glass fiber–reinforced concrete. The house can rise vertically on two interior guideposts up to 12 feet (3.7 meters) as water levels rise. The prototype also contains a butterfly roof to catch water, rooftop solar, and a geothermal heating/cooling system. Although the house did not perform to expectations, the project still served as a precedent for future mass-produced, flood-resilient houses.

The Buoyant Foundation Project also has several amphibious designs that hover, float, and/or rise above the water, including the NRC Amphibious Research Pavilion that was constructed on a stormwater detention pond at the University of Waterloo to test the buoyancy of materials through freeze–thaw cycles and develop flood-resilient structures for vulnerable, First Nation communities. A structural subframe is built with cross beams and buoyant blocks made of expanded polystyrene that resemble a typical, plastic barrel floating dock. The subframe can be affixed to static or telescoping vertical guideposts, the latter allowing the house to rise and fall without floating away from the foundation (BFP, n.d.).

There have also been several conceptual designs for floating cities, including the self-sustainable "Lilypad" by Vincent Callebaut, Wolf Hilbertz's self-forming city via the process of electrochemical collection of sea minerals on a lattice of armatures, naval architect Giancarlo Zema's "Floating House," AT Design's four-square-mile hexagonal floating city, and

FIGURE 12.1 Top: Watervilla in Ijburg, image courtesy of Koen Olthuis/Waterstudio; bottom: the Water Discus underwater hotel by Deep Ocean Technology, image courtesy of Krzysztof Koniuszaniec/Deep Ocean Technology.

several designs and proposals from Koen Olthius's Waterstudio in the Netherlands. Olthius has designed a 36-house floating neighborhood in Ijburg, Amsterdam, and is currently developing a floating city called the "Westland" near The Hague.

Similar to development on land, floating buildings and cities need to take into account their impact on the surrounding environment or, in this case, the marine ecosystem. While these strategies indeed provide "alternative" spaces, their impact on coral reefs, marine life, water currents, and subsequent erosion/sedimentation need to be studied and weighed against their benefits before they are implemented.

Artificial islands

In Jimi Hendrix's song *Voodoo Child,* he talks about chopping down mountains with the edge of his hand, picking up the pieces, and making an island. Although Jimi may not have been thinking about the possibility of all land potentially being submerged under water like in the movie, *Waterworld,* the idea of creating land where there is none may become a viable option not limited to science fiction or Superman villains like Lex Luthor or General Zod.

Little Island, for example, is an artificial island park that opened in 2021 in Hudson River Park in New York City. The park is roughly 2.4 acres (0.97 hectares) and sits above the water on the Hudson with 132, Y-shaped, pre-cast concrete pots (referred to as "tulip pots") that are

FIGURE 12.2 Little Island in New York.

Source: Image courtesy of Hafeez Raji

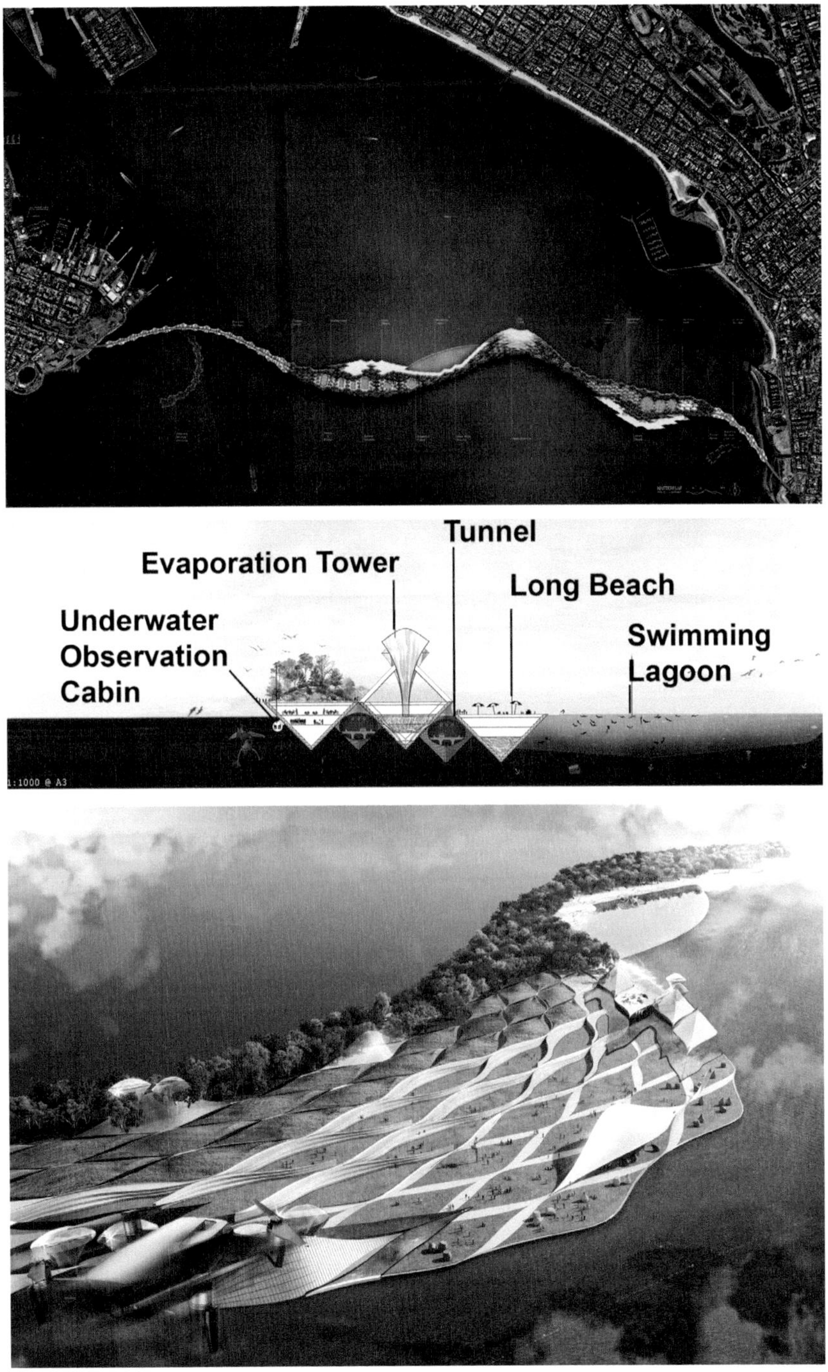

FIGURE 12.3 Sea Line Park. Top: site plan across Hobsons Bay; middle: section of one of the pods; bottom: rendering.

Source: Images courtesy of CX-Landscape

Design Team: George Zhuo, Jing Peng, Simon Zhao, Daniel Dong, Yu Chen supported by 280 concrete piles. The tulip pots vary in height from 15 to 62 feet (4.6–18.9 meters), creating an undulating topography that is designed to protect the park from a 500-year flood where water levels could rise to approximately 11 feet (3.4 meters) (Gannon, 2021; Cubarrubia, 2021). As described by landscape architect Signe Nielsen, the maritime botanical garden consists of 35 species of trees, 65 species of shrubs, and 270 varieties of grasses, perennials, vines, and bulbs (Little Island, n.d.). The park also contains pathways, gathering spaces, activity spaces including dance chimes, spinning discs, vibraphones, and a 687-seat amphitheater.

In Melbourne, Australia, CX-Landscape has also proposed Sea-Line Park as part of a competition for Melbourne Future Park 2050. The park will be 4.3 miles (7 km) long and connect St. Kilda and Williamstown across Hobsons Bay, creating surface-level park space and underwater tunnels. The park will consist of individual, buoyant pods that are produced by 3D printing robots from plastic waste in the ocean. The pods include an underwater observation cabin, a tidal power generation system pod, a seawater purification system pod, an art island pod, and a sports island pod. As described by CX Landscape (2020), "As seawater invades the city, the linear park will be segregated from the land and reform into a self-sustaining island that can cruise away searching a new docking destination or simply float on the sea." The ability of the park to restructure itself in response to unknown water scenarios demonstrates a more "resilient" than "resistant" design strategy.

Underwater colonies

Anyone who has read Jules Verne's classic novel *20,000 Leagues under the Sea* or viewed movies such as *Voyage to the Bottom of the Sea*, *The Abyss*, or *The Spy Who Loved Me* has probably been inspired by the possibility of living underwater. Since the earth is about 70 percent water and humans most likely evolved from the oceans, it would seem only natural for us to seek shelter beneath the surface. Although building underwater provides an alternative to developing on the land, submarine engineering has several challenges, including intense pressure from the water, managing corrosion and erosion from salt water, cold temperatures, using suitable materials, minimal light, and issues relating to the production of food and the disposal of waste.

Bridge construction over water has traditionally used four primary construction methods depending on the depth of the water and characteristics of the waterbed, including battered piles, caissons, coffer dams, and float and lower systems. Battered piles are long, steel tubes that are driven into the bedrock with a pile driver. Once the piles are placed, they are filled with concrete and water is displaced to create a solid pillar. Caissons are watertight, retaining structures that are lowered into the water and anchored to the floor and eventually become the foundation. Box caissons are watertight and open at the top, and pneumatic caissons are sealed at the top but open on the bottom, the latter using compressed air to keep water and mud out of the chamber. In contrast, open caissons have no top and bottom but only vertical walls. They are then lowered into the water, anchored to the floor, pumped dry, and filled with concrete. Coffer dams create a large hole surrounded by water. A geotechnical study is conducted that does a core-penetration test to determine how deep piles must be driven to reach the bedrock. Steel piles are then driven in along with sheet piles that provide an interlocking connection. Two layers of sheet piles filled with sand, gravel, or stone prevent water seepage, and horizontal bracing provides lateral support to resist the inward pressure from the water. A concrete seal course is then attached to the bedrock, and concrete is poured over reinforcing steel to form the foundation. The dam is removed once the concrete settles.

For buildings at extreme depths, the float and lower concept can be utilized, where structures are constructed off-site, lowered to the sea floor, and depressurized. However, working at such depths and traveling to and from the ocean floor can be dangerous due to the hydrostatic pressure of water. In contrast to the decrease in atmospheric pressure as altitude increases, pressure increases at deeper water depths. The air we breathe at sea level comprises 78 percent nitrogen, 21 percent oxygen, 0.94 percent argon, and 0.04 percent carbon dioxide, and the atmospheric pressure is 14.7 psi (1 atmosphere or 1 ATM). The pressure increases about 1 atmosphere for every 33 feet (10 meters) of water depth. So, at 33 feet (10 meters) under the water, a person would experience 29.4 psi of pressure (NOAA, n.d.).

As a person descends deeper underwater, the increased pressure causes more nitrogen and oxygen to be dissolved in the blood. The increased nitrogen pressure causes nitrogen narcosis, which generates symptoms of euphoria, sleepiness, instability, and even unconsciousness. Narcosis usually occurs around 100 feet (30 meters) underwater and requires decompression on the ascent to prevent decompression sickness (DCS), also known as "the bends." Decompression can be done inside a chamber in the case of extremely deep, high-pressure exposure, or in a series of "stops" at preset depths during the ascent to the surface to allow the nitrogen to be released slowly from the body. The severity of DCS and subsequent length of decompression time are directly related to the depth and the amount of time spent underwater. For example, extended underwater habitation experiments, such as the 58-day Tektite I mission in 1969, required approximately 20 hours of decompression therapy before the aquanauts were able to return safely to the surface (Pauli & Cole, 1970).

The first underwater habitat – Conshelf I – was developed in the early 1960s by aquanaut Jacques Cousteau. It was eventually constructed 33 feet (10 meters) below the surface off France's Atlantic coast in 1962. Two aquanauts lived in the habitat for seven days and were among the first people to breathe a mixture of helium and oxygen. Conshelf I was followed by Conshelf II in the Red Sea in 1963 and Conshelf III in the Mediterranean in 1965, in which aquanauts lived at depths of 98–328 (30–100 meters), respectively. Various other underwater living experiments were also conducted in the 1960s and 1970s, including SEALAB off the coast of Bermuda, the Tektite habitat in the U.S. Virgin Islands, Aquarius, Marine Lab, La Chalupa (now Jules' Undersea Lodge), and the Scott Carpenter Space Analog Station. There are also several recreational underwater destinations, including the Hilton Maldives Undersea Restaurant in Singapore, Sea World in Florida, and the Red Sea Star Restaurant in Israel. In addition, several projects are in development, including the Hydropolis Hotel in Dubai, the Poseidon Undersea Resort in Fiji, and aquanaut Dennis Chamberland's Atlantica, which is dubbed the first permanent human underwater colony. Apart from Atlantica, these proposals are at moderate depths of between 16 and 23 feet (5–7 meters) and thus do not require decompression.

Deeper colonies – around the 98-feet (30-meter) mark – are either open to ambient pressure where the interior atmosphere pressure is equal to the surrounding water pressure or closed to the sea with air-locked hatches where interior pressure is less than the surrounding water pressure. Open systems are frequently employed in lab settings where divers are submerged for long periods of time and go between lab and surrounding water frequently. Since the pressure inside the lab is equal to the water pressure, there is no need for decompression each time the diver swims back into the lab.

The Aquarius Undersea Laboratory located off the coast of Key Largo, Florida, is an example of an ambient pressure habitat. Aquarius contains three main components: a life support buoy at the surface that contains generators for power, air compressors, and communication systems; a

120-ton baseplate that is mounted to the sea floor 62 feet (18.9 meters) below the surface; and a habitat cylinder that measures 9 feet in diameter and 43 feet (13 meters) in length. Since the habitat sits on top of the base plate, it is at a depth of roughly 47 feet (14 meters). A wet porch (also known as a moon pool) outside the pressure locks allows aquanauts access to the water. This remains open as the equivalent air pressure inside prevents water from flooding inside. The pressure of the habitat is roughly 2.5 times greater than the atmospheric pressure at sea level. At this depth and pressure, non-saturated visitors have less than 80 minutes to visit and return to the surface before they risk DCS. The mission aquanauts who live in Aquarius can stay there indefinitely, but they must undergo 17 hours of decompression before leaving. At the end of this lengthy decompression session, the aquanauts exit the habitat and swim up to the surface. The decompression chamber is essentially a closed system that maintains sea-level pressure. Since the habitat portion of Aquarius is at only 47 feet, the air used inside is similar to the air we breathe at sea level. However, colonies at greater depths require mixtures of helium and oxygen or mixtures with higher oxygen/lower nitrogen contents (nitrox) to reduce nitrogen uptake in the body's tissues. Helium/oxygen mixtures (also known as heliox environments) were used in the Conshelf missions mentioned earlier.

Aquarius was built in 1986 and is only around 400 square feet (37 square meters), which offers limited space for the aquanauts. Jaques Couteau's grandson, Fabien Cousteau, is currently working with the Proteus Ocean Group on a new project called Proteus that will improve on some of the concepts used on Aquarius. Their goal is to have a network of ambient pressure stations in locations across Europe and the United States. The first unit will be constructed off an island north of Venezuela called Curacao. The station will be about 60 feet deep, 4,000 square feet (372 square meters) to accommodate 12 residents, and will include wet labs, pressurized decompression chambers, a broadcast studio, a hydroponic garden, a sleeping space, a submarine docket hangar, and an umbilical that connect to the surface and to a deeper heliox habitat about 230 feet (70 meters) below the surface (Hu, 2022). Cousteau is also exploring the idea of powering the station by ocean thermal energy conversion.

Closed systems are similar to submarines, where the interior air pressure is maintained around 1 ATM. Such systems require heavy construction to withstand the pressure differential and utilize pressure shafts or submarines to transport people to and from the habitat. Unlike ambient systems, air mixtures in closed systems can be kept closer to those at sea level. The water depth determines the thickness of the walls and openings and the types of connections.

In both types of underwater habitats, oxygen must be replenished, carbon dioxide must be removed, temperature must be maintained, fresh water must be supplied, waste must be recycled, and power must be generated. Oxygen can be supplied from pressurized tanks or through the electrolysis of water, whereby the oxygen molecules are separated from the hydrogen molecules, while carbon dioxide can be removed by scrubbers containing a solution of soda lime. Seawater can be distilled into fresh water via distillation plants that remove the salt. Electrical power can be generated from nuclear reactors, steam turbines, diesel engines, or the sea's currents by using a tidal turbine. Sunlight can be "siphoned" from the surface using solar tubes with advanced optics. Structural consideration must also be implemented to withstand seismic activity on the ocean floor and turbulence from hurricanes.

Brownfield reclamation

The U.S. Environmental Protection Agency defines *brownfields* as "properties where expansion, redevelopment, or reuse may be complicated by the presence or potential presence of a hazardous substance, pollutant, or contaminant" and *land revitalization* "as the process of returning the property to a safe and sustainable use after site assessment and cleanup" (EPA, n.d.a). Brownfields are the result of the by-products and processes used in the manufacturing, distribution, disposal, recycling, or generation of chemical, biological, or physical materials. Contaminants are deposited in the soil and groundwater and can disperse beyond the boundaries of a particular site.

A brownfields cleanup program emerged as a result of the Comprehensive Environmental Response, Compensation and Liability Act (CERCLA) of 1980 – also known as Superfund. CERCLA and its amendments established funds via taxes on chemical raw materials and their associated industries that finance the identification and cleanup of hazardous waste sites that pose a threat to the environment and human health. Hazardous sites are first identified and ranked by severity. The sites with high and immediate threats are placed on the National Priorities List (NPL) for comprehensive cleanup. Over 47,000 Superfund sites have been identified since 1980, and as of 2023, there are 1,336 active and 41 proposed NPL sites (GAO, 2008; EPA, n.d.b).

Prior to CERCLA, other legislation such as the Clean Air Act of 1970, the Clean Water Act of 1972, the Resource Conservation and Recovery Act (RCRA) of 1976, and the Toxic Substances Control Act of 1976 had also imposed limits on the discharge of toxins in the air and water and banned the unsafe and illegal disposal of solid waste on land. The RCRA regulates the construction of landfills and techniques of disposal and requires toxic wastes to be pre-treated via biodegradation or incineration so they will not leach into the soil or groundwater. It also demands stringent cradle-to-grave monitoring of all waste.

In 1995, the U.S. EPA established the Brownfields and Land Revitalization Program. The Brownfields Tax Incentive was also signed into law in 1997 as part of the Taxpayer Relief Act that encouraged the cleanup and reuse of brownfields by making environmental cleanup costs fully deductible in the year they are incurred rather than capitalized and spread over time. In 2002, Congress passed the Small Business Liability Relief and Brownfields Revitalization Act – also known as the Brownfields Law – that provided funds to assess and clean up brownfields, clarified CERCLA liability protections, and provided funds to enhance state and tribal response programs. The 2018 Brownfields Utilization, Investment, and Local Development Act reauthorized the Brownfields Program. Brownfields can also be funded by the 2021 Bipartisan Infrastructure Law and the 2022 Inflation Reduction Act (EPA, n.d.c).

The EPA administers various grants to provide funding for brownfield remediation, including Brownfield Assessment Grants that are used for planning, prioritizing, assessing sites, and developing inventories of brownfields and are funded up to $500,000 over four years; Revolving Loan Fund (RLF) grants, which provide funding for a grant recipient to capitalize a revolving loan fund and make loans and provide sub-grants to carry out cleanup activities at brownfield sites, are funded up to $1 million over five years; Brownfield Cleanup Grants, which are used to fund the technologies required to mitigate the sites, are funded up between $500,000 and $2 million over four years; and Multi-Purpose Grants (MP), Job Training Grants (JT), Technical Assistance, Research, and Training Grants, and State and Tribal Response Program Grants (EPA, n.d.c).

The typical brownfield redevelopment process begins with a Phase 1 site assessment and due diligence that provide an initial screening to determine the extent of the contamination and any possible legal and financial risks. If the site assessment and due diligence reveal no apparent contamination and no significant health or environmental risks, redevelopment activities may begin immediately. If the site seems to contain unacceptably high levels of contamination, a reassessment of the project's vitality may be appropriate. A Phase II site investigation samples the site to provide a comprehensive understanding of the contamination. If this investigation reveals no significant sources of contamination, redevelopment activities may commence. Again, if the sampling reveals unacceptable high levels of contamination, the viability of the project should be reassessed. Should the Phase II site investigation reveal a manageable level of contamination, the next step is to evaluate possible remedial alternatives. If no feasible remedial alternatives are found, the project viability would have to be reassessed. Otherwise, the next step would be to select an appropriate remedy and develop a remedy implementation plan. If additional contamination is discovered after remedy implementation, the entire process is repeated (EPA, 2002).

Converting brownfields into greenspace is often less expensive than similar conversions for residential or other uses because the cleanup requirements are not as stringent (Hellmund & Smith, 2006). Various cleanup technologies can be employed to remediate soil and groundwater. Ex-situ technologies remove contaminants from the site, treat them at a distant location, and return cleaned materials back to the site. In situ technologies treat contamination in place. Choosing the appropriate technology depends on the type, variety, and quantity of contaminants, and the physical conditions of the site. In general, contaminants can be contained, removed, or treated, or some combination of all three. Containment can be accomplished by capping or encasing contaminants via slurry walls, piles, encasements, or grout curtains. Monitoring wells are set up to ensure there is no leakage into adjacent soil and groundwater. Toxins can also be excavated from a site and deposited into a secure landfill at another site, or removed, treated, and deposited back at the original site. Contaminants are often treated via incineration, chemical oxidation/reduction, soil washing, or ultraviolet oxidation. Groundwater can be treated by the vacuum extraction process, where contaminated water is pumped out, treated, and returned to the ground. Other methods used to treat contaminants at the site can be accomplished through bioremediation, phytoremediation, and suction wells. Bioremediation involves the injection of oxygen and microorganisms into the ground to naturally feed on and break down the contaminants. Phytoremediation uses plants and vegetation to extract the contaminants from the soil by direct uptake. The plants also stabilize the soil to prevent further movement of contaminants by erosion. Similar to the vacuum extraction process used for groundwater, water, and detergents can be injected into the ground through a series of peripheral wells and drawn into a central suction well. The soil is cleansed, and the contaminated water is pumped out, treated, and injected back into the site (Wright & Nebel, 2002).

Reinterpreting brownfields

Rather than merely covering up or "hiding the scars" of the past, many former industrial/brownfield sites have simultaneously preserved the history of the site while creating new spaces and functions. As stated by Collins (2000, pp. 461–467):

> Post-industrial space should reveal the legacy of industrialism, not eradicate it or cloak it in nostalgia; create images and stories which reveal both the effect and cause of the legacy, and reveal ecological processes at work in the city, not eradicate them; build infrastructure which embraces ecosystem processes and a philosophy of sustainability.

In 1962, the city of Seattle purchased a 19-acre (7.7 hectare) site once occupied by the Seattle Gas Light Company's coal-gasification plant and repurposed it into a park. The plant closed in

1956 after natural gas became more readily available. Richard Haag was awarded the design and master-planning contract in 1970, and the park officially opened in 1975. Since the site contained the last synthetic gas manufacturing plant in the world, Haag decided to preserve portions of it and incorporate them into the overall design of the park. Soil was remediated through a combination of phytoremediation and bioremediation. The park consists of an earth mound (the Great Earth Mound) constructed from material excavated on-site, a picnic shelter housed in the former boiler house, a children's play barn in the former exhauster – compressor building, and several grassed areas. In consequence, Haag has been hailed as a pioneer in transforming the negative perceptions of a post-industrial site. What was once an abandoned, decrepit, and inaccessible site is now used for concerts, gatherings, recreation, kite-flying, or simply relaxing and taking in the panoramic views of Lake Union.

Such reinterpretation and brownfield adaptive reuse can also be implemented on regional scales. The Ruhr River Valley, for example, is roughly a 200-square-mile (518 square km) site that runs through the state of North Rhine–Westphalia in Germany. In the first half of the twentieth century, iron and steel processing, coal mines, rail lines, and autobahns dominated the region. After World War II, the Ruhr Valley experienced a mass decline and collapse of industry coupled with high rates of unemployment. The local governments found themselves paralyzed by rising social costs, falling tax revenue, and an inability to attract new businesses. In 1989, the Westphalian government commissioned an independent planning corporation, the International Building Exhibition (Internationale Bauausstellung – IBA), to transform the Ruhr's image with innovative architectural and cultural projects. The IBA employed regional planning methods focused on equity, economy, and environment. It established five tenets, including green infrastructure, urban development and renewal, preservation, housing initiatives, and job creation initiatives. Over 120 projects were identified and completed within ten years, ultimately creating a regional network of open space, recreation, and cultural attractions.

After winning the design competition in 1985, Peter Latz became chief designer for the park and had a strong influence on the reinterpretation of the valley. Some of the projects included the reclamation of the Emscher River (which had once carried industrial and human waste), a new stormwater management plan, the construction of pedestrian bridges and interpretive biking routes, displays of experimental forms of energy, new aesthetic lighting, the creation of walkways through the factories, and the 500-acre (202 hectare) Duisburg-Nord Industrial Landscape Park (Landschaftspark Duisburg-Nord). Many of the buildings have also been repurposed for new and interesting functions, including the conversion of an old gasometer into a scuba center, old ore storage bunkers into alpine climbing gardens, a blast furnace into a viewing tower, and a former casting house into a ropes course. At night, multi-colored lighting illuminates the buildings to recreate the life that flourished when the factories were operational.

The concept of "minimal intervention" was employed with the intention of preserving the historical story by leaving certain artifacts of the past in place. Visitors can get a glimpse of the past, the present, and the future as they walk through the park on walkways that follow the routes of old rail lines or water drainages. The success of the Ruhr Valley project lends much hope to decrepit regions throughout the world – and particularly to rust-belt regions in the northeastern United States. The Ruhr can teach such regions how to reinterpret and reclaim space through the fusion of industry and nature, regional planning initiatives, money renegotiation, and the preservation of culture.

Brownfields can also be ideal sites for renewable energy projects. As discussed in Chapter 5, wind turbines and large solar fields are often rejected because of their visual and/or environmental

FIGURE 12.4 Top: Gas Works, Seattle, 1966; bottom: Gas Works Park.

Source: Images courtesy of Seattle Municipal Archives (top) and Seattle Parks and Recreation (bottom)

FIGURE 12.5 Landschaftspark Ubersicht. Top: site overview; bottom: Landschaftspark Lichinstallation (light installation).

Source: Photos by Udo Becker (top) and Horst Zielske (bottom), both courtesy of Claudia Kalinowski/Landschaftspark Duisburg-Nord impacts. Many brownfields are often isolated and/or located in industrialized areas, far away from public use or views. Energy projects can resurrect such forgotten spaces with new life. The old and the new stand side by side, and one can get a glimpse of the past and the future in one vista. Also discussed in Chapter 5 are the concepts of carbon capture and storage (CCS) and carbon capture, utilization, and storage. Often, carbon that is captured is not able to be sequestered on-site and must be transferred to another suitable site. Brownfield sites that contain depleted underground oil and gas fields or inaccessible coal seams can fulfill this need by providing a space for the carbon to be injected.

Utilizing bridges for renewable energy

In addition to the building and rooftop connection strategies discussed in Chapter 9, stand-alone bridges can be used as sites for renewable solar and wind projects that can provide an alternative to placing such systems on undeveloped land. Such a concept was proposed by the design team of Francesco Colarossi, Giovanna Saracino, and Luisa Saracino for the Costaviola Viaduct between Scilla and Bagnara, Italy, in response to a design competition called "Solar Park South." In addition to incorporating wind turbines into the structure of the viaduct, the road surface would be covered in a special type of asphalt produced by the U.S. company Solar Roadways that contains photovoltaic cells. The "Solar Wind" project, as it is called, would generate roughly 40 million kilowatt hours of electricity from the 26 wind turbines and the solar road (Zimmer, 2011). A pedestrian route with an internal carriageway containing solar greenhouses and a farmers' market would run parallel and slightly above the solar road.

Researchers from Spain and the United Kingdom also conducted a study using computational fluid dynamics simulation to test the viability of wind turbines on the Juncal Viaduct in the Canary Islands. The study tested four different configurations, including one turbine 150 feet (46 meters) in diameter, a 150-foot diameter and a 98-foot (30-meter) diameter turbine, two 85-foot diameter (26-meter) discs, and a configuration with 24 small turbines 20 feet (6.1 meters) in diameter. The 46-meter/30-meter combination produced the most power but would

FIGURE 12.6 Conceptual rendering of the Rainbow Bridge between the United States and Canada showing rotational disc turbines in the bridge structure and vertical-axis wind turbines along the bridge road surface.

require more structure to support the higher weights and vibrations; although the 24 smaller turbine design produced about half the gross power than the 46/30-meter combination, the lighter weight would not produce as much vibration or load on the supporting structure (Hernández et al., 2015).

Preserving the ground plane: cliff dwellings and sub-terranean architecture

The Ancestral Pueblo (Anasazi) people who resided in the southwestern United States left behind some of the best examples of cliff dwellings in North America. Many of these examples are visible at Mesa Verda National Park in Colorado and can be dated back to 1,150–1,300 CE. Some of the structures, often referred to as "cliff houses," were built onto the faces of cliffs in levels; other dwellings, known as "cavates," utilized natural alcoves in the cliff face. The dwellings were constructed with hand-hewn stone building blocks and plastered over with an adobe mixture. They offered protection from invaders, while providing access to farming sites on the tops of the mesas.

Cliff sites may provide a viable alternative to consuming "horizontal" land. Architect Laertis Antonios Ando Vassiliou (LAAV Architects) showcased a residential design known as "Casa Brutale" that is carved from the edge of a cliff. The structure consists of three vertical concrete slabs, a glass facade, and a crystalline swimming pool constructed from reinforced glass that is set between the walls along the top of the house. As described by Vassiliou (n.d.), "Casa Brutale is a geometrical translation of the landscape."

Another way to conserve land at the surface is to build under the ground. In 2008, Albert France-Lanord Architects converted a former nuclear shelter into offices and space for computer servers. The project – known as Pionen-White Mountain – is roughly 12,917 square feet (1,200 square meters) and sits almost 100 feet (30 meters) below the granite rocks of the Vita Berg Park in Stockholm, Sweden. The space was inspired by several science fiction films, including *Silent Running*, the *Empire Strikes Back*, and several of Ken Adam's set designs for the James Bond films. Paraphrasing from the architect, one of the design elements was to create contrasting spaces – somewhere the natural rocks dominate and some where the human intervention dominates. To create a pleasant working environment, the center has simulated daylight, plants, waterfalls, and a 2,600-liter saltwater fish tank.

In the Lower East Side of New York City, a concept by James Ramsey of RAAD Studio known as the Lowline has been proposed to repurpose the former Williamsburg Bridge Trolley Terminal into an underground park. To illuminate the space, Ramsey has designed a "remote skylight" where sunlight passes through a glass shield on a parabolic collector dish above-grade and channels the light energy down a "helio tube" via fiber-optic cables to reflective parabolic distributer dishes that are below ground, ultimately allowing vegetation to grow. Although the project is currently paused as of the writing of this book, it showcases a strategy that simultaneously preserves the historic elements of the trolley and creates new, public green space using novel technology (Lowline, n.d.).

Space colonies

In the morbid event that all the earth's open space is consumed, contaminated, or destroyed beyond habitation, the colonization of space might be our last hope. There are three major strategies in colonizing space: orbiting space colonies; the terra-formation of other planets, moons, or asteroids; and interstellar space travel.

FIGURE 12.7 Top images: conceptual renderings of the Lowline, courtesy of James Ramsey, RAAD Studio; bottom: Casa Brutale, Courtesy @laavarchitects/LAAV Architects.

Source: Courtesy @laavarchitects/LAAV Architects

Since the great space race of the mid-twentieth century, there have been significant advances in the exploration of space. In 1926, Robert H. Goddard launched the first liquid-fueled rocket. It reached a height of only 41 feet (12.5 meters) but was soon followed by the first suborbital flight of the German V2 rocket (which ascended to 62 miles/100 kilometers). The Russians put a human in space in 1961, while the United States' Apollo 11 mission was the first to land on the moon in 1969. The 1970s saw probes sent further into space, including the Soviet Union's Venera 7, which landed on Venus in 1970, the United States' Viking 1 and 2, which landed on

Mars in 1976, and the Voyager 1 and 2 missions, which orbited Jupiter in 1979, Saturn in 1980, Uranus in 1986, and Neptune in 1989. The first space shuttle, *Columbia*, was launched in 1981, while the last mission, *Atlantis*, took place in 2011. There were a total of 135 space shuttle flights over those 30 years (Malik, 2011). Other notable achievements include the Near-Earth Asteroid Rendezvous (NEAR), which landed on the Eros asteroid in 2001, the *Spirit* and *Opportunity* rovers, which landed on Mars in 2004, the Cassini probe, which arrived at Saturn in 2004, and the Mars Helicopter in 2021.

In addition to these advances, there has been significant progress in the aerospace engineering of space structures. The Soviet Salyut 1 became the first human-occupied orbital space station in 1971. This was followed by the United States' Skylab in 1973, the Russian Mir station in 1986, and the International Space Station (ISS) in 1998. The ISS is an international joint project between 15 countries and 5 space agencies: the National Aeronautics and Space Administration (NASA), the European Space Agency (ESA), the Russian Federal Space Agency (RKA), the Japanese Aerospace Exploration Agency (JAXA), and the Canadian Space Agency (CSA). The station's total length is 357 feet (109 meters). It is powered by eight solar arrays with a wingspan of 239 feet (72.8 meters) and contains 13,696 cubic feet (388 cubic meters) of habitable volume (NASA, n.d.). Its primary uses are researching into the human exploration of space, monitoring the earth's meteorological and climatic patterns, and studying human and plant biology in a microgravity environment.

NASA's Ames Research Center conducted studies into larger space colonies in the 1970s. The colonies were designed to house 10,000 people and consisted of toroidal, cylindrical, and Bernal sphere designs. All the designs employed air-tight atmospheres rich in oxygen and nitrogen and utilized rotational effects to simulate gravity. The "pseudo gravity" produced by the size and spinning rate of the colony can be designed to simulate Earth-normal gravity of 1g. For example, a colony with a 2,936-feet (895-meter) radius rotating at one rpm (rotations per minute) provides 1 g at the hull (NSS, n.d.). Other space design challenges include power generation, protection from radiation, solar flares, and cosmic rays; methods to recycle resources and waste; and a safe, efficient means of transporting people to and from the colonies. In 2004, the first private spacecraft – SpaceShipOne – was developed by Scaled Composites. Since then, several others have created designs for suborbital travel, including SpaceShipTwo, VMS Eve/VSS Unity by Virgin Galactic, the Lynx from XCOR Aerospace, the Silver Dart from PlanetSpace, Inc., and the Dream Chaser by SpaceDev.

Although space travel has been criticized in recent times as an unnecessary tax and a luxury that will only be afforded by the wealthy, the potential scientific advancements should not be disparaged. Hopefully, by the time the earth is no longer habitable, such technology will be ubiquitous enough that there isn't a class-divide when/if society is reestablished; if not, Hugo Drax's (the villain from the James Bond movie, *Moonraker*) malevolent vision of planned breading to create a society without what he deemed as "imperfection" could become reality.

Terraforming other planets

Colonizing other planets, moons, and asteroids also presents considerable design challenges for architects and engineers. Unlike an orbiting space colony, there really is no way (unless scientists discover gravitons) to simulate the earth's normal gravity of 1 g. For instance:

The lunar surface is at roughly 1/6 g and Mars is at 1/3 g of the earth. Children growing up in such conditions would not be able to develop bones and muscles strong enough to return to the earth, and adults would suffer from a weakening of muscle and bone mass (NASA, 2023).

Other planets/moons would also have to be either terra-formed or self-contained biospheres would have to be constructed to provide an earth-like atmosphere. Schmidt and Zubrin (1996) describe the terra-formation process:

> The essence of the condition is that while Mars's carbon dioxide atmosphere has only about 1 percent the pressure of the earth at sea level, it is believed that there are reserves of carbon dioxide frozen in the south polar cap and adsorbed within the soil sufficient to thicken the atmosphere to the point where its pressure would be about 30 percent that of Earth. To extract this carbon dioxide, the atmosphere would have to be warmed by either pumping artificial greenhouse gases into the atmosphere, angling orbiting mirrors that reflect sunlight at the planet, or by importing ammonia or methane to Mars via asteroids.

Once the planet is warmed, reservoirs of water deep in the crust would have to be released to activate a hydrological cycle, which would in turn release carbon dioxide and water vapor. This would then yield bodies of water, microorganisms, and the subsequent creation of plants, oxygen, and hydrogen. The other possibility is that Mars will eventually warm on its own as

FIGURE 12.8 Space colonies. Top left: toroidal space colony, exterior view; top right: toroidal space colony, interior view; bottom left: cylindrical space colony; bottom right: Bernal sphere space colony.

Source: Images courtesy of Ames Research Center, NASA

the sun continues to expand. In this scenario, it is possible that space colonies orbiting the earth could ultimately migrate to Mars and settle there.

Although some of these "alternative site" designs might seem far-fetched, thinking on this level may be useful if we are truly to engage sustainable design. As previously mentioned, sustainability is defined as meeting our current needs without compromising the needs of future generations. The question remains, however: how far into the future must we plan for? Our universe is almost 14 billion years old, and scientists predict that it will live on for at least another 14 billion years, at which point it will either approach finality in a "Big Crunch" singularity or continue expanding until it becomes complete darkness – the "Big Chill." Our sun is also not infinite: scientists predict it will burn for another five billion years or so. Sustainability, therefore, becomes a relative concept – even before our universe ceases to exist. In addition to natural resource depletion and destruction from extreme storm events that have become more frequent in recent times, our planet will most likely face more extreme, cataclysmic disasters, such as collisions with asteroids, nuclear holocaust, terrorism, pandemics, and biological/genetically engineered pathogens, to name a few. Maybe our fate is inevitable and such events are simply too large for us to comprehend, or are simply "out of sight, out of mind." But maybe thinking on a grander scale will provide solutions to the more imminent dangers our planet faces.

References

Buoyant Foundation Project (BFP). (n.d.). *Waterloo, Ontario: NRC Pavilion*. www.buoyantfoundation.org/waterloo-ontario-nrc-research-pavilion

Collins, T. (2000). Interventions in the rust belt: The art and ecology of post-industrial public space. *British Urban Geography Journal, Ecumene*, *7*(4), 461–467.

Cubarrubia, E. (2021). *Petals, pots, and pilings produce pier-park hybrid Little Island*. ENR New York. www.enr.com/articles/51343-petals-pots-and-pilings-produce-pier-park-hybrid-little-island

CX Landscape. (2020). *Sea line park*. www.cx-landscape.com/sea-line-park

Danford, G. S. (1994). *Spring 1994 syllabus: Environmental behaviour and design: The Nike Jesus Project*. University at Buffalo, State University of New York.

Federal Emergency Management Administration (FEMA). (2005). *National flood insurance program: Floodplain management requirements*. www.fema.gov/sites/default/files/documents/fema-480_floodplain-management-study-guide_local-officials.pdf

Gannon, D. (2021). *NYC's newest park Little Island finally opens on the Hudson River. 6 Square Feet New York City*. www.6sqft.com/nycs-newest-park-little-island-finally-opens-on-the-hudson-river/

Hellmund, P. C., & Smith, D. S. (2006). *Designing greenways: Sustainable landscapes for nature and people* (p. 230). Island Press.

Hernández, O. S., Volkov, K., Mederos, A. C. M., Padrón, J. F., & Lorenzo, A. E. F. (2015). Power output of a wind turbine installed in an already existing viaduct. *Renewable and Sustainable Energy Reviews*, *48*, 287–289. https://doi.org/10.1016/j.rser.2015.03.097

Hu, C. (2022, March 4). Jaques Cousteau's grandson is building a network of ocean floor research stations. *Popular Science*. www.popsci.com/technology/fabien-cousteau-proteus-underwater-international-space-station/

Intergovernmental Panel on Climate Change (IPCC). (2018). *Special report on the ocean and cryosphere in a changing climate*. www.ipcc.ch/site/assets/uploads/sites/3/2022/03/01_SROCC_SPM_FINAL.pdf

Lindsey, R. (2022, April 19). *Climate change: Global sea level*. NOAA. www.climate.gov/news-features/understanding-climate/climate-change-global-sea-level

Little Island. (n.d.). *Design and construction*. https://littleisland.org/design-construction/

Lowline. (n.d.). *About the lowline project*. http://thelowline.org/about/project/

Malik, T. (2011). *NASA's space shuttle by the numbers: 30 years of a spaceflight icon*. Space.com. www.space.com/12376-nasa-space-shuttle-program-facts-statistics.html

National Aeronautics and Space Administration (NASA). (2023). *Counteracting bone and muscle loss in microgravity*. www.nasa.gov/missions/station/iss-research/counteracting-bone-and-muscle-loss-in-microgravity/

National Aeronautics and Space Administration (NASA). (n.d.). *Station facts*. www.nasa.gov/international-space-station/space-station-facts-and-figures/
National Oceanic and Atmospheric Administration (NOAA). (n.d.). *Atmospheric and hydrostatic pressure*. www.omao.noaa.gov/sites/default/files/documents/Atmospheric%20and%20Hydrostatic%20Pressure_slides121316.pdf
National Space Society (NSS). (n.d.). *Space settlement basics*. https://nss.org/settlement/nasa/Basics/wwwwh.html
Pauli, D. C., & Cole, H. A. (Eds.). (1970). *Summary report on Project Tektite I*. Office of Naval Research. https://ntrs.nasa.gov/api/citations/19710016053/downloads/19710016053.pdf
Phelan, J. (2022). *Italy's plan to save Venice from sinking*. BBC. https://www.bbc.com/future/article/20220927-italys-plan-to-save-venice-from-sinking
Schmidt, S., & Zubrin, R. (Eds.). (1996). *Islands in the sky: Bold new ideas for colonizing space* (p. 126). John Wiley & Sons.
Tognin, D., Finotello, A., D'Alpaos, A., Viero, D. P., Pivato, M., Mel, R. A., Defina, A., Bertuzzo, E., Marani, M., & Carniello, L. (2022). Loss of geomorphic diversity in shallow tidal embayments promoted by storm-surge barriers. *Science Advances*, *8*(13), 8446. http://doi.org/10.1126/sciadv.abm8446
U.S. Environmental Protection Agency (EPA). (2002). *Technical approaches to characterizing and cleaning up brownfield sites: Municipal landfills and illegal dumps*. https://nepis.epa.gov/Exe/ZyPDF.cgi/30004IE6.PDF?Dockey=30004IE6.PDF
U.S. Environmental Protection Agency (EPA). (n.d.a). *Brownfields*. www.epa.gov/brownfields/about
U.S. Environmental Protection Agency (EPA). (n.d.b). *Superfund: National priorities list (NPL)*. www.epa.gov/superfund/superfund-national-priorities-list-npl
U.S. Environmental Protection Agency (EPA). (n.d.c). *Overview of EPA's brownfields program*. www.epa.gov/brownfields/overview-epas-brownfields-program
U.S. Government Accountability Office (GAO). (2008). *Superfund: Funding and reported costs of enforcement and administration activities*. GAO-08-841R. www.gao.gov/products/gao-08-841r
Vassiliou, L. A. A. (n.d.). *Casa Brutale*. LAAV Architects. www.laav.nl/#/casa-brutale-lyrical-brutalism/
Venezia Autentica. (n.d.). *How Venice was built: An engineering marvel*. https://veneziaautentica.com/how-was-venice-built/#:~:text=The%20buildings%20in%20Venice%20were,hold%20up%20the%20buildings%20above
Wright, R. T., & Nebel, B. J. (2002). *Environmental science: Toward a sustainable future* (8th ed., p. 498). Pearson Education.
Zanchettin, D., Bruni, S., Raicich, F., Lionello, P., Adloff, F., Androsov, A., Antonioli, F., Artale, V., Carminati, E., Ferrarin, C., Fofonova, V., Nicholls, R. J., Rubinetti, S., Rubino, A., Sannino, G., Spada, G., Thiéblemont, R., Tsimplis, M., Umgiesser, G., . . . & Zerbini, S. (2021). Sea-level rise in Venice: Historic and future trends (review article). *Natural Hazards and Earth Systems Science*, *21*, 2643–2678. https://doi.org/10.5194/nhess-21-2643-2021
Zimmer, L. (2011). *Wind turbine bridge transforms Italian viaduct into public space*. Inhabitat. https://inhabitat.com/solar-wind-turbine-bridge-repurposes-viaduct-for-public-space/

13

IMPLEMENTING REGREENING STRATEGIES

The current approach to the built environment is fragmented due to the compartmentalization of the design disciplines – particularly planning, urban design, engineering, architecture, and landscape architecture. In any given project, civil engineers design the roads, parking, site grading, utilities, and stormwater systems; architects focus on the design of buildings, often with little or no site analysis; mechanical and electrical engineers design heating, ventilation, air conditioning, electrical, and plumbing systems; landscape architects are often left to design any spaces that remain after the roads, parking, utilities, and buildings are complete. Planning and zoning regulations are often viewed as obstacles by all the designers, causing animosity and forcing compromises to the design, for better or worse. Engineers feel architects are not technical enough; architects view engineers as number crunchers with no design or aesthetic sensibility; developers, project managers, and property owners are often most concerned with meeting a budget and/or maximizing profit, which means sacrifices are often made to all the systems and designs – even if they are good for the environment. Analysis of building or site performance after a project is completed is rare, and even if it is undertaken, it is usually conducted by academics whose ideas are often viewed as idealistic, over-conceptual, and impractical by those in practice. With this amount of disjunction, it is easy to see why changing the culture of design and the makeup of the built environment is so difficult. It is also why the built environment is often out of sync and in conflict with the natural environment.

Such siloed practices are placing the design professions in jeopardy of becoming extinct in higher education and displaced in professional practice by more specialized fields in the sciences and humanities that are better equipped to design solutions to complex problems that our planet faces. Engineers, biologists, and ecologists are starting to assume more design roles, dubbing themselves "eco-designers" or "bioengineers," and consuming projects and conducting research that should be done by architects, landscape architects, and urban designers. Given the specialized technical aptitude of these other disciplines, one might question the need for the design generalist at all – especially given the complexity of some of the infringing global environmental problems. According to Manley and Claydon (2000),

> The belief in the powers of the generalist is not well founded given the complexity of today's development industry and the diverse environment in which development takes place . . .

DOI: 10.4324/9781032668666-13

> there is ample evidence of other specialists taking over roles to the extent that the architect is no longer considered a leader in the development process.

Steiner and Thompson (1997) also expressed concern about the relevancy of landscape architecture:

> If the discipline embraces ecological design and planning, then it has a leadership role in contemporary society throughout the world. If landscape architecture, however, turns inward and ignores its larger responsibility to the public good, then it will become marginalized and less relevant.

New academic pedagogies

In the academic setting, architecture/landscape architecture faculty holding professional degrees are often displaced as the result of proprietary and confusing degree structures. In the United States, the National Architectural Accrediting Board (NAAB) accredits three different degrees for architecture (B.Arch., M.Arch., D.Arch.) and five different tracts leading up to these degrees. The degrees are not earned in sequence like the typical bachelors-masters-doctorate path ubiquitous in other disciplines, and although they are designed for professional practice, they still are not considered adequate to enable graduates to sit for the licensing exam directly out of school. In essence, the degrees are over-conceptual for practice but not scholarly enough for academia. Disciplines outside of architecture also do not understand how these professional degrees fit into their research and teaching mission. The result has been the inability of an architecture faculty to teach and/or conduct research across disciplines or to secure adequate research funding. This is detrimental to both universities and architecture faculty since there has been a recent surge in design-based programs and research that are offered under schools outside of architecture.

One possible strategy to sustain the design programs is to merge architecture, landscape architecture, and urban design into one, comprehensive discipline. Landscape architecture contains courses in ecological design, land analysis, site design, and environmental science that architecture lacks; architecture contains courses in building technology, energy systems, and structural analysis that landscape architecture lacks; and urban design contains courses in urbanism, housing, transportation, sociology, equity, planning, and zoning that both architecture and landscape architecture lack. The fusion of the three fields could produce a new generation of designers with a more comprehensive set of social and environmental knowledge that can provide less insular and more holistic solutions to problems.

Such a merger would require a coordinated effort between the regulating bodies that govern the professions and organizational changes at the schools offering the programs. Given the amount of bureaucracy, such collaboration and/or the creation of a new, single accreditation agency, professional body, and registration exam that encompasses all three fields would be daunting, to say the least, not to mention the fact that the increased content would most likely extend the duration of design programs. Despite this complexity, some schools have shown that such a task can be accomplished. The University of the West of England (UWE), for example, had success in combining architecture and planning in the early 2000s. The BA (Hons) Architecture and Planning is currently accredited by the Royal Institute of British Architects (RIBA), the Royal Town Planning Institute (RTPI), and the Architects Registration Board (ARB). The degree promotes ideas of understanding the physical, social, and economic context of development before engaging in design and established a new way of educating people to work as architect-planners (Manley & Claydon, 2000).

Many schools only offer one of the disciplines, and the schools that do offer all three programs often offer them in separate colleges. For example, landscape architecture is often part of an agricultural or horticultural college, architecture is often its own school or part of the humanities or fine arts, and urban design can sometimes be included in geography or sociology programs. Some programs do offer some combination of the degrees under the same academic college, which has made cross-disciplinary collaboration somewhat smoother, and in some cases, has fostered the creation of dual degree programs, such as the Master of Architecture/Master of Landscape Architecture (M.Arch./M.L.A.), the Master of Urban Planning/Master of Landscape Architecture (M.U.P./M.L.A.), or the Master of Urban Planning/Master of Landscape Architecture/Master of Architecture (M.U.P./M.L.A./M.Arch.). While such programs are indeed a step in the right direction and provide precedents for a new merged degree, they are still optional and relatively rare, which means students are still graduating from single-style programs with too narrow of a focus to compete in the future marketplace. Both architecture and landscape architecture also have "non-pre-professional plus" tracts where a student completes an undergraduate degree in a non-architectural/landscape architecture discipline, then completes the professional degree at the graduate level. Such programs can produce architects with diverse backgrounds since the completed undergraduate degree can be in any discipline. However, depending on the undergraduate degree earned, this still does not ensure adequate training in environmental, ecological, or social issues. In addition, such "non pre-professional plus professional" programs are outnumbered by five-year or four plus two "pre-professional plus" programs where exposure to other academic topics is not as prevalent.

In addition to offering a variety of degree options, both research and professional programs should be offered at the doctorate level, such as the D.Arch./PhD or D.L.A./PhD, rather than the ambiguous "terminal" master's degrees (M.Arch., M.L.A., M.U.P., MFA) that are often misunderstood by other fields in the sciences and humanities and by society as a whole. The professional Doctor of Architecture was first accredited by the NAAB in 2004. Although only one school has adopted it (University of Hawaii, Manoa), the new degree has the potential to rebrand the discipline by adding curricular enhancements, shortening the intern/licensing process, and transforming the building/form-centric perception of architecture to one that is more about healing ecological and social fragmentation.

Outside of new degree programs, schools can also foster a more immediate impact by revamping the culture of the design studio and integrating more research into their programs. As previously mentioned, the current structure of the design studio tends to overemphasize signature design, form-making, and the culmination of star architects (starchitects) rather than engaging in scientific research methods or quantitative/qualitative analysis. Ostentatious displays are often encouraged while concepts grounded in science, ecology, geospatial analysis, or environmental impact are often deprecated and/or accused of lacking creativity or originality. As Van der Ryn and Cowan (2007, p. 28) point out:

> Architecture is sometimes taught and envisioned as though sites were interchangeable background slides projected behind the human (*sic*) object . . . and unfortunately, at most universities, it is still entirely possible to earn a master's degree in architecture without knowing how the sun moves through the sky, without being aware of energy or resource use in a building, and without taking a course in environmental science.

Given the fact that the design studio makes up about one-third of the credits, more of the time could be spent on scholarly research into the problem rather than on the dogmatic arts and crafts form-making exercises that have dominated architecture programs since they were first formalized in seventeenth-century France with the establishment of the Academie Royal d'Architecture and the Ecoles des Beaux-Arts.

Van der Ryn and Cowan (2007, pp. 171–172) go on to say:

> Studios as research/design clinics could begin to build a cumulative knowledge base that is lacking in design fields such as architecture. The present work of design studios is wasted on one-time exercises that are rarely implemented outside of the classroom. Studios based on analyzing particular issues – megacities, urban wildlife corridors, the future of suburbs, constructed wetlands – would provide a continuous cycle of information.

Although many studios have specific themes, academic transcripts do not reflect the details, and the breadth of what was examined in a specific studio may only be evident if a student/graduate presents a portfolio of their work. Certainly, this is something that could be fixed clerically with the registrar's office by adding the title of the studio, but the fact remains that what architecture students do in the studio remains a mystery to people on the "outside."

Some schools have been successful in making the studio more research- and less arts-and-crafts-based. The Design Research Lab (DRL) at the Architectural Association in London offers a 16-month program that leads to a Master of Architecture (M.Arch.) degree. Rather than engaging in several design projects and studios customary in most architecture curricula, students at the DRL partake in one comprehensive design project that incorporates research, knowledge, and skills acquired throughout the course of study. A design agenda is established each year and used to inform the studio, workshop, and seminar courses. Such coordination ensures students apply what they learn in the technical and theoretical courses to their design work. In what the DRL calls the Open-Source Studio, self-organized teams of students, faculty, and consultants work collaboratively on projects and share information across networks. Work and research results are carried from one year to the next, further promoting a non-proprietary, collaborative approach to design research.

The integration of environmental research centers and labs within schools of architecture/landscape architecture is also a way to transform pedagogies by allowing more experimentation and testing of ideas and technologies. Carnegie Mellon University (CMU) has adopted a position of environmental accountability in architectural design that led to the creation of the Center for Building Performance and Diagnostics and a 7,000-square-foot workspace that is used to test building subsystems, including materials, electrical/mechanical components, energy, air quality, acoustics, lighting, and structure. Like CMU, the University of Minnesota established the Center for Sustainable Building Research, which examines sustainable design, energy efficiency, innovative building components, and post-occupancy evaluation of human – space interaction. Texas A&M University College of Architecture has five research centers – the Center for Health Systems and Design, the Center for Heritage Conservation, the Center for Housing and Urban Development, the Center for Leadership and Management in the Design and Construction Industry, and the Hazard Reduction and Recovery Center – investigate topics around various themes, including community management of environmental hazards, sustainability of human and natural ecosystems at various spatial scales, the social and economic viability of neighborhoods and communities, healthy neighborhoods and cities, transportation mobility, safety, access, and physical forms, building information modeling, architectural preservation and conservation, healthcare facilities design, and energy-efficient buildings (Texas A&M, n.d.).

K–12 education

In addition to college-level design programs, ecological design, sustainability, and environmental science should be integrated into K–12 curricula. Many states have already acted on this matter. In the state of Washington, for instance, environmental education was made mandatory in all grade levels in 2009. Instruction about conservation, natural resources, and the environment is taught in an interdisciplinary manner throughout the sciences, social studies, and humanities. The program includes a full set of learning standards, a literacy plan to ensure students have ample opportunities and funding to engage in project-based learning and a state-wide, interdisciplinary sustainable design project that brings industry, business, higher education, and community members together with students to design solutions to real-world problems. The Colorado Alliance for Environmental Education has also developed academic standards that integrate environmental literacy in the curriculum by teaching students to connect content across disciplines related to natural and human systems; understand relationships between people, places, and the environment; demonstrate processes and skills with critical thinking and research; and participate in proactive decision-making in the community and the school. Similar to Washington, such standards can be integrated with existing courses or an entirely new course on a topic can be created. In 2004, the Green Schools Initiative was founded with the mission of making schools more environmentally sensitive and teaching environmental literacy and stewardship to students. It adopted the "7 Steps to a Green School" program from Eco-Schools International, which includes establishing a green team or eco-committee; adopting an environmental vision statement; conducting a school environmental/energy audit; creating a green school action plan; monitoring and evaluating progress; integrating greening into the curriculum; and communicating achievements. These steps provide an effective means to change the culture from the inside out by "transforming schools into models of sustainability for communities and engaging youth in hands-on, inquiry-based learning" (GSI, n.d.).

International and national policies

At the global level, governments have collaborated and pledged their commitments to several international accords. The Montreal Protocol of 1987 regulated the production and consumption of nearly 100 ozone-depleting substances (ODS), including the complete elimination of hydrochlorofluorocarbons in developed countries in 2020 and in developing countries by 2030. In 2016, the Kigali Amendment to the Montreal Protocol added hydrofluorocarbons to the list and approved a timeline for their gradual reduction through the mid-twenty-first century. The protocol has phased out 98 percent of ODS globally compared to 1990 levels, and between 1990 and 2010, it is estimated the treaty reduced greenhouse gas emissions by the equivalent of 135 gigatons of carbon dioxide (UNEP, n.d.a).

With the focus on HCFCs and HCFs in the Montreal Protocol, the Kyoto Protocol was adopted in 1997 to limit and reduce greenhouse gas (GHG) emissions. After a complex ratification process, it entered into force in 2005 and adopted the Doha Amendment in 2012. As of the writing of this book, there are 192 parties that have signed the agreement (UNEP, n.d.b). The Paris Agreement is an international treaty on climate change that was developed at the 2015 United

Nations Climate Change Conference. The agreement sets long-term goals that guide nations to substantially reduce global greenhouse gases to hold global temperature increase to below 2 degrees Celsius above pre-industrial levels and pursue efforts to limit it to 1.5 degrees Celsius above pre-industrial levels; periodically assess their progress toward achieving the goals; and provide financing to developing countries to mitigate climate change, strengthen resilience, and enhance abilities to adapt to climate impacts (UN, n.d.). It also aims to reach global net-zero emissions in the second half of the century. To date, 195 parties have joined the agreement.

A technical report from the first global stock take was released in September 2023 to assess the progress of the agreement. Seventeen key findings were noted, and although progress was evident, an urgent warning was issued that more ambitious targets and mitigation measures are needed to reduce global GHG emissions below target levels (UN, 2023). Developing countries have also been critical of the agreement because developed countries were able to grow their economies before emission targets were put into place. This was addressed at the COP27 conference with the establishment of the Loss and Damage Fund that provides financial assistance to developing countries that often have the lowest emissions but are most vulnerable to climate disasters.

According to Climate Action Tracker, there is a 50 percent chance that warming associated with the current pledges and target scenarios will exceed 2.1 degrees Celsius in 2100, with estimates in the range of 2.5 degrees Celsius to 2.9 degrees Celsius (Climate Action Tracker, n.d.). Since participation in the Paris Agreement is voluntary and can change with leadership (i.e., the United States withdrawing in 2020), a more effective strategy might be carbon pricing that places a fee on emissions. This can come in the form of an emissions trading system (or cap and trade) where emissions are capped, and entities need to hold one emission credit for each ton of GHG emitted but can buy and sell units. This strategy is specified in Article 6.2 of the Paris Agreement by establishing the potential of trading emission credits across borders. Other forms of carbon pricing include emission reduction funds where a government can buy credits for emission reduction projects and/or a carbon tax that directly taxes fossil fuels burned by specific entities, including industries, transportation, and residential heating/cooling (UN, n.d.c). British Columbia passed a carbon tax in 2008 that started at Canadian $10 per ton of carbon dioxide emitted and has grown to $50 per ton in 2022 (World Bank, 2023).

In addition to Climate Action Tracker, other agencies have recognized that the Paris Agreement may fall short of what is required to reduce GHG emissions. The International Energy Agency published Net Zero by 2050 that calls for ramping up renewables, improving energy efficiency, cutting methane emissions, and increasing electrification. The World Green Building Council's Advancing Net Zero program also focuses on specific strategies that can be done at the building level to achieve net-zero carbon buildings. Many countries have also adopted national policies around climate change to achieve or surpass their commitment to the Paris Agreement. In 2015, France passed a national law that requires all new commercial buildings to be partially covered by either green roofs or solar panels. In the United States, the creation of a National Climate Change Task Force is working on goals to reduce greenhouse gas emissions 50–52 percent below 2005 levels in 2030, reach 100 percent carbon-pollution-free electricity by 2035, achieve net-zero emissions by 2050, and deliver 40 percent of the benefits from federal investments in climate and clean energy to disadvantaged communities (White House, n.d.). In addition, the passage of the bipartisan Inflation Reduction Act of 2022 issued tax provisions that will accelerate the deployment of clean energy with the goal of creating a clean energy economy that lowers energy costs, creates jobs in renewable energy and advances environmental justice.

Federal tax credits can also be earned for certain practices. The Residential Renewable Energy Tax Credit allows a taxpayer to claim a credit of 30 percent of qualified expenditure for a renewable energy system that serves a residence (IRS, n.d.). Tax credits can also be earned for Energy Star appliances. In 2019, the United Kingdom passed a law for domestic residences to achieve net-zero greenhouse gas emissions by 2050. Likewise, Sweden's Climate Act of 2017 sets out policies for the country to have zero net emissions of greenhouse gases by 2045.

The challenge with international and national policies is an effective means of enforcement. In addition, changes in leadership may cause some countries to renege on their pledge and/or drop out of the treaty altogether. As far as politics is concerned, global climate change should not be a partisan issue, nor a divisive cultural or socio-economic issue. It should not matter if you are liberal or conservative, rich or poor, educated or uneducated, urban or rural, a hermit or an international traveler – as stated throughout this book, this is something we all need to accept and work together to mitigate.

State, municipal, and local policies

Since international cooperation and enforcement is a complex task and national ratification is often stuck in stalemate because of partisan politics, establishing policies at the state or local level might be *somewhat* easier to achieve. As discussed in Chapter 9, although some Green Building Rating Systems (GBRSs) programs such as Japan's CASBEE are mandated, most GBRSs are still voluntary, which begs the question if the standards would have more impact if they were integrated with building codes or zoning regulations. The International Code Council's International Green Construction Code (IgCC), for example, is designed to be incorporated into a jurisdiction's codes and ordinances and function as an overlay to other International Code Council codes adopted by the jurisdiction. Likewise, many jurisdictions have also adopted the American Society of Heating, Refrigeration, and Air-Conditioning Engineers' ANSI/ASHRAE/USGBC/IES Standard 189.1 standard, the Americans with Disabilities Act (ADA), or universal/inclusive design standards into their building codes. Jurisdictions that do adopt standards also have the option of making the code more stringent.

As discussed in Chapter 4, Vermont's Act 250 is an example of a state regulation that is designed to safeguard the environment and the character of communities throughout the state. The state of California approved a measure banning the sale of new gas-powered cars, SUVs, and pick-up trucks by 2025. City and local municipalities have also had success in adopting green infrastructure and land conservation into their zoning ordinances by regulating lot coverage limits, subdivisions, setback distances, parking standards, building heights/sizes/scale, open-space conservation, tree removal/replacement, erosion control, and stormwater management, to name a few. As part of their Growing Greener codes, the London Grove Township in southern Chester County, Pennsylvania, requires developers to designate half the acreage in all new subdivisions as permanent conservation lands (NLT, 2009). Similar regulations were implemented in the city of Topeka, Kansas, where ordinances for stream buffers, landscape, and open space were enacted that require minimum buffer widths around streams, best management practices utilizing green stormwater infrastructure, and require developers to set aside five acres of open space per 1,000 residents, respectively (Topeka, 2002). Stormwater fees can also be used to fund and/or incentivize green infrastructure. Since more runoff would equate to higher fees, many developers may seek to decrease impervious surfaces and/or capture the stormwater runoff on-site through the use of green infrastructure and detention ponds.

In 2017, the city of San Francisco required all new commercial and residential buildings up to ten stories high to have rooftop solar systems for heat and/or electricity. This legislation, which is part of San Francisco's goal of transitioning to 100 percent renewable energy, is an extension of an earlier statute that requires all buildings of ten floors or less to utilize solar systems on at least 15 percent of the rooftops. Santa Monica, California has a similar but slightly more stringent mandate that requires new single-family homes to install a minimum of 1.5 watts of solar per square foot of the structure. In 2009, the city of Toronto, Canada, adopted a by-law that requires green roofs on any new commercial, institutional, and residential development with a minimum gross floor area of 2,000 square meters (~21,528 square feet); the percentage of coverage is then weighted on the basis of the size of the building. Expediting the permit process for green infrastructure can also incentivize property owners to incorporate sustainable, low-impact designs into their projects. Chicago's Green Permit Program, for example, expedites the permit process and provides permit-fee discounts to qualifying projects that incorporate green infrastructure. Similar to federal tax credits, states can also issue tax credits for renewable energy projects, energy-efficient appliances, and best management practices.

Marketing green infrastructure

Just saying something is good for the environment is not necessarily sufficient to change the culture and influence policymakers and the professions concerned with the built environment. The consequences of not implementing green infrastructure and the benefits of implementing it need to be quantifiable and communicated effectively to the public. For example, the pair of aerial images that appear at the beginning of Chapter 7 show development from the past to the present; a third image showing the future could be generated using annual average land consumption and development rates. Three-dimensional modeling software and animation can also be used to provide effective visual demonstrations of the impact of development. Several traffic-simulation software packages can generate animations of traffic-flow patterns using data projections for a proposed transit system. Such visuals and animations make it easy for the layperson to see what traffic will be like under a variety of future scenarios.

Simply seeing the loss of open space may be enough to convince environmentally conscious citizens but might be insufficient to convince capitalistic land developers and companies. Since money is often the only thing that garners certain people's attention, economic forecasting is also needed. For example, cost of community service (CCS) studies can be conducted to estimate the fiscal impact of different land uses within a municipality. Expenditures and revenues from a municipality's budget are examined across residential, commercial/industrial, and agricultural/open-space land uses to yield an expenditure-to-revenue ratio. Ratios greater than one indicate that the land use costs more in municipal expenses than it collects in revenue; a ratio of less than one means the land use generates more revenue than it incurs in expenses (i.e., a "net profit"). Although many variables can affect the ratio – including geography, school budgets, and population density – most of the studies indicate that residential developments contribute less in revenue than they require in government expenditure, while commercial/industrial and agricultural/open-space lands contribute more in revenue than they require in expenditure, with agricultural/open space generating the lowest ratios and highest revenues (Harrison & French, n.d.). Although CCS studies are only used for current land use patterns, they can certainly inform future decisions. They can also highlight the notion of natural capital, which is often, for want of a better expression, often "lost in the woods" in economic analysis.

The notion of conducting energy audits in buildings to show the owner the potential economic savings if certain measures are implemented has been common practice in the energy industry. In 2015, Google created Project Sunroof, which utilizes the company's database of aerial imagery and maps to determine if a rooftop is suitable for solar panels. The app factors in geography, local climatic data, roof size/angle, and shadowing from adjacent structures/trees along with federal and state tax credits, utility rebates, and renewable energy credits to determine solar energy production, investment costs, and payback. A similar audit or app can also be applied to green infrastructure. A study by the Natural Resource Defense Council used this approach when developing an algorithm by compiling data from existing projects to show the potential economic benefits of hypothetical green infrastructure retrofits to owners of office buildings, apartment complexes, and retail centers in the Philadelphia area. Using a set of pre-building conditions and specific sets of green infrastructure improvements, the projections showed the owners line by line how much money could be saved by reducing energy usage, reducing maintenance needs, lowering stormwater fees, and increasing property values. For example, in a typical four-story, 32-unit/33,740-square-foot apartment building, the installation of a green roof, 12 strategically planted trees, and a series of bioswales and rain gardens were projected to save $1,780 in annual energy costs, $128,160 from reduced roof maintenance, and $1,230 in annual stormwater fees. The retrofitted building could also receive $52,720 in tax credits, a $77,720 increase in rental income, and a $37,500 increase in property value (Clements & Juliana, 2013).

Quantifying the environmental benefits of natural capital is also becoming more efficient due to advances in computer software and databases. For example, in 2010, the Central Vermont Regional Planning Commission (CVRPC, 2010) quantified the benefits of low-impact development using GIS (geographic information systems) software. Impermeable surfaces, average rainfall data, soil types, land cover, and build-out scenarios were modeled in the software and analyzed using the Natural Resources Conservation Service Runoff Equation to determine the volume of water runoff from impermeable and permeable surfaces, and the LID Technique Equation to determine the volume of water runoff when a LID technique was utilized. Their results showed that with a projected 2.5 percent increase of impermeable surface area, there would be an 8 percent increase in runoff during a typical, 2.2-inch storm. These models were compared to ones using LID techniques, whereby depending on the size, type, and quantity of techniques, showed proportionate decreases in stormwater runoff. Building information modeling software can also be used to quantify and project energy usage and costs of different building systems.

Although data and numbers motivate some people, others are more influenced by personal stories. To satisfy this need, the Energy Action Network (n.d.) in Vermont created a Community Energy Dashboard that helps communities understand energy at the local level. The dashboard provides interactive tools, such as a timeline to track a community's progress toward meeting 90 percent of local energy needs through efficiency and renewable energy sources by 2050, a place to add actions centered on efficiency, an interactive map that identifies locations of renewable energy projects throughout the state, statistics that track renewable energy production data so communities can see how they rank against other towns, and links to personal stories, analysis, and resources that provide a level of transparency about technology cost and performance. Real people – not companies trying to sell you services or products – can relay what worked best for them in terms of planning, financing, and installation. Meanwhile, Efficiency Vermont is a non-profit organization that provides education, services, and incentives for energy efficiency for homeowners and businesses. The organization conducts energy assessments and provides financial assistance, advice, and planning for renovation and new construction, project support, and education and awareness (Efficiency Vermont, n.d.). Similar to the Vermont Energy Dashboard, one service it provides is the "neighbor comparison," which shows homeowners how much energy they use compared to nearby homes. This is often all that is needed to motivate people to reduce their energy consumption.

Changing the culture

Although new academic pedagogies and ratifying public policies can be contentious and complex, to say the least, changing culture, behavior, and rituals may prove to be equally difficult. In an interview for a course in civics, one student spoke with Liz Gamache (the Mayor of St. Albans, Vermont, and director of Efficiency Vermont), who identified three reasons why people resist change: they don't get it; they don't like it; or they don't like you (Gamache, 2017). Educating people (as discussed earlier) and persuading them to participate in the change (discussed in the following text) can address the first two issues; as far as getting people to like you, the key is to become a better listener, be more open-minded, and respect different ideas and opinions, no matter how far out you think they are.

Transparency, engaging community input, and ensuring everyone in a community has equal access to the decision-making process are critical to the success of any project. Cross-sectoral collaboration through public–private–non-profit partnerships is a way to unify stakeholders and centralize resources to achieve mutually beneficial solutions. Kania and Kramer (2011) indicate that to have success in cross-sectoral collaboration, participants must have a common agenda with a clear understanding of the problem and an approved set of solutions; a shared measurement system to report on the program's success or failure; clearly defined activities that are mutually enforcing; continuous communication; and a separate organization to serve as the backbone of the entire initiative. The authors cite the example of the Elizabeth River Project in Virginia in 1993, when 100 stakeholders – including the city governments of Chesapeake, Norfolk, Portsmouth, and Virginia Beach, the Virginia Department of Environmental Quality, the U.S. Navy, local businesses, schools, community groups, environmental organizations, and universities – came together to develop an 18-point plan that, to date, has conserved or restored 1,000 acres (405 hectares) of watershed.

Forming partnerships between schools and public and private organizations is also an effective way to foster change in communities. Such partnerships provide real-world learning opportunities for students and a cost-effective way for communities to complete work/research on an initiative that might otherwise be too expensive. The University at Buffalo's Urban Design Project (now part of the UB Regional Institute) forged a symbiotic relationship between research and practice by connecting faculty members and urban design students with local governments, community-based organizations, and citizens to both understand places, sites, and neighborhoods and to devise solutions to real-world issues in communities. The student work provided a plethora of design options that ranged from ardent, long-term visions to more short-term, pragmatic solutions. The studio included both architecture and urban design students and fostered inter-discipline collaboration and learning. Community design allows students, design professionals, and community members to come together to voice, write, and or sketch their ideas in open-forum workshops known as charrettes. In their book *Placemaking*, Schneekloth and Shibley (1995) refer to such open space for dialogue as the "dialogic space":

> The dialogic space is the context in which hopes, fears, ideas, and frustrations about a place and the people who live there are discussed. Through the process of sharing within this space,

> new insights, and knowledge about what is and what could be are constructed. "Knowledges" brought from the outside, such as the professional knowledge of an architect, interior designer, teacher, or office manager, are offered within the discussion and transformed as part of the developing collective world that will eventually frame any placemaking action.

Unfortunately, in the twenty-first century, we still see communities that are socially, culturally, and economically divided. This division perpetuates the "us versus them" mentality that hinders any progress, resolution, or common ground that is essential for peaceful and sustainable community development. The dialogic space seeks to erase this fragmentation by giving everyone in the community an equal voice in a democratic, civil fashion. Although such a forum will produce a variety of ideas and concerns, the simple act of engaging citizens can generate support for an initiative, raise awareness, and increase media coverage. Inclusiveness, giving equal consideration to all concerns, and transparency in the decision-making process can also keep community participation active throughout the project.

Indeed, diversity, inclusiveness, and a variety of viewpoints can ultimately make a project relevant and successful. Despite someone's occupation, level of education, income bracket, ethnicity, race, religion, or gender, we can all learn something from one another. In the twenty-first century, we need to put our differences aside when it comes to environmental protection and converge around the benefits of clean air, clean water, open space, renewable energy, sustainable design, and green infrastructure. As President Barack Obama said in a 2014 commencement address at the University of California, Irvine:

> We need scientists to design new fuels. We need farmers to help grow them. We need engineers to invent new technologies. We need entrepreneurs to sell those technologies. We need workers to operate assembly lines that hum with high-tech, zero-carbon components. We need builders to hammer into place the foundations for a clean energy age. We need diplomats and businessmen and women, and Peace Corps volunteers to help developing nations skip past the dirty phase of development and transition to sustainable sources of energy. In other words, we need you.
>
> *(Obama, 2014)*

References

Central Vermont Regional Planning Commission (CVRPC). (2010). *Quantifying the impacts of low impact development: A GIS analysis of impervious surfaces and informational design charrette workshop*. http://centralvtplanning.org/wp-content/uploads/2012/03/Final-LID-Report-CVRPC1.pdf

City of Topeka (Topeka). (2002). *Topeka: Sustainable development: Moving toward a greener community*. www.fs.usda.gov/nac/assets/documents/morepublications/topekasustainable.pdf

Clements, J., & Juliana, A. S. (2013). *The Green edge: How commercial property investment in green infrastructure creates value*. Natural Resources Défense Council. www.nrdc.org/sites/default/files/commercial-value-green-infrastructure-IB.pdf

Climate Action tracker. (n.d.). *The CAT thermometer*. https://climateactiontracker.org/global/cat-thermometer/

Efficiency Vermont. (n.d.). *Energy solutions for all Vermonters*. www.efficiencyvermont.com/

Energy Action Network. (n.d.). *Vermont energy dashboard*. https://eanvt.org/tracking-progress/community-energy-dashboard/

Gamache, L. (2017). Interview conducted for final project for Leadership Champlain.

Green Schools Initiative (GSI). (n.d.). *7 steps to a green school*. www.greenschools.net/article.php-id=70.html

Harrison, S., & French, C. (n.d.). *An introduction to cost of community services studies*. UNH Extension. https://extension.unh.edu/resources/files/Resource002693_Rep3977.pdf

Internal Revenue Service (IRS). (n.d.). *Residential clean energy credit.* www.irs.gov/credits-deductions/residential-clean-energy-credit
Kania, J., & Kramer, M. (2011). *Collective impact.* Stanford Social Innovation Review, Leland Stanford Jr. University.
Manley, S., & Claydon, J. (2000). Achieving richness and diversity: Combining architecture and planning at UWE, Bristol. In D. Nicol & S. Pilling (Eds.), *Changing architectural education: Towards a new professionalism.* Spon Press.
Natural Lands Trust (NLT). (2009). *Growing greener: Conservation by design.* www.greenerprospects.com/PDFs/growinggreener.pdf
Obama, B. (2014, June 14). *UC Irvine commencement address.* www.whitehouse.gov/the-press-office/2014/06/13/remarks-president-university-california-irvine-commencement-ceremony
Schneekloth, L., & Shibley, R. (1995). *Placemaking: The art and practice of building communities* (p. 7). John Wiley & Sons.
Steiner, F., & Thompson, G. (1997). *Ecological design and planning.* John Wiley & Sons.
Texas A&M University College of Architecture (Texas A&M). (n.d.). *Academics.* www.arch.tamu.edu/academics/#research
United Nations (UN). (n.d.). *The Paris agreement.* www.un.org/en/climatechange/paris-agreement
United Nations Climate Change (UN). (n.d.c). *About carbon pricing.* https://unfccc.int/about-us/regional-collaboration-centres/the-ciaca/about-carbon-pricing#What-does-the-Paris-Agreement-say-on-carbon-pricin
United Nations Environment Programme (UNEP). (n.d.a). *About Montreal protocol.* www.unep.org/ozonaction/who-we-are/about-montreal-protocol
United Nations Environment Programme (UNEP). (n.d.b). *What is the Kyoto Protocol*? https://unfccc.int/kyoto_protocol
United Nations Framework on Climate Change (UN). (2023). *Technical dialogue of the first global stocktake*. https://unfccc.int/documents/631600
Van Der Ryn, S., & Cowan, S. (2007). *Ecological design* (10th anniversary ed., pp. 28, 171–172). Island Press.
The White House. (n.d.). *National climate task force.* www.whitehouse.gov/climate/
World Bank. (2023). *Climate policies with real-world results.* www.worldbank.org/en/news/feature/2023/09/19/climate-policies-with-real-world-results

14

CONCLUSION

In the first Democratic presidential primary debate in 2016, Anderson Cooper asked the candidates what they thought was the greatest national security threat to the United States. Senator Bernie Sanders of Vermont answered emphatically, "climate change." He went on to say, "The scientific community is telling us, if we do not address climate change, transform our energy system away from fossil fuels, the planet we'll be leaving our kids and our grandchildren may well not be habitable." Although it was my intent to keep politics out of this volume, love him or hate him, Bernie's statement hit the nail on the head, in my opinion. And despite a person's stance on climate change, it seems making the planet a better place should be something that unites us, not divides us.

However, it is also recognized that uniting people around a central cause may be overly idealistic and somewhat unrealistic. The regreening principles discussed in this book are moot if they do not embrace the specific priorities and needs of communities. Listening to all viewpoints – whether they are or are not congenial with current urban design or architectural theory – is critical to embracing the social "equity" arm of sustainability. In addition to merely listening to ideas, it is imperative that the composition of decision-making bodies is diverse to ensure there is equal representation from all races, ethnicities, socio-economic groups, religions, ages, genders, sexual orientations, and people with varying degrees of physical and mental abilities. As stated in a recent post by the Immigration Coalition, "Distance breeds assumption and fear, proximity breeds empathy and compassion."

It is also important to recognize that the notions of sustainability and resilience are relative and are subject to adapt and/or change to emerging challenges, technologies, or social movements. Recognizing the limitations of sustainability may also soothe climate or eco-anxiety that seems to have become more commonplace – especially among younger generations. My niece – who is just finishing up eighth grade as I complete this book – came home the other night on the verge of tears. She said something along the lines of,

> When my generation gets older, we are not going to be able to swim in lakes, drink clean water, breathe the air, or go outside . . . many plant and animal species will be extinct . . . all the ice will be melted, it will never snow again, and all the beaches will be under water . . .

DOI: 10.4324/9781032668666-14

> and then if that's not bad enough, I'm worried about world war, nuclear bombs, AI taking over society, another irrational world leader, another pandemic, or worse yet, the sun burning out and life on Earth ending!

Given the dire messages of some political leaders and environmental activists, the plethora of stories in the news, new curriculum in K–12 regarding climate change, and the abundance of movies and television shows dealing with the annihilation of the earth and/or the machines taking over, it is easy to see why all people (not just younger generations) may experience anxiety, fear, or guilt about the fate of our planet. However, fear mongering, condemnation, or finger-pointing are not productive strategies to solve this problem, in my opinion. If people who read this book experience anxiety or guilt concerning climate change or the planet or feel my tone is sanctimonious, then I have failed as an author. Although climate change is an imminent matter, it is not something that should be feared but rather something that can present new opportunities to restore natural balance, mitigate social and ecological fragmentation, and establish a new, green economy. And opposed to popular culture where bullying, berating, gaslighting, and cancelling people have become commonplace, it is ok to make mistakes. In fact, owning and learning from our past mistakes is the only way to advance knowledge and evolve as a society.

After reading this book, it should be apparent that there is no magic bullet that will solve all of the earth's issues but rather a combination of big and small strategies working together at multiple scales of development. Comprehensive plans and zoning ordinances can guide future development, preserve open space, and catalyze mixed-use development and mixed-income housing. Public transit that links regional services and attractions can reduce vehicle miles traveled, which in turn can foster the reduction, retrofitting, or complete removal of roads and large surface parking, ultimately reconnecting fragmented ecosystems and neighborhoods. Stepping-stone green spaces on rooftops and in alleys can be linked to community and/or regionally scaled greenways. Adaptive reuse can preserve historically significant architecture and provide an alternative to developing on open space. Making interior spaces more comfortable by bringing nature inside can foster health, wellness, and productivity. Alternative sites on the surface of water, under water, on brownfields, cliffsides, bridges, or in outer space can be used to make infrastructure more sustainable and/or resilient to both imminent and existential environmental problems. Public policy for environmental protection, integrating green building rating systems into the building codes, tax incentives, and tax credits for green infrastructure and energy-efficient retrofits can increase sustainable and resilient development.

It is also important to keep things in perspective. In the preface of this book, I mentioned the passing of my father and my brother-in-law and several health problems I am dealing with as I enter middle age. As I said, sometimes it takes tragedy (or a few) for you to realize what is important in life. All the regreening strategies discussed in this book are pointless if they sacrifice health, well-being, safety, personal comfort, quality of life, or equity. Balancing the needs of the planet and the needs of people is indeed a delicate process, but seeing the empathy, open-mindedness, inclusiveness, inquisitiveness, awareness, and general concern for our planet expressed by younger generations, I am hopeful that there are solutions on the horizon.

Our planet and humanity are inherently resilient. Throughout history, both have survived ice ages, meteor collisions, natural disasters, plagues, extinctions, diseases, wars, ozone depletion, terrorism, oil spills, fascism, and pandemics. We have evolved from agrarian lifestyles to industrial ones to the current digital age of instantaneous information. Change, growth, and development are inevitable and have consequences that can alter space and time. Our planet is also finite, and nothing is eternal. However, we owe it to the earth, the universe, and humanity to do the best we can with our available knowledge and technology to preserve natural resources and sustain health, wellness, and equity for future generations.

INDEX

Note: Page numbers in *italics* indicate a figure and page numbers in **bold** indicate a table on the corresponding page.

For Product Safety Concerns and Information please contact our EU representative GPSR@taylorandfrancis.com Taylor & Francis Verlag GmbH, Kaufingerstraße 24, 80331 München, Germany

Batch number: 10722104

Printed by Printforce, the Netherlands